AF757470

ANGIOTENSIN II RECEPTORS

VOLUME 2

Medicinal Chemistry

Edited by
Robert R. Ruffolo, Jr.

CRC Press
Boca Raton Ann Arbor London Tokyo

Library of Congress Cataloging-in-Publication Data

(Revised for vol. 2)
Angiotensin II receptors / edited by Robert R. Ruffolo, Jr.
p. cm. — (Pharmacology and toxicology)
Includes bibliographical references and index.
1. Angiotensin—Receptors. I. Ruffolo, Robert R. II. Series.
III. Series: Pharmacology & toxicology (Boca Raton, Fla.).
[DNLM: 1. Receptors, Angiotensin. 2. Angiotensin II.
QU 68 A5878 1994]
QP572.A54A53 1994
615′.71—dc20
ISBN 0-8493-8380-3 (v. 1)
ISBN 0-8493-8545-8 (v. 2)
DNLM/DLC
for Library of Congress 93-36648
CIP

This book contains information obtained from authentic and highly regarded sources. Reprinted material is quoted with permission, and sources are indicated. A wide variety of references are listed. Reasonable efforts have been made to publish reliable data and information, but the author and the publisher cannot assume responsibility for the validity of all materials or for the consequences of their use.

Neither this book nor any part may be reproduced or transmitted in any form or by any means, electronic or mechanical, including photocopying, microfilming, and recording, or by any information storage or retrieval system, without prior permission in writing from the publisher.

All rights reserved. Authorization to photocopy items for internal or personal use, or the personal or internal use of specific clients, may be granted by CRC Press, Inc., provided that $.50 per page photocopied is paid directly to Copyright Clearance Center, 27 Congress Street, Salem, MA 01970 USA. The fee code for users of the Transactional Reporting Service is ISBN 0-8493-8545-8/94/$0.00+$.50. The fee is subject to change without notice. For organizations that have been granted a photocopy license by the CCC, a separate system of payment has been arranged.

CRC Press, Inc.'s consent does not extend to copying for general distribution, for promotion, for creating new works, or for resale. Specific permission must be obtained in writing from CRC Press for such copying.

Direct all inquiries to CRC Press, Inc., 2000 Corporate Blvd. N.W., Boca Raton, Florida 33431.

© 1994 by CRC Press, Inc.

No claim to original U.S. Government works
International Standard Book Number 0-8493-8545-8
Library of Congress Card Number 93-36648
Printed in the United States of America 2 3 4 5 6 7 8 9 0
Printed on acid-free paper

Pharmacology and Toxicology: Basic and Clinical Aspects

Mannfred A. Hollinger, Series Editor
University of California, Davis

Published Titles

Inflammatory Cells and Mediators in Bronchial Asthma, 1990, Devendra K. Agrawal and Robert G. Townley
Pharmacology of the Skin, 1991, Hasan Mukhtar
In Vitro *Methods of Toxicology*, 1992, Ronald R. Watson
Basis of Toxicity Testing, 1992, Donald J. Ecobichon
Human Drug Metabolism from Molecular Biology to Man, 1992, Elizabeth Jeffreys
Platelet Activating Factor Receptor: Signal Mechanisms and Molecular Biology, 1992, Shivendra D. Shukla
Biopharmaceutics of Ocular Drug Delivery, 1992, Peter Edman
Beneficial and Toxic Effects of Aspirin, 1993, Susan E. Feinman
Preclinical and Clinical Modulation of Anticancer Drugs, 1993, Kenneth D. Tew, Peter Houghton, and Janet Houghton
Peroxisome Proliferators: Unique Inducers of Drug-Metabolizing Enzymes, 1994, David E. Moody
Angiotensin II Receptors, Volume 1: Molecular Biology, Biochemistry, Pharmacology, and Clinical Perspectives, 1994, Robert R. Ruffolo, Jr.

Forthcoming Titles

Alcohol Consumption, Cancer and Birth Defects: Mechanisms Involved in Increased Risk Associated with Drinking, Anthony J. Garro
Animal Models of Mucosal Inflammation, Timothy S. Gaginella
Brain Mechanisms and Psychotropic Drugs, A. Baskys and G. Remington
Development of Neurotransmitter Regulation and Function: A Pharmacological Approach, Christopher A. Shaw
Direct Allosteric Control of Glutamate Receptors, M. Palfreyman, I. Reynolds, and P. Skolnick
Drug Delivery Systems, V. V. Ranade
Genomic and Non-Genomic Effects of Aldosterone, Martin Wehling
Human Growth Hormone Pharmacology: Basic and Clinical Aspects, Kathleen T. Shiverick and Arlan Rosenbloom
Immunopharmaceuticals, Edward S. Kimball
Neural Control of Airways, Peter J. Barnes
Pharmacological Effects of Ethanol on the Nervous System, Richard A. Deitrich
Pharmacology in Exercise and Sport, Satu M. Somani
Pharmacology of Intestinal Secretion, Timothy S. Gaginella
Phospholipase A2 in Clinical Inflammation: Endogenous Regulation and Pathophysiological Actions, Keith B. Glaser and Peter Vadas
Placental Pharmacology, B. B. Rama Sastry
Placental Toxicology, B. B. Rama Sastry
Rational Drug Design, David B. Weiner and William V. Williams
Receptor Characterization and Regulation, Devendra K. Agrawal
Ryanodine Receptors, Vincenzo Sorrentino
Serotonin and Gastrointestinal Function, Timothy S. Gaginella and James J. Galligan
Stealth Liposomes, D. D. Lasic and F. J. Martin
Targeted Delivery of Imaging Agents, Vladimir P. Torchilin
TAXOL: Science and Applications, Matthew Suffness
Therapeutic Modulation of Cytokines, M.W. Bodner and Brian Henderson

DEDICATION

This book is dedicated to my wife, Stephany, with all my love. Her patience and support mean so much to me and have inspired me to fullfill my goals in life. For that I am eternally grateful.

EDITOR

Robert R. Ruffolo, Jr., Ph.D., is currently Vice President and Director of Pharmacological Sciences (U.S., U.K., Europe, and Australia), at SmithKline Beecham Pharmaceuticals, King Of Prussia, Pennsylvania.

Dr. Ruffolo is a graduate of The Ohio State University receiving his B.S. degree in Pharmacy, *summa cum laude* (with distinction), in 1973. He received his Ph.D. in Pharmacology from The Ohio State University in 1976. Thereafter, he received his postdoctoral training at the National Institutes of Health under the direction of Dr. Marshall Nirenberg who received the Nobel Prize in Medicine in 1968.

Dr. Ruffolo joined Eli Lilly and Company as Senior Pharmacologist in 1978 and was appointed Chairman of the Cardiovascular Research Committee. Dr. Ruffolo subsequently joined SmithKline Beecham Pharmaceuticals in 1984 as Director of Cardiovascular Pharmacology, and was appointed to his current position in 1992.

In 1988, Dr. Ruffolo was awarded the coveted John Jacob Abel Award in Pharmacology by the American Society for Pharmacology and Experimental Therapeutics. He was recognized for his contributions in the fields of receptor theory and adrenergic and cardiovascular pharmacology. His work has had significance in the understanding of catecholamine receptors and the design of new generations of α-adrenoceptor agonists with therapeutic potential in cardiovascular disease and other therapeutic areas. In 1989, Dr. Ruffolo received the Distinguished Alumni Award from The Ohio State University, and in 1994 he received the award of Distinguished Visiting Professor from the Department of Pharmacological and Pharmaceutical Sciences at the University of Houston.

Dr. Ruffolo has authored nearly 400 full-length publications, and has lectured extensively throughout the world. He is currently Editor-in-Chief of *Pharmacology Communications*, and is Editor of *Critical Reviews in Pharmacology* as well as an Editor for Cardiovascular and Renal Drugs for *Current Opinion in Investigational Drugs*. He was previously an Editor for *The Journal of Pharmacology and Experimental Therapeutics* and he currently serves on 17 editorial boards. Dr. Ruffolo has edited 7 books.

PREFACE

Major advances have been made recently in our understanding of the angiotensin II receptors. At least two subtypes of the angiotensin II receptor are now known, designated as AT_1 and AT_2, which have been identified through biochemical and pharmacological techniques. Both of these angiotensin II receptor subtypes have been cloned and expressed, and consequently, a more detailed understanding of the molecular structures and biochemical functions of the angiotensin II receptor subtypes is being realized.

The elucidation of new functions mediated by the angiotensin II receptors at the cellular, tissue, and organ level has led to an increased understanding of the roles that angiotensin II receptors play in physiological and pathophysiological processes, and these advances have fueled excitement in this area and have driven biological research at a rapid pace. The biological aspects of research on the angiotensin II receptors, from the gene to the human, are reviewed in detail in the accompanying volume to this book, *Angiotensin II Receptors, Volume I: Molecular Biology, Biochemistry, Pharmacology and Clinical Perspectives.* In spite of the dramatic progress made in understanding the biology of the angiotensin II receptors, the most important developments of the past decade relating to angiotensin II receptors involves the medicinal chemistry and the structure–activity relationships for peptide angiotensin II receptor agonists and antagonists, and more importantly the nonpeptide angiotensin II receptor antagonists. The chemistry of the angiotensin II receptor agonists and antagonists is the subject matter for the present volume, *Angiotensin II Receptors, Volume II: Medicinal Chemistry.*

With the identification of AT_1 and AT_2 receptors for angiotensin II, structure–activity relationships for activation and inhibition of these receptors have expanded dramatically and have resulted in the identification of potent and highly selective angiotensin II receptor agonists and antagonists which serve as important "pharmacological tools" that have enabled investigators to characterize the existing angiotensin II receptor subtypes, and may aid in the possible identification of additional, but presently unknown, angiotensin II receptor subtypes. Indeed, several new and exciting compounds that have arisen from these recent advances in the structure–activity relationships for angiotensin II receptor antagonists, especially the nonpeptide antagonists, are finding their way into clinical trials for the treatment of diseases, such as hypertension, congestive heart failure, and renal failure.

What we believe distinguishes this book on the medicinal chemistry of angiotensin II receptors, along with the accompanying volume on the molecular biology, biochemistry, pharmacology, and clinical perspectives related to angiotensin II receptors, from other books on the subject is the broadness of the subject matter covered. We have compiled and reviewed the most recent and important advances made in relation to the angiotensin II receptors, which span from the intracellular level of the nucleus, including the messenger RNA that encodes for these receptors, through the biochemical processes activated or inhibited by these receptors, up to and including the functional effects that angiotensin II receptors mediate in intact organs, animals, and ultimately in humans, both in normal and disease states (Volume I). Furthermore, we have also reviewed in detail all of the major developments relating to the medicinal chemistry of angiotensin II receptor agonists and antagonists, including most importantly the structure–activity relationships for nonpeptide angiotensin II receptor antagonists (Volume II). To my knowledge, a comprehensive review dealing with angiotensin II receptors from the standpoint of molecular biology, biochemistry, pharmacology and clinical applications (Volume I), and medicinal chemistry (Volume II) has not been attempted previously, and I believe that it is the integration of these disciplines, and the exciting findings that have been generated recently from them, that make these two volumes entitled *Angiotensin II Receptors, Volume I: Molecular Biology, Biochemistry, Pharmacology and Clinical Perspectives* and *Angiotensin II Receptors, Volume II: Medicinal Chemistry* most unique.

Medicinal chemists have played a pivotal role in facilitating research on angiotensin II receptors. Most notably, medicinal chemists are responsible for the identification of important new chemical tools that have enabled the studies emanating from molecular biology and pharmacology of the angiotensin II receptors to proceed. As additional information is uncovered on existing angiotensin II receptors, or as additional subtypes of the angiotensin II receptor are identified, medicinal chemists will continue to design new and selective agonists and antagonists for these receptors, which will allow molecular biologists, biochemists, pharmacologists, physiologists, and physicians to explore further the physiological and pathophysiological roles of the angiotensin II receptors.

These two volumes on angiotensin II receptors are constructed in a manner similar to previous volumes that we have compiled on the molecular biology, biochemistry, medicinal chemistry, and pharmacology of α-adrenoceptors and β-adrenoceptors. It may appear presumptuous for an entire volume to be written by investigators from a single institution where, clearly, the majority of significant contributions occurring in this field could not possibly have originated. However, it should be noted that the authors of these two volumes on angiotensin II receptors have significant experience in this field and have worked extensively on the discovery as well as the preclinical and clinical development of novel, nonpeptide angiotensin II receptor antagonists that are currently in clinical trials. In total, the authors of these two volumes have published in excess of 100 papers relating to the renin-angiotensin system, and as such, they have established their credibility in the field of angiotensin II receptors. In particular, the chemists that have worked on Volume II which deals with the medicinal chemistry of angiotensin II receptors have worked for over one decade on the structure–activity relationships of angiotensin II receptor agonists and antagonists, have synthesized hundreds of peptide and nonpeptide angiotensin II analogs and have published extensively on the subject. The research of the medicinal chemists who have contributed to this volume has resulted in the identification of nonpeptide angiotensin II receptor antagonists that are currently undergoing clinical trials. Because of the pivotal role that medicinal chemists have played over the past decade in research on angiotensin II receptors, it is appropriate to devote an entire volume to the medicinal chemistry of angiotensin II receptor agonists and antagonists.

Robert R. Ruffolo, Jr.

CONTRIBUTORS

Major contributions to this volume have been written by investigators from SmithKline Beecham Pharmaceuticals, King of Prussia, Pennsylvania. Listed below are the names of those contributing individuals followed by their respective position within the organization attesting to their significant experience in the field of angiotensin II receptors.

John Gleason, Ph.D.
Vice President and Director
Department of Medicinal Chemistry

Judith C. Hempel, Ph.D.
Senior Investigator
Department of Physical and Structural Chemistry

Richard M. Keenan, Ph.D.
Assistant Director
Department of Medicinal Chemistry

Catherine Peishoff, Ph.D.
Senior Investigator
Department of Physical and Structural Chemistry

Domenico Regoli, M.D.*
Professor
Department of Pharmacolgy

Robert R. Ruffolo, Jr., Ph.D.
Vice President and Director
Department of Pharmacological Sciences

James M. Samanen, Ph.D.
Associate Director
Department of Medicinal Chemistry

Joseph Weinstock, Ph.D.
Senior Fellow
Department of Medicinal Chemistry

*Currently associated with University of Sherbrooke, Sherbrooke, Quebec, Canada.

TABLE OF CONTENTS

Chapter 1

INTRODUCTION: MEDICINAL CHEMISTRY OF ANGIOTENSIN II RECEPTORS

John G. Gleason and Robert R. Ruffolo, Jr.

The renin-angiotensin system plays a central role in the regulation of blood pressure and fluid and electrolyte balance, and as a result, plays a major role in the pathophysiology of a number of disease states.[1] The predominant active component of this system, angiotensin II, one of the most potent vasoconstrictors known, is formed by the sequential enzymatic proteolysis of the circulating tetradecapeptide prohormone, angiotensinogen (Figure 1). In the kidney, and in a number of other tissues, renin, an aspartyl protease, acts on angiotensinogen to form angiotensin I, a relatively inactive decapeptide. Angiotensin I is further processed by the zinc metalloprotease, angiotensin converting enzyme (ACE), which is present on endothelial cells to form the biologically active octapeptide, angiotensin II.

Angiotensin II has a number of important physiologic actions which relate, either directly or indirectly, to regulation of blood pressure and sodium retention.[2] Among its most important actions in the peripheral circulation are vascular smooth muscle contraction and potentiation of norepinephrine release from sympathetic nerves.[3] Angiotensin II also stimulates release of aldosterone from the adrenal cortex. In the kidney, angiotensin II stimulates sodium reabsorption, both by a direct action on the proximal tubule and by an indirect action on the distal nephron *via* aldosterone. In the brain, angiotensin II increases sympathetic outflow and stimulates the thirst mechanism.

Considering the importance of the renin-angiotensin system in the regulation of blood pressure and electrolyte balance, it is not surprising that this area has been, and remains, a fruitful area for drug discovery. ACE inhibitors[4] have proven to be highly successful in a number of disease states and are well tolerated in patients.[5-8] However, ACE is identical to kininase II, a protease which plays a major role in kinin degradation[9] and the metabolic degradation of substance P.[10] Thus, inhibition of ACE may result in an increase in both substance P and bradykinin, which has been proposed as the cause of the dry cough and angioedema, the predominant side effects of ACE inhibitor therapy.[11] Inhibition of renin, while an attractive clinical strategy, remains elusive. Potent inhibitors have been described,[12] but development of this class as drugs has been limited by pharmacokinetic difficulties.

Most, if not all, of the pharmacological effects of angiotensin II are mediated via interaction with angiotensin II receptors present throughout the body (Chapter 2). To date, there are two receptor subtypes which have been described, AT_1 and

0-8493-8545-8/94/$0.00+$.50
© 1994 by CRC Press, Inc.

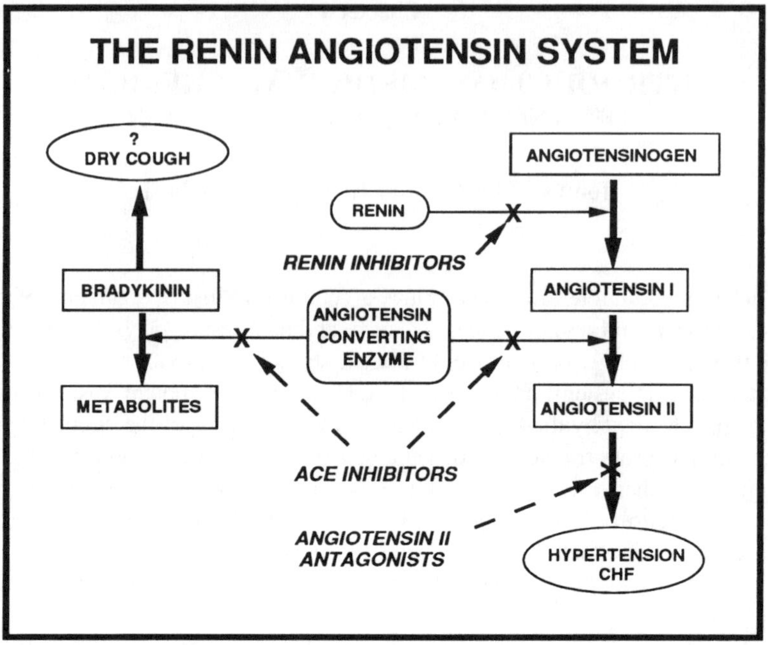

FIGURE 1. The renin-angiotensin system.

AT_2. The AT_1 receptor, through which all of the known pharmacological effects of angiotensin II appear to be mediated, is a G-protein coupled receptor. The discovery of potent and selective AT_1 receptor antagonists offers an attractive alternative to ACE inhibitor therapy, and is the focus of the chapters in this volume. Although the functional significance of the AT_2 receptor remains to be determined,[13] antagonists of this receptor subtype have been identified (Chapter 4) and should provide the requisite tools to unravel further the role of angiotensin II in disease.

The discovery of nonpeptide, orally active angiotensin II receptor antagonists represents a major achievement in medicinal chemistry, albeit one which has spanned several decades. The foundations for this discovery were laid with the extensive empirical exploration of the structure–activity relationships of peptide and semi-peptide analogs of angiotensin II (Chapter 2). These studies resulted in the identification of peptide angiotensin II receptor antagonists, such as saralasin, and recognition of the importance of the carboxyl terminal phenylalanine (Phe^8) and a tyrosine residue (Tyr^4) in agonist function (Figure 2).[14] Appropriate modification of either residue could switch the function of the molecule from agonist to antagonist, observations which are rationalized by Marshall[15] in terms of a conformational requirement for agonist activity which places the Phe8 side chain over the Pro-Phe peptide bond. This recognition of a key conformational requirement for the carboxyl terminal region of angiotensin II subsequently played a

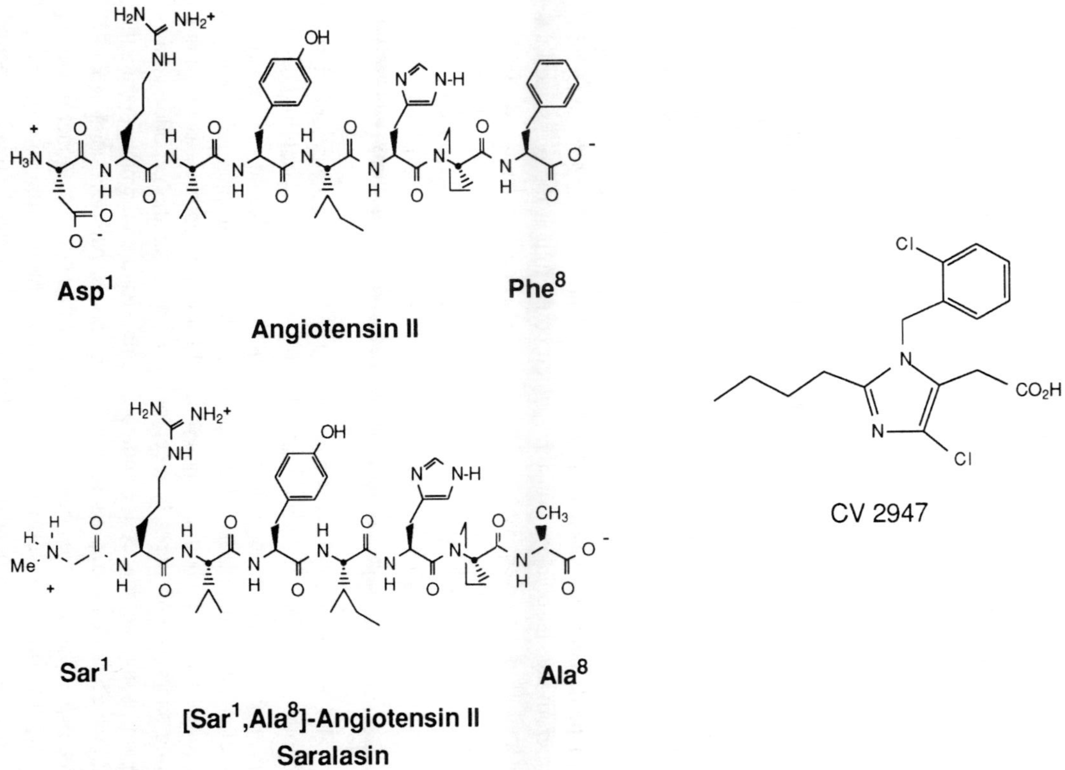

FIGURE 2. The structures of angiotensin II and the competitive peptide receptor antagonist, [Sar[1],Ala[8]]-Angiotensin II (Saralasin). CV 2947 is the first nonpeptide angiotensin II receptor antagonist identified.

critical role in the discovery and development of high affinity nonpeptide angiotensin II receptor antagonists. The amino terminal region of angiotensin II is also of considerable importance for receptor binding; for example, replacement of the amino terminal aspartic acid residue with sarcosine, particularly in the antagonists, provides a significant increase in affinity.[16] Thus, the critical residues for receptor binding and activation span the full length of the octapeptide, resulting in an apparently large functional pharmacophore (Figure 2). While providing valuable chemical, pharmacological, and clinical knowledge about angiotensin II receptor antagonists, these early peptides, however, possessed significant liabilities which limited their therapeutic value. All retained weak, but significant, partial agonist activity which precluded their use as antihypertensives. In addition, these peptide antagonists lacked oral bioavailability, and the large functional pharmacophore seemingly required for peptide antagonists prevented significant reduction in the size and peptide nature which would be necessary to circumvent these limitations.

The apparently large pharmacophore required for peptide antagonist affinity would suggest that the design of a small molecule, nonpeptide angiotensin II receptor antagonist based on the structure (and conformation) of the peptide antagonists would be difficult. Thus, the discovery of such a nonpeptide angiotensin II receptor antagonist, albeit with low affinity, CV 2947 (Figure 2), by Furakawa and colleagues[17–19] represented a major breakthrough in the angiotensin II receptor antagonist field. This antagonist is a weak, but competitive receptor antagonist with a micromolar dissociation constant, which could inhibit the pressor effects of angiotensin II in normotensive rats and lower blood pressure in renin-dependent hypertensive rats.[20,21] Most notably, unlike the peptide antagonists, this nonpeptide is devoid of agonist activity. This discovery provided the basis for a major effort over the past decade which has resulted in the discovery of several high affinity nonpeptide angiotensin II receptor antagonists which are currently undergoing clinical evaluation.

Although CV 2947 represented a potential starting point for the rational design of nonpeptide angiotensin II receptor antagonists, to apply a rational drug design approach, it was necessary to relate the structure of CV 2947 to the important structural elements of angiotensin II as they are displayed in the pharmacologically relevant conformation. It should be noted that this design approach assumed that the binding site(s) within the AT_1 receptor overlapped for the peptide agonists and antagonists and the nonpeptide antagonists, an assumption which, despite the remarkable success of the approach, remains unproven. Although angiotensin II is a linear peptide, and hence could have considerable conformational mobility, utilizing spectroscopic and physical chemical data in combination with structure–activity data from a number of analogs, several models of potential receptor-bound conformations of angiotensin II have been developed. The genesis of these models and a discussion of the relative merits of the models is presented in the appendix to Chapter 3. Utilizing the model of Fermandjian et al.,[22] and focusing attention on the carboxyl terminal region for the reasons described above, Duncia et al.[20]

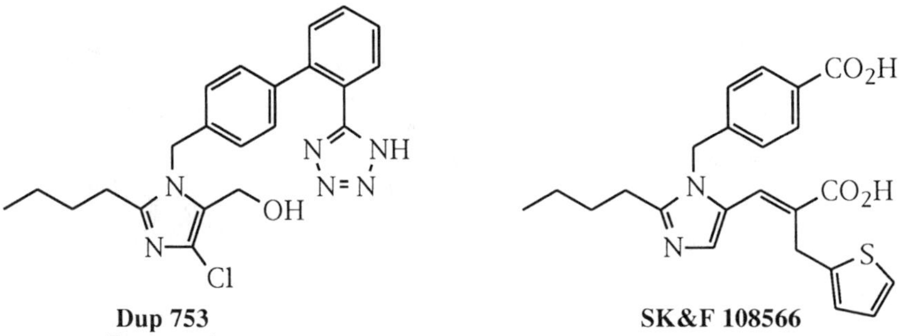

Dup 753
(Losartan)

SK&F 108566

FIGURE 3. Chemical structures of the nonpeptide angiotensin II receptor antagonists, DuP 753 (Losartan) and SK&F 108566.

developed a model which overlaid the carboxylate group of CV 2947 with the C-terminal acid group of angiotensin II, the imidazole of CV 2947 with His[6] of angiotensin II, a residue in angiotensin II which is important for receptor affinity, and oriented CV 2947 in such a manner that the *ortho*-chlorobenzyl moiety pointed toward the amino terminal residue, Asp1, and specifically toward the carboxyl side chain of this residue. This model directed Duncia et al.[20] through an extensive medicinal chemical effort to the discovery of the first high affinity, nonpeptide, orally active angiotensin II receptor antagonist, Dup 753, which has subsequently been given the name, losartan (Figure 3).[20] The elegant medicinal chemistry which allowed the progression from CV 2947 to losartan is described in detail in Chapter 4. Although the carboxylic acid of CV 2947, which was critical to the overlay model with angiotensin II, is no longer present in losartan, the corresponding molecule in which the hydroxymethyl group is replaced by a carboxylate is a major metabolite of losartan and is approximately one order of magnitude more potent than losartan. This metabolite appears to contribute significantly to the observed pharmacological effects of losartan.

Losartan represents the first of a growing class of angiotensin II receptor antagonists possessing in common the biphenyltetrazole moiety linked to a nitrogen-containing heterocycle, most often an imidazole. Chemical structures of several members of this class which have progressed to clinical development are depicted in Figure 4.[23] The structure–activity relationships of this class of compounds is discussed extensively in Chapter 4.

Utilizing a different model for the receptor bound conformation of angiotensin II derived from consideration of the structures of conformationally restricted analogs of angiotensin II, accessible low energy conformations of the peptide, and the agonist hypothesis developed by Marshall,[15] Weinstock et al.[21] developed a different, but equally successful design strategy. In this model, the overlay of the carboxyl of CV 2947 with the carboxyl terminal Phe[8] of angiotensin II and the

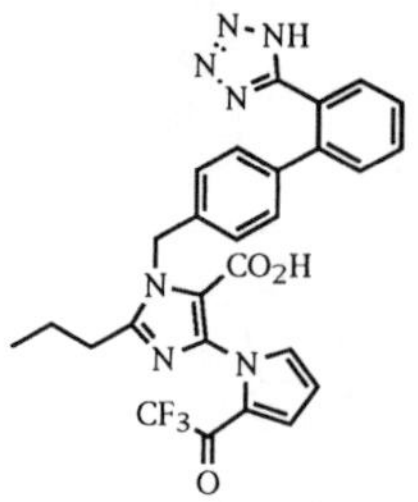

CI-996

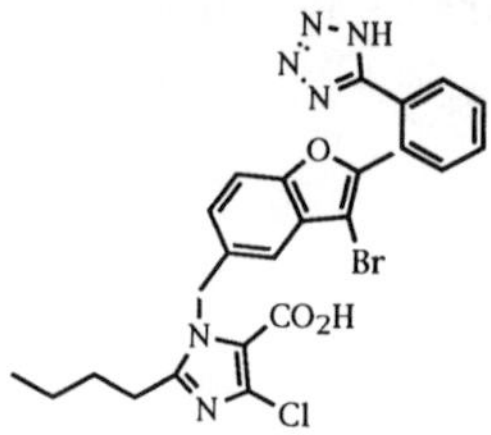

GR-117289

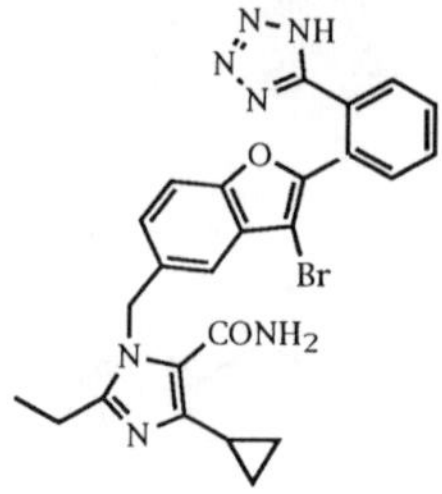

GR-138950

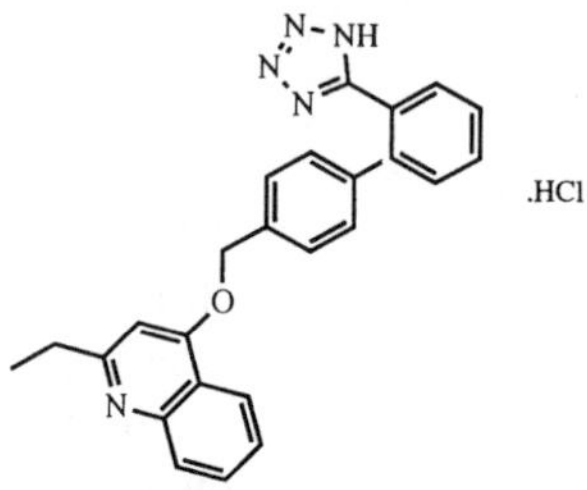

ICI-D6888

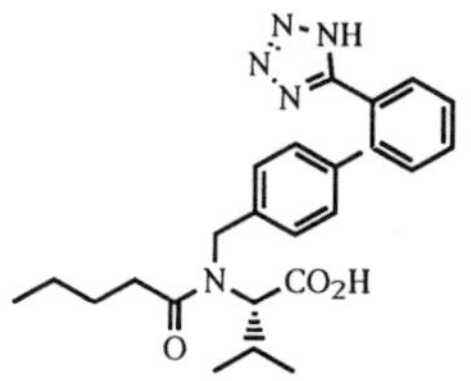

Valsartan

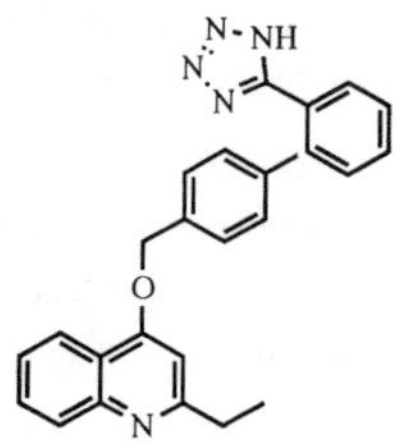

ICI-D8731

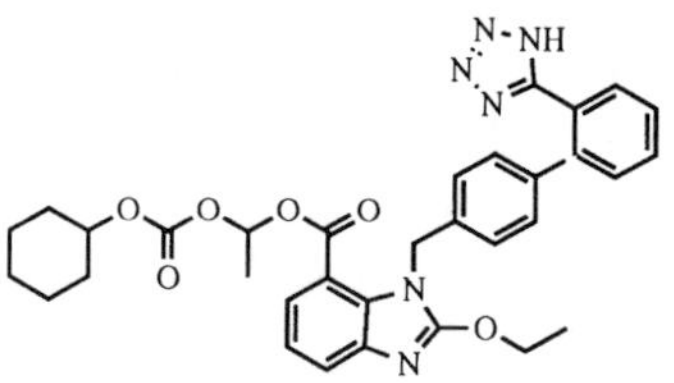

TCV-116

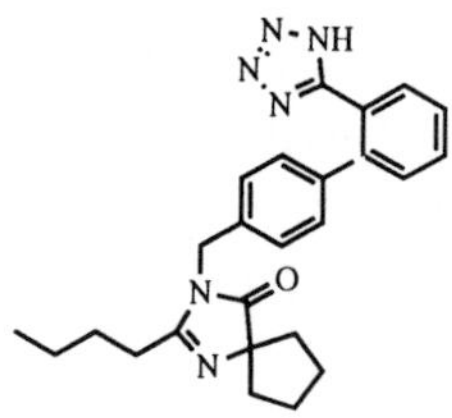

SR-47436

FIGURE 4. Chemical structures of some angiotensin II receptor antagonists in clinical development.

ortho-chlorobenzyl of CV 2947 with Tyr4, provided a model which led to the discovery of the imidazoleacrylate, SK&F 108566.[21] In SK&F 108566 (Figure 3), the thienymethyl sidechain is proposed to mimic the sidechain of Phe8, while the toluic acid residue may function as a tyrosine isostere. A stereo view of the proposed overlay of SK&F 18566 with angiotensin II is illustrated in Figure 5. The structure–activity relationships which guided the discovery of SK&F 108566 are described in Chapter 4. As with losartan, SK&F 108566 is a high affinity, potent, AT_1 selective, orally active angiotensin II receptor antagonist and is currently in clinical development.

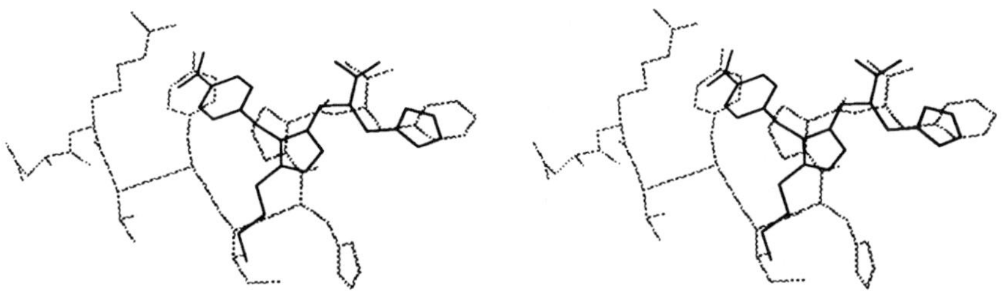

FIGURE 5. Stereo overlay of SK&F 108566 (solid lines) with angiotensin II (Dotted lines); the conformation of angiotensin is as described in Reference 21.

As a class, these novel angiotensin II receptor antagonists have been shown to antagonize angiotensin II-induced effects in animals, and more importantly, do not exhibit the partial agonist properties characteristic of the peptide antagonists. They have demonstrated efficacy in a number of disease models, including renin-dependent hypertensive dogs[24] and rats,[24,25] spontaneously hypertensive dogs[26] and rats,[27–30] congestive heart failure in rats,[31] and vascular damage in rats following endothelial injury.[32,33] Losartan, the most advanced clinical candidate, inhibits the pressor response of angiotensin II in normotensive subjects,[34] reduces proteinuria in patients with nondiabetic renal disease,[35] and lowers blood pressure in patients with essential hypertension.[36,37]

While the major focus of activities within the angiotensin II receptor antagonist area has been on the AT_1 receptor, both peptide and nonpeptide antagonists which are selective for the AT_2 receptor subtype have been discovered. These include the peptide, CGP 42112,[38] and the nonpeptides, PD 123319 and WL 19.[39] The chemistry of these latter two compounds is discussed in Chapter 4. More recently, compounds have been described which possess high affinity for both the AT_1 and AT_2 receptor subtypes.[40] While the AT_2 subtype selective antagonists have provided valuable tools for the characterization of angiotensin II receptors, all of the known functions of angiotensin II appear to be mediated by the AT_1 receptor subtype. The physiological role of the AT_2 receptor subtype remains unknown. Thus it remains unclear at this time what therapeutic benefit the AT_2 or AT_1/AT_2 receptor antagonists will have.

Since the initial discovery of renin in 1898, a tremendous effort has been devoted to the investigation of the renin-angiotensin system and its physiological and pathophysiological roles. The development of ACE inhibitors which block the production of the primary active component of the renin angiotensin system, angiotensin II, has provided a significant therapeutic advance, and the subsequent development of selective nonpeptide angiotensin II receptor antagonists, and possibly nonpeptide renin inhibitors, may provide for further refinement in the regulation of the renin-angiotensin system.

REFERENCES

1. **Garrison, J. C. and Peach, M. J.,** Renin and angiotensin, in *Goodman and Gilmans The Pharmacological Basis of Therapeutics,* Gilman, A., Rall, T., Nies, A., et al., Eds., Pergamon Press, New York, 1990, 749.
2. **Peach, M. J. and Dostal, D. E.,** The angiotensin II receptor and the actions of angiotensin II, *J. Cardiovasc. Pharmacol.,* 16, (Suppl. 4), S25, 1990.
3. **Starke, K.,** Regulation of noradrenaline release by presynaptic receptor systems, *Rev. Physiol. Biochem. Pharmac.,* 77, 1, 1977.
4. **Wyvratt, M. J. and Patchett, A. A.,** Recent developments in the design of angiotensin-converting enzyme inhibitors, *Med. Res. Rev.,* 5, 483, 1985.
5. **Kostis, J. B.,** Angiotensin-covering enzyme inhibitors, *Am. J. Hypertension,* 2, 57, 1989.
6. **Williams, G. H.,** Converting-enzyme inhibitors in the treatment of hypertension, *N. Engl. J. Med.,* 319, 1517, 1988.
7. **Packer, M., Kukin, M. L., Neuberg, G. W., Pinsky, D. J., Penn, J., and Abittan, M. H.,** The current status of angiotensin converting enzyme inhibitors in the management of patients with chronic heart failure, *J. Hypertension,* 7 (Suppl. 5), S33, 1989.
8. **Riegger, A. J. G.,** ACE inhibitors in congestive heart failure, *Cardiology,* 76 (Suppl. 2), 42, 1989.
9. **Regoli, D. and Barabe, J.,** Pharmacology of bradykinin and related kinins, *Pharmacol. Rev.,* 32, 1, 1980.
10. **Cascieri, M. A., Bull, H. G., Mumford, R. A., Patchett, A. A., Thornberry, N. A., and Liang, T.,** Carboxyl-terminal tripeptidyl hydrolysis of substance P by purified rabbit lung angiotensin-converting enzyme and the potentiation of substance P in vivo by captopril and MK-422, *Mol. Pharmacol.,* 25, 287, 1984.
11. **Levens, N. R., de Gasparo, M., Wood, J. M., and Bottari, S. P.,** Could the pharmacological differences observed between angiotensin II antagonists and inhibitors of angiotensin converting enzyme be clinically beneficial?, *Pharmacol. Toxicol.,* 71, 241, 1992.
12. **Greenlee, W. J.,** Renin inhibitors, *Med. Res. Rev.,* 10, 173, 1990.
13. **Timmermans, P. B. M. W. M., Wong, P. C., Chiu, A. T., and Herblin, W. F.,** Nonpeptide angiotensin II receptor antagonists, *Trends Pharm. Sci.,* 12, 55, 1991.
14. **Moore, A. F. and Fulton, R. W.,** Angiotensin II antagonists-saralasin, *Drug Dev. Res.,* 4, 331, 1984.
15. **Marshall, G. R.,** Determination of the receptor bound conformation of angiotensin, *Dtsch. Apoth. -Ztg.,* 85, 2783, 1986.
16. **Regoli, D. and Park, W. K.,** The pressor and myotropic effects and the antagonist properties of several analogs of angiotensin II, *Can. J. Physiol. Pharmacol.,* 50, 99, 1972.
17. **Furukawa, Y., Kishimoto, S., and Nishikawa, K.,** U.S. Patent Application 4,355,040, 1982.
18. **Furukawa, Y., Kishimoto, S., and Nishikawa, K.,** U.S. Patent Application 3,340,598, 1982.
19. **Furukawa, Y. and Nishikawa, K.,** Patent Application 4,582,847, 1986.
20. **Duncia, J. V., Carini, D. J., Chiu, A. T., Johnson, A. L., Price, W. A., Wong, P. C., Wexler, R. R., and Timmermans, P. B. M. W. M.,** The discovery of DuP 753, a potent, orally active nonpeptide angiotensin II receptor antagonist, *Med. Res. Rev.,* 12, 149, 1992.
21. **Weinstock, J., Keenan, R. M., Samanen, J., Hempel, J., Finkelstein, J. A., Franz, R. G., Gaitanopoulos, D. E., Girard, G. R., Gleason, J. G., Hill, D. T., Morgan, T. M., Peishoff, C. E., Aiyar, N., Brooks, D. P., Fredrickson, T. A., Ohlstein, E. H., Ruffolo, R. R. Jr., Stack, E. J., Sulpizio, A. C., Weidley, E. F., and Edwards, R. M.,** 1-(Carboxybenzyl) imidazole-5-acrylic acids: potent and selective angiotensin II receptor antagonists, *J. Med. Chem.,* 34, 1514, 1991.
22. **Fermandjian, S., Lintner, K., Haar, W., Fromageot, P., Khosla, M. C., Smeby, R. R., and Bumpus, F. M.,** Conformation-function relationship of angiotensin II and analogues, in *Peptides, 1976, Proc. 14th Eur. Peptide Symp.,* Loffet, A., Ed., University of Brussels, 1976, 339.

23. **Prous, J., Castaner, J., and Graul, A.,** ICI-D8731, angiotensin II antagonist, *Drugs of the Future,* 18, 428, 1993.
24. **Brooks, D. P., Fredrickson, T. A., Weinstock, J., Ruffolo, R. R., Jr., Edwards, R. M., and Gellai, M.,** Antihypertensive activity of the nonpeptide angiotensin II receptor antagonist, SK&F 108566, in rats and dogs, *Naunyn-Schmiedeberg's Arch. Pharmacol.,* 345, 673, 1992.
25. **Wong, P. C., Price, W. A., Reilly, T. M., Duncia, J. V., and Timmermans, P. B. M. W. M.,** Antihypertensive mechanism of captopril in renal hypertensive rats: studies with a nonpeptide angiotensin II receptor antagonist and an angiotensin II monoclonal antibody, *J. Pharmacol. Exp. Ther.,* 250, 515, 1989.
26. **Bovee, K. C., Wong, P. C., Timmermans, P. B. M. W. M., and Thoolen, M. J. M. C.,** Effects of the nonpeptide angiotensin II receptor antagonist DuP 753 on blood pressure and renal functions in spontaneously hypertensive PH dogs, *Am. J. Hypertension,* 4, 327S, 1991.
27. **Fenoy, F. J., Milici, I., Smith, R. D., Wong, P. C., Timmermans, P. B. M. W. M., and Roman, R.,** Effects of DuP 753 on renal function of normotensive and spontaneously hypertensive rats, *Am. J. Hypertension,* 4, 321S, 1991.
28. **Camargo, M. J. F., von Lutterotti, N., Pecker, M. S., James, G. D., Timmermans, P. B. M. W. M., and Laragh, J. H.,** DuP 753 increases survival in spontaneously hypertensive stroke-prone rats fed a high sodium diet, *Am. J. Hypertension,* 4, 341S, 1991.
29. **von Lutterotti, N., Camargo, M. J. F., Mueller, F. B., Timmermans, P. B. M. W. M., and Laragh, J. H.,** Angiotensin II receptor antagonist markedly reduces mortality in salt-loaded Dahl S rats, *Am. J. Hypertension,* 4, 346S, 1991.
30. **Bunkenburg, B., Schnell, C., Baum, H-P., Cumin, F., and Wood, J. M.,** Prolonged angiotensin II antagonism in spontaneously hypertensive rats, *Hypertension,* 18, 278, 1991.
31. **Raya, T. E., Fonken, S. J., Lee, R. W., Daugherty, S., Goldman, S., Wong, P. C., Timmermans, P. B. M. W. M., and Morkin, E.,** Hemodynamic effects of direct angiotensin II blockade compared to converting enzyme inhibition in rat model of heart failure, *Am. J. Hypertension,* 4, 334S, 1991.
32. **Kauffman, R. F., Bean, J. S., Zimmerman, K. M., Brown, R. F., and Steinberg, M. I.,** Losartan, a nonpeptide angiotensin II (ANG II) receptor antagonist, inhibits neointima formation following balloon injury to rat carotid arteries, *Life Sci.,* 49, PL223, 1991.
33. **Laporte, S. and Escher, E.,** Neointima formation after vascular injury is angiotensin II mediated, *Biochem. Biophys, Res. Comm.,* 187, 1510, 1992.
34. **Christen, Y., Waeber, B., Nussberger, J., Lee, R. J., Timmermans, P. B. M. W. M., and Brunner, H. R.,** Dose-response relationships following oral administration of DuP 753 to normal humans, *Am. J. Hypertension,* 4, 350S, 1991.
35. **Gansevoort, R. T., de Zeeuw, D., de Jong, P. E.,** Antiproteinuric effect of the angiotensin-II (A-II) receptor antagonist MK-954 in non-diabetic renal disease, *JASN,* 3, 282, 1992.
36. **Nelson, E., Arcuri, K., Ikeda, L., Snavely, D., and Sweet, C.,** Efficacy and safety of losartan in patients with essential hypertension, *J. Hypertension,* 10 (Suppl. 4), S122, 1992.
37. **Weber, M. A.,** Clinical experience with the angiotensin II receptor antagonist losartan, *Am. J. Hypertension,* 5, 247S, 1992.
38. **Whitebread, S., Mele, M., Kamber, B., and de Gasparo, M.,** Preliminary biochemical characterization of two angiotensin II receptor subtypes, *Biochem. Biophys. Res. Comm.,* 163, 284, 1989.
39. **Blankley, C. J., Hodges, J. C., Klutchko, S. R., Himmelsbach, R. J., Chucholowski, A., Connolly, C. J., Neergaard, S. J., Van Nieuwenhze, M. S., Sebastian, A., Quin, J. III, Essenburg, A. D., and Cohen, D. M.,** Synthesis and structure-activity relationships of a novel series of nonpeptide angiotensin II receptor binding inhibitors specific for the AT_2 subtype, *J. Med. Chem.,* 34, 3248, 1991.
40. **De Laszlo, S. E., Quagliato, C. S., Greenlee, W. J., Patchett, A. A., Chang, R. S. L., Lotti, V. J., Chen, T.-B., Scheck, S. A., Faust, K. A., Kivlighn, S. S., Schorn, T. S., Zingaro, G. J., and Siegl, P. K. S.,** A potent, orally active, balanced affinity angiotensin II AT_1 antagonist and AT_2 binding inhibitor, *J. Med. Chem.,* 36, 3207, 1993.

Chapter 2

STRUCTURE–ACTIVITY RELATIONSHIPS OF PEPTIDE ANGIOTENSIN II RECEPTOR AGONISTS AND ANTAGONISTS

James Samanen and Domenico Regoli

TABLE OF CONTENTS

0-8493-8545-8/94/$0.00+$.50
© 1994 by CRC Press, Inc.

I. THE FIRST PEPTIDE ANTAGONISTS

An important milestone in the development of antihypertensive therapy was reached with the discovery of potent antagonist analogs of angiotensin II, such as saralasin [Sar^1,Ala^8]-angiotensin II and [Sar^1,Ile^8]-angiotensin II.* Both compounds owe their antagonist activities to modifications of the position eight side-chain, from phenylalanine in angiotensin II to alanine and isoleucine in [Sar^1,Ala^8]-angiotensin II and [Sar^1,Ile^8]-angiotensin II, respectively.

Angiotensin II

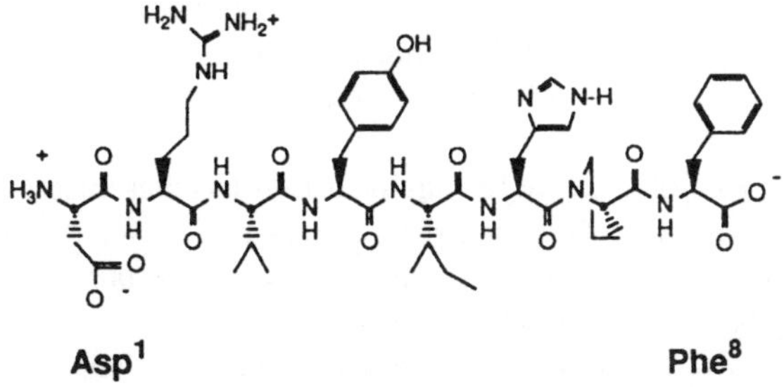

Saralasin [Sar¹,Ala⁸]- Angiotensin II

[Sar¹,Ile⁸]- Angiotensin II

Studies of these early octapeptide antagonists in humans[2,3] clarified the issues in angiotensin II antagonist design:

1. The octapeptide antagonists retained sufficient agonist activity to prevent a reduction in blood pressure in all but those individuals with high renin

*IUPAC nomenclature[1] for peptide structures will be followed in Chapters 2 and 3, in which the native amino acids are abbreviated by three letter code, and analogs of AII are denoted as shown for saralasin, with the modifications enclosed in square brackets. Abbreviations employed in this text, which have not been described by IUPAC convention and which are described in Table 19, include ε-Ahx, γ-Abu, Aib, ε-Ape, Apb, Apib, Abu, (αMe)Ala, (αMe)Phe, (αMe)Tyr, (αMe)Val, Cha, Chg, Cit, Cle, ΔAla, ΔPhe, Dtc, ε-Ghx, γ-Gbu, (γ-Im)Abu, hArg, hCys, Ima, Ind, L,L-lactam-Phe, t-Mpr, (NMe)Ala, (N-3Me)His, (N-1Me)His, (NMe)Ile, (NMe)Phe, (4-NH_2)Phe, (OMe)Thr, (OMe)Tyr, (OEt)Tyr, Peg, (pF)Phe, Pgl, (βMe)Phe, (SAcm)Phe, (SMe)Pen, Pza, Tal, Thiq.

levels. ACE inhibitors, however, which lacked angiotensin II activity, were capable of lowering blood pressure even in humans who displayed "normal" renin levels.[4] Thus, clinically useful angiotensin II antagonists must be devoid of agonist activity.

2. The octapeptide antagonists lacked oral activity. In general, peptides and semipeptides (peptides that contain some mimetic elements of peptide structure) do not display adequate oral bioavailability. The recent report of adequate oral bioavailability in certain semipeptide renin inhibitors[5,6] is an exciting counter example. Until more experience is gained in the development of peptides and semipeptides with adequate oral bioavailability, the pursuit of high affinity small nonpeptide ligands will continue to be the surest route toward oral drugs.

As a molecular target, angiotensin II has presented a third challenge to medicinal chemists, namely that the residues critical for receptor binding span the entire peptide structure of saralasin. Thus, the peptide antagonists present a very large pharmacophore to the receptor.

On the basis of these issues, medicinal chemists have pursued two approaches toward the search for oral angiotensin II antagonists: (1) modification and removal of peptidic elements of structure and (2) discovery and enhancement of novel nonnative ligands to the angiotensin II receptor. The first approach would give rise initially to semipeptides, which contain some mimetic elements of structure. Further effort would lead to nonpeptides, in which the native peptide structure is no longer recognizable. The second approach short-cuts the first approach by screening compound databases and biological samples for nonpeptide leads that can then be enhanced by classical empirical modifications. It is the second approach that gave rise to the first reported nonpeptide angiotensin II antagonists. A benzylimidazole with weak angiotensin II antagonist activity was discovered by Furukawa et al. among a group of diuretic agents.[7,8] The group at Dupont modified that lead into the biphenylimidazoles, e.g., Dup 753, which spearheaded the dramatic proliferation of nonpeptide angiotensin II antagonists that is described in detail in Chapter 4, Volume II.

Although most angiotensin II semipeptide and nonpeptide analogs have been designed empirically, it has long been recognized that ligand design could be enhanced significantly from an understanding of the bioactive or receptor-bound conformation. This chapter describes empirical peptide-based approaches toward clinically useful angiotensin II antagonists and the quest for analogs that could help define the receptor-bound conformation. Chapter 3, Volume II reviews molecular models of angiotensin II, including theoretical models, models relating to physical studies of conformation, models relating to conformationally constrained analogs of angiotensin II, and finally, template models that relate the benzylimidazole nonpeptide angiotensin II antagonists to the peptides. Chapter 3, Volume II discusses the role of the template model in angiotensin II antagonist design that led to SK&F 108566. Chapter 4, Volume II focuses upon the structure–activity studies of nonpeptide angiotensin II antagonists that have led to a variety of potent, orally active clinical candidates.

II. CONSTITUTIONAL ELEMENTS OF ANGIOTENSIN II CRITICAL FOR AGONIST AND ANTAGONIST ACTIVITY

The structure–activity relationships found among angiotensin II agonists and antagonists have been reviewed previously.[9,10] This chapter discusses the important concepts that have been developed from structure–activity relationship studies of angiotensin II agonists and antagonists, especially those that have been important in nonpeptide analog design and that add new data since the early reviews. Since the angiotensin II agonist and antagonist structure–activity relationships hold much in common, both types of activities are discussed together. It has only been in the last few years that receptor binding assays have been employed routinely in the assay of angiotensin II analogs. For this reason, binding assay results are discussed in a separate section (see Section V). It is important to note that many of the concepts that have been developed in the absence of binding data will need to be reexamined. This section focuses upon constitutional elements of structure, i.e., structural elements that do not necessarily define conformation. Conformational elements of structure, which constrain or promote peptide conformation, are discussed in Section III.

For ease of reference, data tables are organized at the end of the chapter. They do not follow the order of discussion but begin with position one modifications, move through to position eight modifications, followed by multiple substitutions, cyclic peptides, angiotensin I analogs, angiotensin II fragments, and end with two comparison tables. The data tables contain many new analogs from our laboratories that have never been published.* Although the tables contain over 300 analogs, they do not include every analog in the literature. The authors believe that exhaustive data tables would overwhelm the reader and detract from their relevance to the text. The authors have attempted, however, to be encyclopedic with regard to the most important aspects of structure–activity relationship and have attempted to note areas that are underexplored. The data tables, furthermore, do not contain all of the data generated on an analog. An attempt has been made to include the data most relevant to the text.

The structure–activity studies on angiotensin II antagonists are summarized in Figure 1. The figure highlights binding groups, and agonist "switches," etc. Notice that, as noted above, the binding groups span the entire structure of angiotensin II and its analogs and are positioned on every other amino acid residue.

*All analogs presented for the first time in this chapter, denoted by the reference,[11] have been prepared by solid phase peptide synthesis by the methods described in previous references, e.g., Samanen et al.,[12] and as in such previous work, all peptides were purified to ≥95% by column chromatography or semipreparative hplc and gave satisfactory amino acid analyses (within 10% of theoretical values).

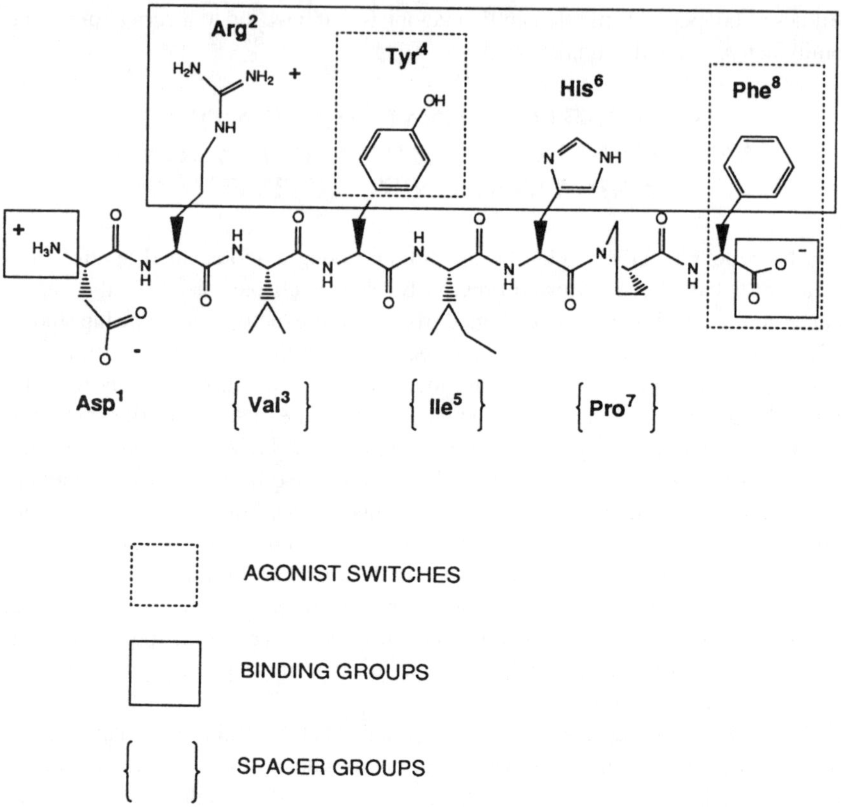

FIGURE 1. Summary of the structure–activity relationship of angiotensin II. Superimposed upon the structure of angiotensin II are boxes delineating functional groups responsible for agonist activity (agonist switches shown in dotted boxes), receptor binding (shown in solid boxes), and sidechains that appear to have a steric role (spacer groups shown in parentheses).

Notice also that, with the exception of phenylalanine Phe^8, the binding groups are relative polar functional groups. In Figure 1, the binding groups all appear to be on one face of the molecule. In Chapter 3, Volume II it will be seen that this may not be the case in the receptor-bound conformation.

Of the many constitutional elements in angiotensin II, the most important are the two ‘‘switches’’ in angiotensin II, which, if modified, turn angiotensin II into an antagonist: tyrosine Tyr^4 and phenylalanine Phe^8 (see Figure 1). Among the first reported antagonists of angiotensin II,[13] was the analog [Phe^4,Tyr^8]-angiotensin II *72*[14,15] (see Table 4)[2,4,16–21] which contained an alteration of the tyrosine in position four. Other weak antagonists were reported shortly after: [(pF)Phe^4]-angiotensin II[16] *73* and [Phe^4]-angiotensin II *1*[13] *71* (see Table 4). In that same period, antagonists of much greater potency were discovered by alteration of the eight position: [Ala^8]-angiotensin II *206,* [Leu^8]-angiotensin II *208,*

[Ile[8]]-angiotensin II *207,* and [Cys[8]]-angiotensin II *209* (see Table 8). In this book, angiotensin II agonists are those that contain Tyr[4] and Phe[8], while angiotensin II antagonists are those analogs that contain a modification to either Tyr[4] or Phe[8] that imparts antagonist activity.

Another critical modification is the replacement of aspartic acid, Asp[1], with sarcosine, Sar. This modification gives both agonists and antagonists with enhanced affinity and potency.[22] Sar[1] antagonists typically display a slower off-rate than angiotensin II.[20] Such slowly reversible antagonists depress both the slope and the maximum of the angiotensin II dose-response curve *in vitro* at high doses but not at low doses. The pA_2 values (i.e., $-\log K_B$) typically reported for Sar[1] analogs are determined at low doses in the range of competitive inhibition and consequently may be overestimated.[20] This type of antagonist activity has come to be described as "insurmountable,"[23] as opposed to "noncompetitive," since increasing doses of an agonist in both cases cannot overcome the antagonist blockade. An insurmountable antagonist, however, binds to the agonist receptor binding site, whereas a noncompetitive antagonist binds in an allosteric mode, i.e., outside the agonist receptor binding site.

Although the development of [Sar[1],X[8]]-angiotensin II antagonists continued,[9] investigations of [Sar[1],X[4]]-angiotensin II antagonists were not reported until recently. This may have been due to the lower potency of the [X[4]]-angiotensin II antagonists (see Table 4) compared to the [X[8]]-angiotensin II antagonists (Table 8).

A. THE PHE[8] AGONIST HYPOTHESIS

1. Modifications to the C Terminus of Angiotensin II That Give Agonists, Mixed Agonists, and Antagonists

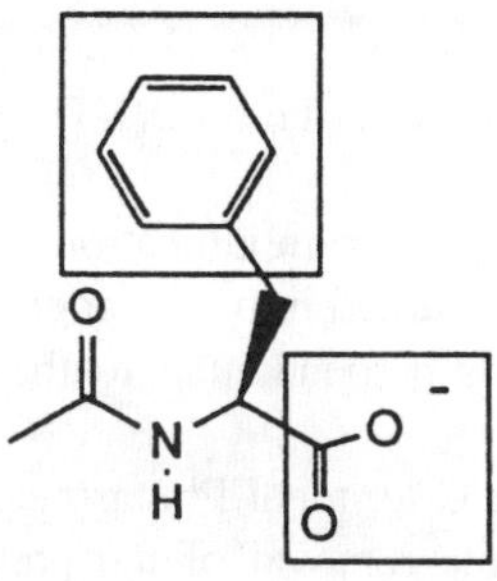

One of the most fascinating amino acid residues in any peptide is the C terminal phenylalanine in angiotensin II. The elegance of the C terminal switch was most fully delineated by Marshall in 1986,[24] in what will be referred to as the "agonist hypothesis." The agonist hypothesis proposes that

1. An angiotensin II octapeptide analog will act as an agonist if the position four residue is tyrosine and the position eight residue contains a phenylalanine of the L-configuration.

2. Any modification to the C terminal Phe that prevents the positioning of the benzyl sidechain over the Pro-Phe peptide bond, shown above and in Figure 2, will yield an antagonist analog.

The structure–activity relationship that supports this agonist hypothesis is given in Table 8. As seen in the table and Figure 2, modifications to Phe that retain full agonist activity include

1. the addition of a methyl group to the alpha carbon to give α-methylphenylalanine, (αMe)Phe, in analog *186*
2. dehydrogenation to give dehydrophenylalanine, ΔPhe, in analog *187*
3. creation of a cyclic structure through attachment of a methylene group to the α amine and ortho carbon on the phenyl ring to give the *L*-tetrahydroisoquinoline-2-carboxylic acid, *L*-Thiq, in analog *188*
4. intercalation of a methylene group between the α carbon and carboxylate groups to give β-aminophenylbutyric acid, Apb in analog *189*

These modifications do not prevent the positioning of the benzyl sidechain over the Pro-Phe amide. Modifications that disrupt this agonist orientation and give potent angiotensin II antagonists include

1. the inversion of configuration to *D*-Phe in analogs *205* and *94*
2. the addition of a methyl group to the Pro-Phe amide to give (NMe)Phe in analog *217*
3. intercalation of a methylene group between the α carbon and the ortho carbon on the phenyl ring to give 2-aminoindane-2-carboxylic acid, Ind, in analog *219*
4. deletion of the β carbon to give phenylglycine, Pgl, in analog *220*
5. inversion of configuration of *L*-Thiq to *D*-Thiq in analog *224*

Perhaps the most elegant of these modifications is the Ind substitution. With this modification, an agonist is turned into an antagonist by freezing the aromatic ring into an orientation that is perpendicular to the desired agonist orientation (Figure 2).

Many modifications to the C terminal Phe aromatic ring can give potent antagonists (see Table 8). Simple removal of the phenyl group gives alanine in analog *206.* A wide variety of alkyl groups can substitute for alanine, to give antagonists. Interestingly, substitution of heteroaromatic rings, such as thienylalanine (Tal), gives partial agonists *195* and *196.* Donating substituents on the aromatic group tends to increase agonist activity, while withdrawing substituents tends to decrease agonist activity, as described by Hsieh et al.[25] Bulky substituents tend to reduce agonist activity.[25] Less important are substituents that are lipophilic. Aromatic character is required in a sidechain for maximal agonist activity. Aliphatic sidechains impart a fair degree of agonist activity, although noncyclic

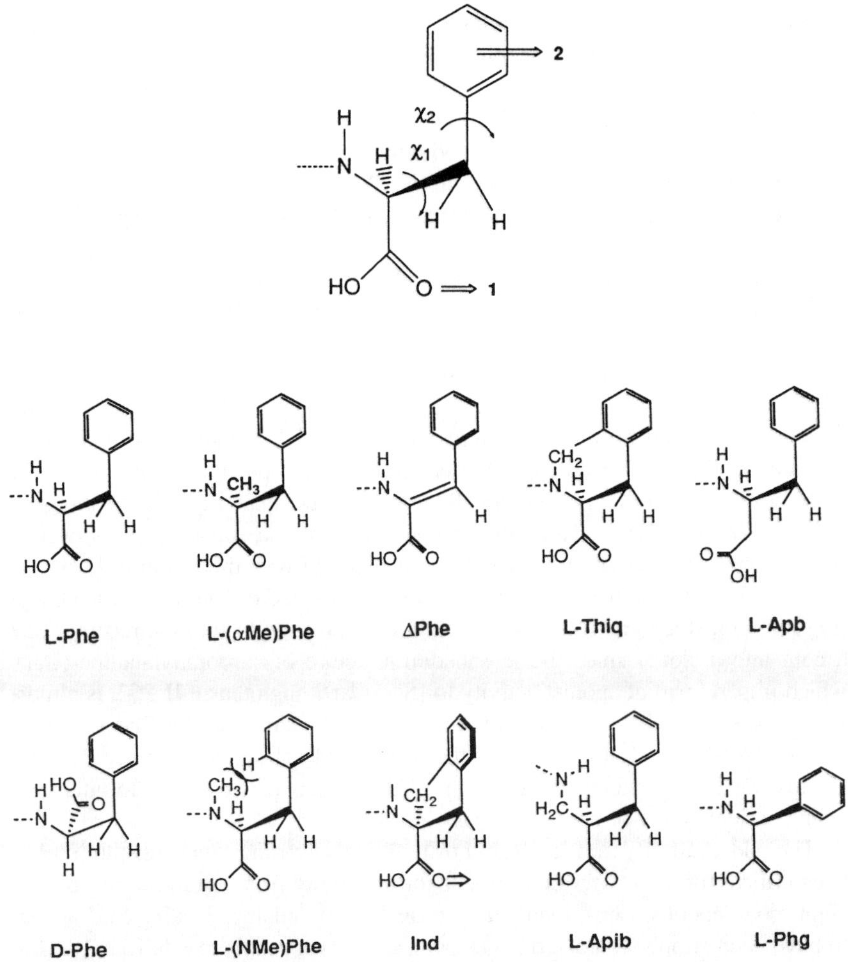

FIGURE 2. The agonist hypothesis. Besides binding sites within the rest of the angiotensin II sequence, the C-terminal carboxylate group interacts with a receptor group **1**, and the phenyl group of phenylalanine interacts with a second group **2.** If the phenyl group is positioned over the Pro-Phe amide group ($\chi_1 = \sim 0$ deg) with the edge of the phenyl ring pointing towards the amide group ($\chi_2 = \sim 0$ deg), interaction of angiotensin II in this orientation with the receptor will produce an agonist response. If either the carboxylate group or the phenyl ring are moved out of these orientations, analog binding to the receptor will not produce an agonist response.

branched aliphatic sidechains totally abrogate an agonist response. Thus, Cha provides 20% agonist activity in analog *198,* while Leu provides <0.1% agonist activity and gives rise to an antagonist analog *208.* Abu provides only 0.5% agonist activity, whereas Nle provides 7.3% agonist activity.[25] ΔOal, which bears a long nonbranched aliphatic sidechain held in the agonist conformation by the double bond, provides potent partial agonist activity in [Sar^1,ΔOal^8]-angiotensin II *199* (55% agonist activity, with a pA_2 of 7.0 in the rat adrenal glomerulosa cell). Electron-rich substituents increase antagonist activity in analogs bearing branched aliphatic sidechains, e.g., (OMe)Ser in [Sar^1,(OMe)Ser^8]-angiotensin II *232* provides an antagonist with exceedingly high antagonist activity (pA_2 9.66). The lipophilic sidechains in [Sar^1,Phe(Br$_5$)^8]-angiotensin II *221,* [Sar^1,Car^8]-angiotensin II *202* and [Sar^1,Hfv^8]-angiotensin II *222* appear to be responsible for the prolonged duration of these analogs (>6 h for the Phe(Br$_5$)^8 analog *221 in vitro*).[26] Intercalation of a methylene group between the α amine and the α carbon gives aminophenylisobutyric acid, Apib, which imparts partial agonist activity to analog *204.*

Critical to the agonist hypothesis is the fact that the agonist activities of the ΔPhe^8 and L-Thiq^8 analogs *187* and *188* are retained, since the only phenyl group orientation allowed by these structures is the eclipsed orientation over the Pro-Phe amide (Figure 2). An eclipsed orientation of the β-phenyl group ($\chi 1 = \approx 0°$) with the Pro-Phe amide is energetically less favorable than a gauche relationship ($\chi 1 = -60$ to $-100°$). Thus, the receptor may play a role in promoting this orientation, i.e., agonists may undergo an induced fit with the receptor. Escher et al.[27] reasoned that the receptor ultimately requires a fixed orientation of the phenyl ring for agonist activity (''the sidechain has therefore to fit into the receptor like a coin into a slot'') since the icosahedral structure of carboranylalanine (Car), which imparts partial agonist activity to [Sar^1,Car^8]-angiotensin II *202,* is almost spherical, essentially containing almost the same volume as the Phe sidechain rotating about its 1′-4′ axis and lacks the slim key-like structure of a phenyl group. Clearly, analogs in which the position eight aryl group is not able to attain the agonist orientation will be antagonists.

The added affinity derived from position one sarcosine (Sar) substitution has been critical for the observation of antagonist activity from a number of position eight modifications. Thus, many early position eight analogs lacking Sar^1 appear to have been simply characterized as poor agonists. For this reason [Sar^1]-angiotensin II analogs bearing Ind, Trp, or Apib should be examined as angiotensin II antagonists.

2. D-Phe^8 Antagonist Structure–Activity Relationship: Topological Relationship to L-Phe^8 Agonists

Studies of the sidechain of D-Phe^8 antagonists have shown, in general, that the sidechain requirements that give rise to potent agonist activity in the L-Phe^8 series give rise to potent antagonist activity in the D-Phe^8 series (see Table 8).[28] There are two types of analogs that are important exceptions to this trend, the diphenylalanine (Dip) analogs and biphenylalanine (Bip) analogs,[29] which will be discussed below. The data in Table 8 allows one to expand upon the agonist hypothesis to consider the geometry of receptor binding groups for both D-Phe^8 and L-Phe^8 analogs (Figure 3). The agonists bearing L-Thiq^8 and ΔPhe^8 define

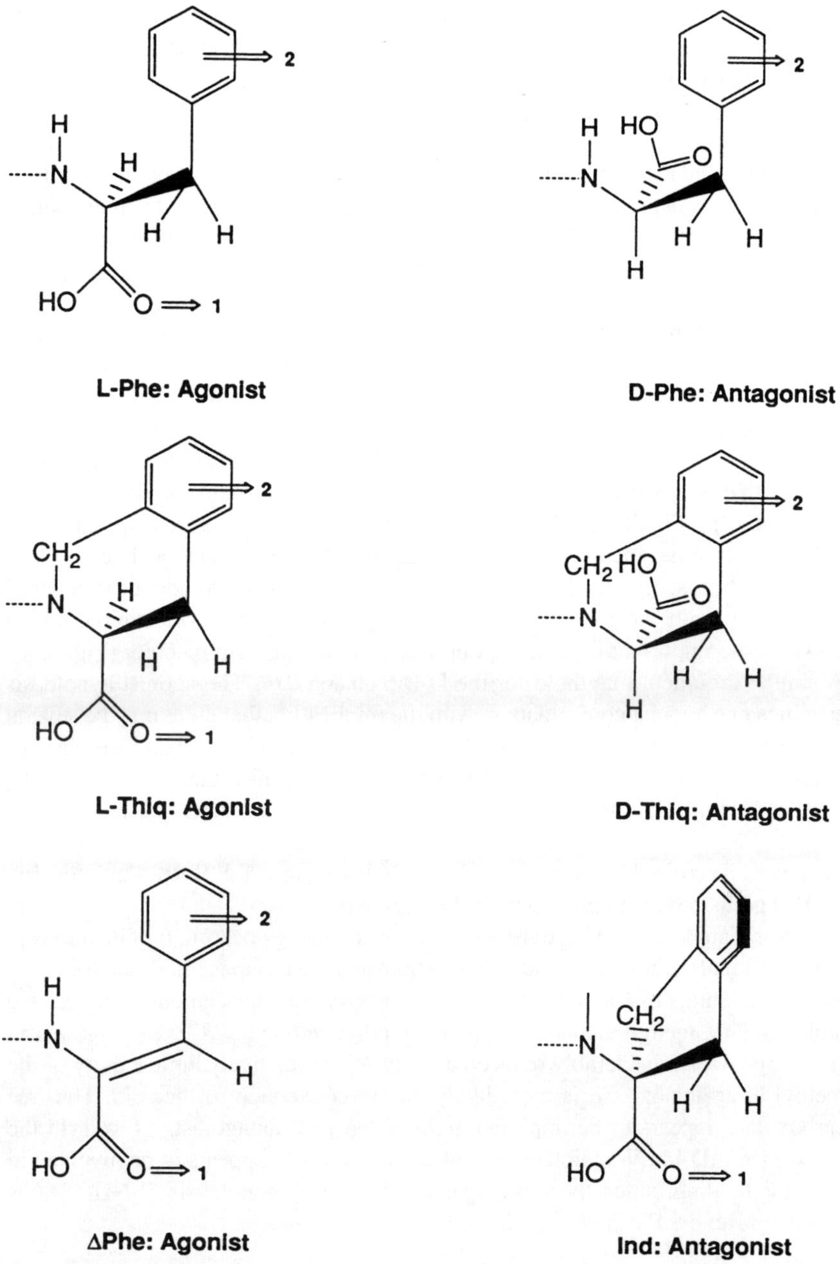

FIGURE 3. The relationship between [L-Phe[8]-angiotensin II] agonists and D-Phe[8]-angiotensin II antagonists, see legend in Figure 2.

the agonist conformation as one in which the β-phenyl group is nearly eclipsed by the α-amino group and *trans* to the C terminal carboxylate (see Figure 3). The D-Phe[8] antagonist shows that disrupting this relationship eliminates receptor stimulation but not high affinity interaction with the receptor. The D-Thiq[8] antagonist alters the orientation of the D-AA[8] carboxylate group, while the Ind[8] antagonist alters the orientation of the β-phenyl group, thereby disrupting receptor stimulation. Hence, as depicted on the right in Figure 3, antagonists arise from disrupting either the orientation of the β-phenyl group or the C terminal carboxylate group. The super-agonist activity of the L-Dip[8] analog *191*[29] suggests that a third agonist group should be considered, another β-phenyl group, as in Figure 4. Difficult to interpret is the reduced agonist activity of the D-Dip[8] analog *192*.[29] One might expect antagonist activity based upon Figure 4. The observation of reduced agonist activity could be indicative of partial agonist activity, which may be difficult to distinguish from pure agonist activity. In either event, the model in Figure 4 suggests that the second β-phenyl group in D-Dip salvages some of the agonist activity that was lost upon reorientation of the C terminal carboxylate group relative to L-Dip. Finally, the L-Bip[8] analog *213* must be considered a special *para*-substituted L-Phe analog. Its potent antagonist activity could be interpreted a number of ways; e.g., the biphenyl group may be too bulky to adopt the eclipsed relationship with the Pro-Phe amide but may adopt another orientation, which prevents the important β-phenyl group from interacting with receptor (Figure 4). A similar argument may hold for the D-Bip analog *214*. The agonist hypothesis specifies one agonist conformation, with the implication that there may be several high affinity antagonist conformations. Not shown in Figure 3 are the various skewed configurations of the β-aryl groups in the antagonists that may be possible. The reader may wish to draw them out to verify that the model still holds.

3. The C Terminal Carboxylate Group is Important for Both Angiotensin II Agonists and Angiotensin II Antagonists

Several angiotensin II agonist and antagonist analogs bearing modifications to the C terminal carboxylate group demonstrate that the group is important for either agonist or antagonist activity (Table 8). Carboxylate replacements, such as the amide in *240*, methyl alcohol in *236*, and methyl ketone in *237*, give angiotensin II analogs with considerably reduced activity.[30] The relatively high activity of the methyl ester analog *238* is most likely due to conversion to the acid. The free carboxylate appears to be important in angiotensin II antagonists as well. In the case of [Sar[1],D-Phe[8]]-angiotensin II *94*, the carboxylate appears to be involved in receptor binding, since the affinity of [Sar[1],D-Phe[8]]-angiotensin II-NH_2 *241* is reduced (Table 8.B).

4. Des-Phe[8] Analogs

Removal of the C terminal Phe from angiotensin II (i.e., angiotensin II (1–7) *351*) markedly decreases activity *in vitro*[31] and in humans.[32] Bovy et al.[33] have found that [Sar[1],des-Phe[8]]-angiotensin II *352* displays moderate antagonist

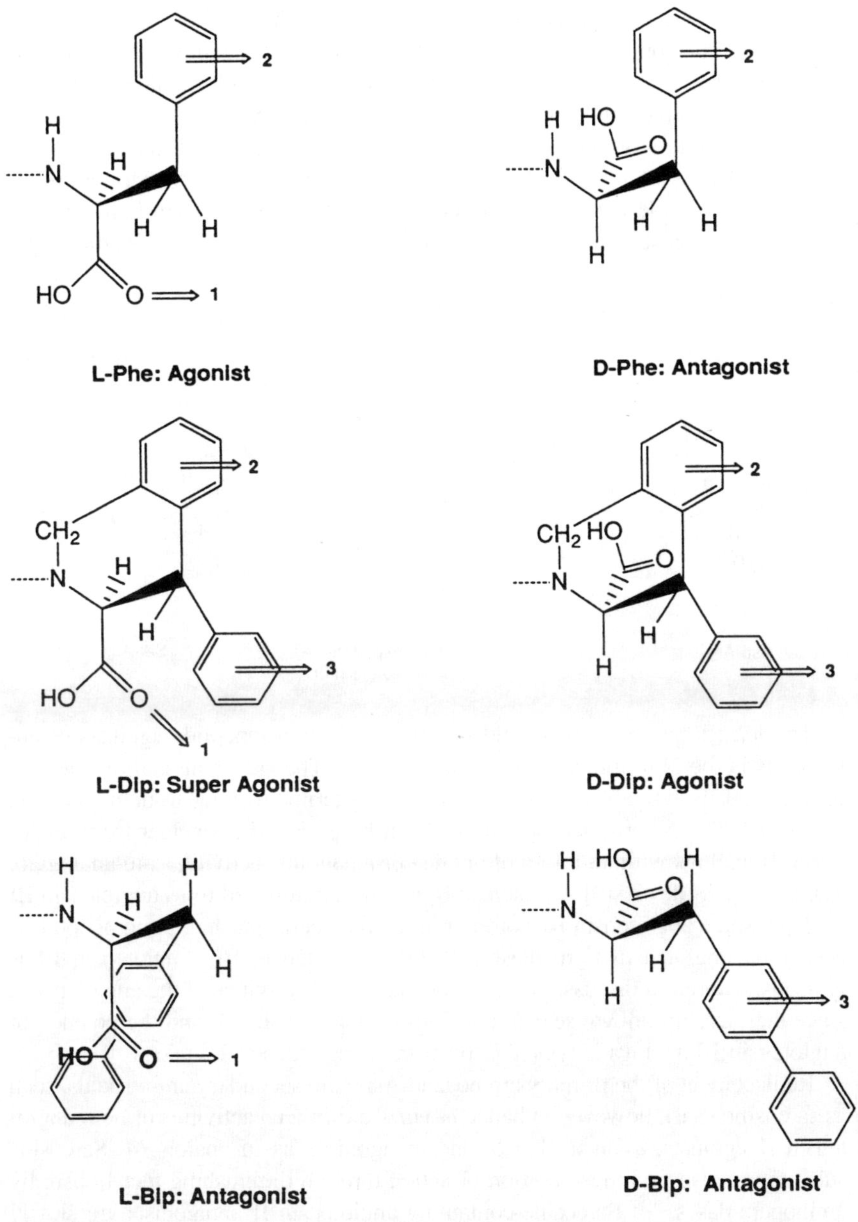

FIGURE 4. The relationship between [L-Bip and L-Dip]^{8}-angiotensin II agonists and [D-Bip and D-Dip^{8}]-angiotensin II antagonists. The nonnative amino acids Bip and Dip potential add a third group **3,** that may interact with the receptor group responsible for agonist activity (e.g., the butyl group in the side chain of a lysine), giving rise to superagonist activity with [L-Dip^{8}]-angiotensin II, and diminished but continued agonist activity with [D-Bip^{8}]-angiotensin II (perhaps due to maintenance of an interaction with the receptor group responsible for agonist activity). The long biphenyl sidechain in [L-Bip^{8}]-angiotensin II may be incapable of interacting with receptor group **2** preventing an agonist orientation of the biphenyl group, yet [D-Bip^{8}-angiotensin II is capable of a different high affinity interaction with the receptor.

activity[33] (see Table 15). Bovy et al. showed that the Pro^7 carboxylate is not involved in antagonist activity, since the Pro^7 amide *353* is equipotent. In fact, the role of proline at least in this series is to constrain the conformation of the preceding His^6 residue, since proline can be replaced with dimethylamine, in analog *356,* with little loss of antagonist activity. Somewhat surprising is the fact that the des-Phe analogs continue to display the residual agonist activity that commonly plagues the peptide antagonists. It should be noted that the pyrrolidine ring in octapeptide agonists may be involved in receptor interaction (see Section II.E).

B. N-TERMINAL ASPARTIC ACID

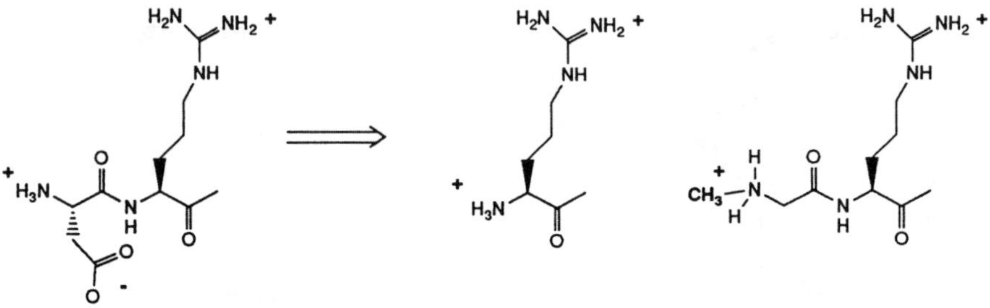

Asp-Arg-

(des-Asp)-Arg-
= Angiotensin III

Sar-Arg-

The least important amino acid in angiotensin II octapeptide agonists or antagonists is the N terminal amino acid (Table 1). The sidechain carboxylate can be removed, as in the Ala^1 analog *5,* as can the N terminal amine as in the succinyl analog *12.* The N terminal aspartic acid can be removed altogether from angiotensin II analogs without a loss of agonist or antagonist activity, as in analog *20,* Table 2. Such [des-Asp^1]-angiotensin II analogs are referred to as angiotensin III analogs, since angiotensin III is found in abundance in plasma. Aminopeptidase acts upon angiotensin II in plasma to give angiotensin III. Angiotensin III is typically ignored in discussions of the physiological activities of the angiotensins, since there are no known selective actions of angiotensin III, and the potency of angiotensin III analogs is typically lower than angiotensin II analogs.

Replacement of the primary amino acid Asp with secondary amino acids, such as sarcosine (Sar), however, enhance *in vitro* and *in vivo* activities of both angiotensin II agonists, as in analog *8,* and antagonists, as in analog *14.* Sarcosine substitution also prolongs duration of action through diminishing metabolism by aminopeptidases.[20,34] Sarcosine-containing angiotensin II antagonists are slowly reversible antagonists, depressing both the slope and maximum of the angiotensin II dose-response curve *in vitro* at high doses but not at low doses.[35] The "pA_2" values typically reported for [Sar^1]-antagonists are determined at low doses in the range of competitive antagonism and may be overestimated. But such pA_2s are of value for comparative purposes. The activities of angiotensin II analogs bearing modifications in other positions are typically enhanced through Sar[1] substitution. For example, the antagonist activity of [D-Phe[8]]-angiotensin II is barely measurable, whereas the antagonist activity of [Sar[1],D-Phe[8]]-angiotensin II is readily measurable. Despite the importance of sarcosine, the literature is devoid of sarcosine analogs. We found that [D-(NMe)Ala[1] and (NEt)Gly[1],Ile[8]]-angiotensin II analogs *17* and *18* did not display enhanced antagonist activities. Thus, position one can be summarized as nonessential, except that sarcosine in position one is an important modification.

The instability of Asp[1] or Asn[1] analogs must be noted. Over time, these analogs undergo rearrangement to the β-Asp analogs, e.g., *2, 3,* or *15,* a fact first observed by Riniker and co-workers.[36] The β-Asp substituent is not recognized by plasma aminopeptidases; hence N terminal Asp removal to give the less active angiotensin III analog is not observed. Thus, Asp[1] or Asn[1] analogs may appear to gain potency over time, since the β-Asp analogs are more potent *in vivo* than their Asp[1] or Asn[1] precursors. Sar[1] analogs do not suffer this problem. Within the angiotensin sequence, this is the only position that is a point of chemical "instability."

C. TYR[4] ANTAGONISTS

1. Modifications to the *p*-Hydroxy Group of Tyr[4]

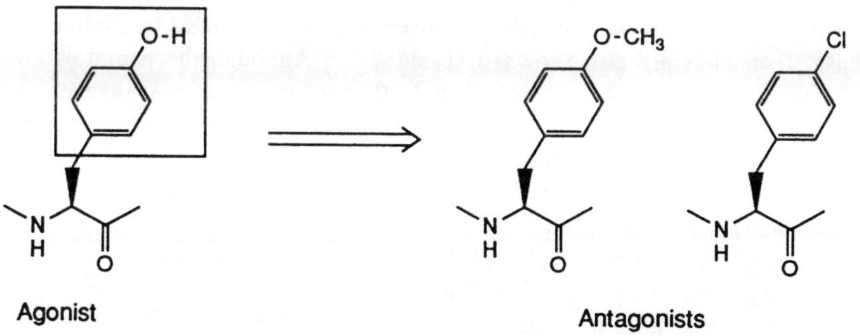

The tyrosine in position four of angiotensin II constitutes one of the agonist "switches," such that a number of modifications to that group will give rise to analogs that display high affinity for the angiotensin II receptor but will display only residual agonist activity. Table 4.A lists the early [Asp[1],(pX)Phe[4]]-angiotensin II analogs that were shown to be angiotensin II antagonists. As seen in the table, none of these analogs displayed comparable potency to the early [Asp[1],X[8]]-angiotensin II analogs. For this reason, the development of [(pX)Phe[4]]-angiotensin II antagonists lagged behind the [X[8]]-angiotensin II, until the more recent discovery that [Sar[1],(pX)Phe[4]]-angiotensin II analogs were more potent.[18,37] Table 4.B contains a number of [Sar[1],(pX)Phe[4]]-angiotensin II analogs rank-ordered with potent agonists at the top, potent antagonists at the bottom, and analogs with mixed agonist/antagonist activities in the middle. Table 4.C contains a number of [Sar[1],(pX)Phe[4],Ile[8]]-angiotensin II analogs, with potent antagonists at the top and weak antagonists at the bottom. These analogs have been reported previously by several laboratories.[17,37,38,39] The pX substituent inductive effects[40] σ_p, lipophilicities[40] π, and hydrogen bonding characteristics H^+ and H^-, are also listed in Table 4.C. Hydrogen bonding and the inductive effects (electronegativity) of the *para*-substituent have been shown to be important for optimal *agonist* activity of position four analogs of [Sar^1]-angiotensin II.[17,38] The same correlation between pX structure in [Sar^1,(pX)Phe^4]-angiotensin II and *antagonist* activity cannot be made (Table 4.B). The data suggests, however, a general trend:[39] position four *para*-substituents that are hydrophilic and capable of donating a hydrogen atom in a hydrogen bond promote agonist activity, while *para*-substituents that are hydrophobic and incapable of donating a hydrogen atom promote antagonist activity. These antagonist properties were found to be optimal in the *p*-chloro substituent.[39] The resulting analog [Sar^1,(pCl)Phe^4]-angiotensin II *85* is a potent angiotensin II antagonist *in vivo.*

The pX substituents which promote antagonist activity in the [Sar^1,(pX)Phe^4]-angiotensin II series were unfavorable in [Sar^1,(pX)Phe^4,Ile^8]-angiotensin II analogs, Table 4.C.[39] Angiotensin II analogs that are antagonists by virtue of an alteration in position eight require a position four agonist side chain, Table 4.D. Concurrent modifications of position four and eight do not give rise to potent antagonists with reduced partial agonist activity.[39]

The *in vivo* partial agonist activities of some of the [Sar^1,(pX)Phe^4,Ile^8]-angiotensin II analogs are also shown in Table 4.C.[26] Although the partial agonist activities are lower in the analogs *87* and *89* to *92,* the antagonist potencies of these analogs are correspondingly lower, both *in vitro* and *in vivo.* Partial agonist activity will be discussed further in a later section.

Potent antagonists to angiotensin II have also been obtained with D-amino acid substitutions in position eight.[28] Table 4.D contains a series of [Sar^1,(pX)Phe^4,D-AA^8]-angiotensin II analogs. The (pF)Phe^4 modification increases *in vitro* and *in vivo* antagonist activity in [Sar^1,(pF)Phe^4,Phe^8]-angiotensin II *83* but reduces *in vitro* and *in vivo* antagonist activity of the corresponding Ile^8 analog *91.* It also reduces *in vitro* and *in vivo* antagonist activity in the corresponding D-Phe^8 analog *97.* It appears, therefore, that concurrent antagonist modifications of position four and eight do not give rise to potent angiotensin II antagonists with reduced partial agonist activity. Interestingly, it will be seen that this negative effect of concurrent modifications in angiotensin II analogs has been observed in other analogs.

The apparently different receptor requirements for position four agonist and antagonist analogs suggests that a position four agonist side chain (pX^h, where h = hydrophilic) interacts with receptor functionality that is different from that which interacts with a position four antagonist side chain (pX^l, where l = lipophilic). This hypothesis may be useful in rationalizing the activities of benzylimidazole nonpeptide antagonists (Chapter 3, Volume II). In the present study, the octapeptide sequence [Sar^1,(pX^l)Phe^4]-angiotensin II appears to reach maximal antagonist activity when X = Cl. It remains to be seen whether alternate sequences bearing a chlorophenylalanine group might provide angiotensin II antagonist activity with greater potency.

D. ARG², HIS⁶: CRITICAL HYDROPHILIC RECEPTOR BINDING GROUPS

For both angiotensin II agonists and antagonists,[41] the arginine Arg² guanidinium and the histidine His⁶ imidazole groups are important for high affinity receptor interactions.

1. Position Two Arginine

As seen in Table 2, either the primary amine of lysine or the isosteric but neutrally charged urea of citrulline are poor replacements for guanidine in angiotensin II agonists[42,43] or in angiotensin II antagonists.[41]

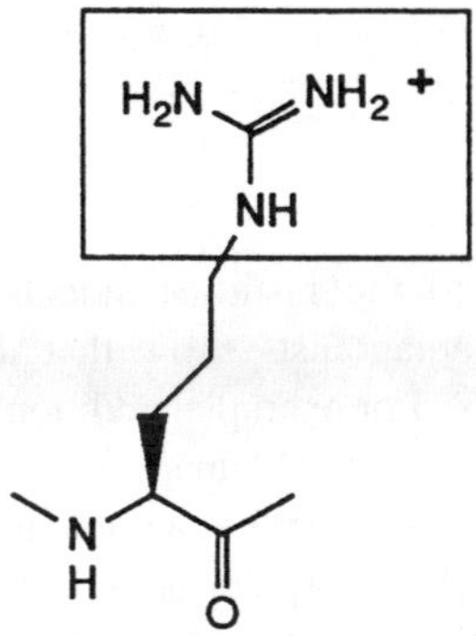

Similarly, lysine is a poor replacement for arginine in the [des-Asp¹]-angiotensin II agonist *22*,[44] but D-Lys is a good replacement in analog *23*. In fact, simple D-amino acids, such as D-Ala or D-(NMe)Ala, may be substituted for Arg in [des-Asp¹]-angiotensin II or [des-Asp¹,Ile⁸]-angiotensin II.[41] The D-(NMe)Ala substitution in the corresponding agonist *33* and antagonist *48* retain more *in vivo* activity than *in vitro* activity. These results were ascribed to greater resistance of D-(NMe)Ala to aminopeptidase activity than other substitutions.[41] Bovy et al. studied the cationic chain length requirements in a series of [des-Asp¹,desNH$^{\alpha}$-AA²,Ile⁸]-angiotensin II analogs *34* to *38*.[45] Antagonist activity was optimal with ε-Ape (ε-aminopentanoic acid) in analog *35*.

Finally, Soffer et al.[46] have described a series of potent octapeptides bearing Lys-Gly in place of the N terminal Asp-Arg and Ser or Ala in place of Pro in position seven, e.g., *49* and *50*. Although the origin of these peptides was not described, one could imagine how the Lys¹ ε-amine could reach the same region of space occupied by the distal amine on Arg² in angiotensin II.

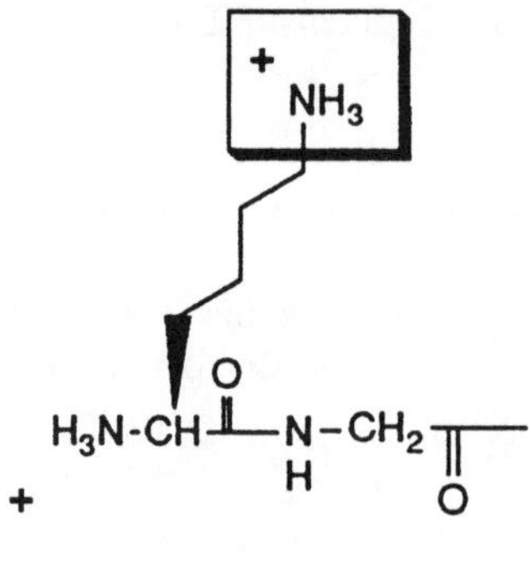

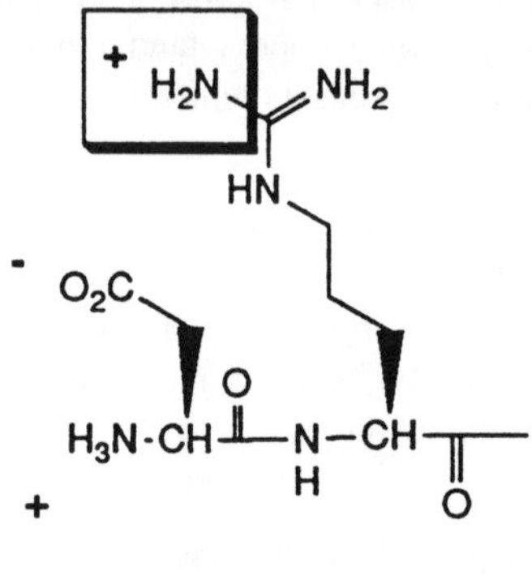

Newer cationic amino acids, such as (*p*-aminomethyl)phenylalanine[47] or the 4-aminoalkylprolines[48] have not been incorporated into position two of angiotensin II agonists or antagonists.

2. Position Six Histidine

Studies of modifications to the histidine sidechain in both angiotensin II agonists and angiotensin II antagonists show that the His^6 sidechain is quite sensitive to alteration (Table 6). For example, addition of a methyl group to either imidazole nitrogen gives angiotensin II analogs with reduced agonist activity: [(N-πMe)His^6]-angiotensin II* *144* 0.05% angiotensin II pressor activity, [(N-τMe)His^6]-angiotensin II *143* 5.0% pressor activity.[49] These data would suggest that the π^{im}-N is involved either in interaction with the receptor or in an intramolecular hydrogen bond that stabilizes the receptor bound conformation.

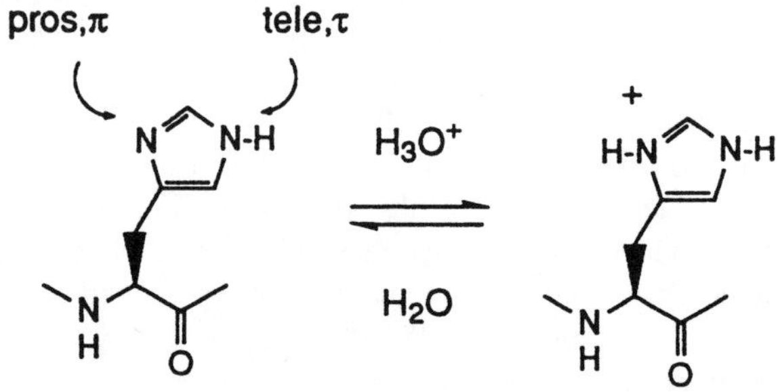

[His^6]-Angiotensin II pKa 6.3-6.4

*By IUPAC convention, the two imidazole nitrogens have been designated pros, π, or "near" to the connecting His beta carbon, and tele, τ, or "far" since biochemists had defined the position one nitrogen as the π nitrogen even though by the chemist's nomenclature the π nitrogen is position three.

Among the aryl group modifications in Table 6, summarized below, pyrazole is the next best heteroaryl group. [Pza[6]]-angiotensin II *142* displays 20.7% relative agonist activity in the rat uterus and 49.6% relative activity in the rat blood pressure assay.[50] With a pKa of 2.2 the pyrazole substitution is not protonated at neutral pH. The relatively high activity of the Pza analog suggests, therefore, that the imidazole sidechain is not protonated in the receptor bound conformation. The fact that the most potent analogs with heterocyclic amino acids in position six all contain a π-nitrogen or a corresponding π-heteroatom further supports the importance of the π-nitrogen.

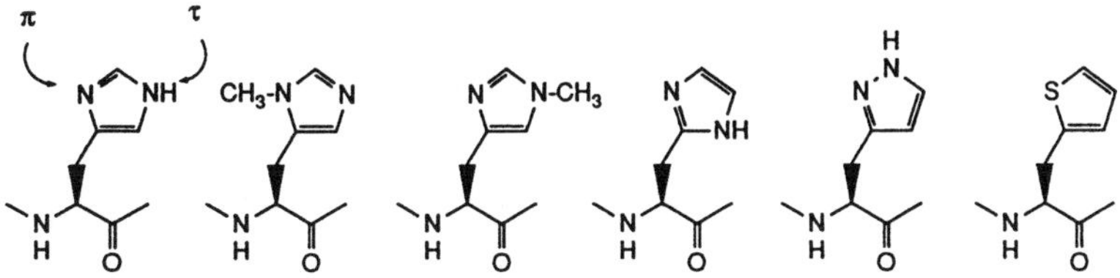

	His	(πMe)His	(τMe)His	Ima	Pza	Thi
[Asn[1]]-AII Pressor:	100	0.05	5.0	0.4	49.6	19.3

Juliano and Paiva[51] found that the histidine pKa in [Asn[1]]-angiotensin II-NH_2 (6.14) is lower than either [Asn[1]]-angiotensin II (6.30) or angiotensin II (6.40), which the authors took as evidence that the His[6] sidechain interacts with the C terminal carboxylate. Both angiotensin II agonists and antagonists display the same rank ordering of modifications: His ≫ Tal > Phe, (N-3Me)His ≫ Lys (Table 6).[41]

Pottorf et al.[52] found that Gln[6] substitution was less deleterious than Asn[6] substitution in a series of angiotensin II analogs, e.g., analogs *150* and *151* (Table 6). Speculating that Gln and Asn substitutions test the role of the τ and π nitrogens in His respectively, they conclude that the τ-NH is important for both receptor binding and for agonist activity in angiotensin II analogs. Despite the high affinity of [Gln[6]]-angiotensin II *151* (IC_{50} 67 n*M*) and low agonist activity (90% contraction at 10 μ*M*), there was no mention of potential antagonist or partial agonist activity.

The fact that antagonist activity has not been described in previous position two or position six analogs of angiotensin II suggests that neither residue is involved in receptor stimulation.[41] The histidine residue could serve one or both of two roles in agonist and antagonist receptor binding to receptor: (1) direct interaction with receptor ligands or (2) intramolecular interactions to maintain a proper conformation for receptor binding. Bioassay results cannot distinguish between the two roles. Conformational studies of angiotensin II in solution have suggested possible interactions between either the tyrosine sidechains[53] or the C terminal carboxylate group with the histidine sidechain.[51] Such possibilities will be examined in greater detail in Chapter 3, Volume II.

E. VAL[3], ILE[5], PRO[7]: LIPOPHILIC SPACER GROUPS

None of these amino acids appear to be involved in primary interactions with receptor, since removal of any of these sidechains is significantly less detrimental than removal of the sidechains in positions two, four, six, or eight. This group of amino acids may be rank ordered in importance as follows: $Val^3 < Ile^5 < Pro^7$.

1. Position Three Valine

The fact that alanine may be substituted into position three in both angiotensin II and [Sar^1,Ile^8]-angiotensin II with only modest losses in activity (Table 3), suggests that position three is of minimal importance.

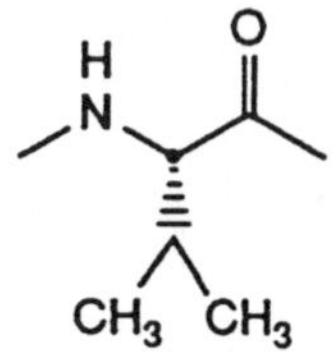

A number of lipophilic amino acids may be substituted into position three in [Sar^1,Ile^8]-angiotensin II without a loss of *in vivo* potency. The polar amino acids Glu and Lys, however, reduce *in vitro* activity in analogs *69* and *70* (Table 3.C).

2. Position Five Isoleucine

Amino acids with lipophilic sidechains that contain more than one group on the β-carbon, i.e., a β-branched hydrocarbon moiety, are required in position five of angiotensin II analogs with potent agonist activity (Table 5.A).[54,55]

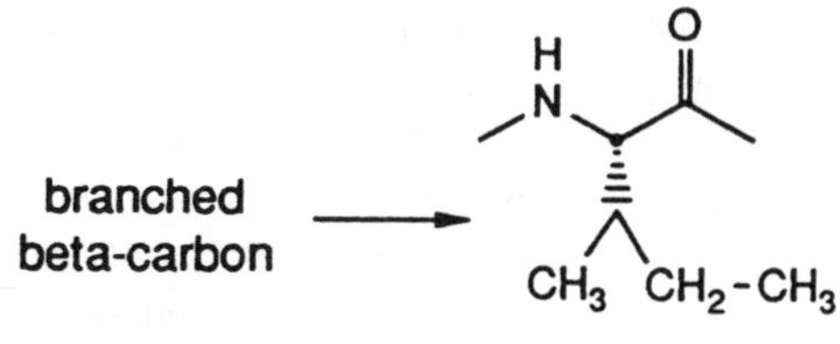

This requirement for agonist activity does not follow for angiotensin II analogs with potent antagonist activity. Straight chain amino acids may be substituted into position five of [Sar^1,X^5,Ile^8]-angiotensin II with retention or enhancement of antagonist activity, e.g., (X,pA_2 rabbit aorta) Phe, 9.15; Tyr, 9.6; His 9.0; Arg, 9.0; Glu, 9.0; Nle, 8.85, in analogs *117, 118, 120, 122, 123, 124* (Table 5.C) compared to Ile, 9.1 in analog *16.*[56] Nonetheless, β-branched sidechains can still enhance the antagonist activities of [Sar^1,X^5,Ile^8]-angiotensin II analogs, e.g., X^5 = (βMe)Phe, pA_2 9.3 in analog *130,* or X^5 = (OMe)Thr, pA_2 9.7 in analog *127.* Other β-MeO amino acids would be worthy of investigation.

3. Position Seven Proline

The proline pyrrolidine ring plays an important conformational role in angiotensin II, which will be discussed in Section III.A. The importance of the pyrrolidine ring is implied by the reduced activities of [Ala[7]]-angiotensin II *160* (Table 7.A).

For this reason, structure–activity studies in position seven have focused upon amino acids that constrain conformation. Table 7.D contains some [Sar[1],X[7],Ile[8]]-angiotensin II analogs that have never been published but extend the scope of our knowledge about position seven in angiotensin II antagonists. As seen in the table, [Sar[1],X[7],Ile[8]]-angiotensin II analogs *175* and *176,* bearing amino acids with cationic (Arg) or anionic (Glu) sidechains, are essentially inactive. [Sar[1],X[7],Ile[8]]-angiotensin II analogs *180-183,* bearing amino acids with aromatic sidechains, are also inactive, while [Sar[1],X[7],Ile[8]]-angiotensin II analogs *177, 178,* and *184,* bearing amino acids with aliphatic sidechains, are almost as potent as the native Pro[7] analog. We have no explanation for this extreme selectivity for aliphatic over aromatic sidechains. Nonetheless, these results support the hypothesis that position seven prefers lipophilic amino acids. Thioproline in position seven of [Sar[1],Tpr[7]]-angiotensin II gives a superagonist *165 in vitro,* (Table 7.B) and a more potent antagonist *171 in vitro* in [Sar[1],Tpr[7],Ile[8]]-angiotensin II.

F. ANGIOTENSIN I ANTAGONISTS

Angiotensin I, the precursor to angiotensin II, is generally considered to be inactive. This inactivity is difficult to demonstrate, however, due to rapid cleavage to angiotensin II in plasma by angiotensin converting enzyme.

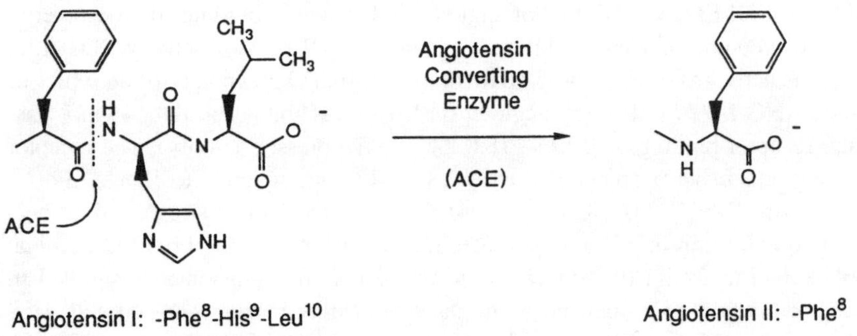

Angiotensin I: -Phe^{8}-His^{9}-Leu^{10} Angiotensin II: -Phe^{8}

Thus, angiotensin I displays high levels of agonist activity in the rat, (compound *319* in Table 14). Angiotensin I *in vitro,* however, displays lower levels of activity.

With the advent of radioreceptor assays for angiotensin II analogs, it has been possible to demonstrate that angiotensin I displays genuine affinity for the angiotensin receptors, albeit at lower affinity, IC_{50} 67 n*M* (rat mesentery), compared to 3.2 n*M* for angiotensin II (Table 14). Unlike angiotensin II analogs, Sar^1 substitution in [Sar^1]-angiotensin I *320* does not display enhanced activity or affinity (Table 14). Another logical extension from the angiotensin II structure–activity relationship would be the consideration of angiotensin I analogs bearing modifications to position eight. Such modifications, of course, could give angiotensin II antagonists. An early study with [D-Phe^8]-angiotensin I *321* found that this analog was devoid of agonist, antagonist, or angiotensin converting enzyme inhibitory activity.[57] [Sar^1,Cys(SMe)^8]-angiotensin I is a potent angiotensin II antagonist, both *in vitro* and *in vivo.*[58] Its activity appears to derive from the action of angiotensin converting enzyme upon the compound to give [Sar^1,Cys(SMe)^8]-angiotensin II *230* (Table 8.A), since the angiotensin converting enzyme inhibitor, captopril, completely abolishes antagonist activity both *in vitro* and *in vivo.* Jackson recently reported an investigation of the inhibitory effects of [X^8]-angiotensin I analogs on angiotensin I and angiotensin II challenge.[59] [Ile^8, Cys^8, Val^8, and Leu^8]-angiotensin I *327* to *330* inhibited the pressor response to either angiotensin I or angiotensin II, and to a lesser extent they inhibited the pressor response to angiotensin II in the presence of captopril (Table 14). None of the compounds appeared to inhibit ACE. The differences in IC_{50}s between non-treated and captopril-treated rats that were reported for these compounds varied considerably. [Val^8 and Leu^8]-angiotensin I *329* and *330* displayed the lowest variation, suggesting that these compounds had a direct action on the angiotensin II receptor. These results led Jackson to propose the investigation of corresponding mutant angiotensinogens (which would yield angiotensin I antagonist analogs) as a potential gene therapy for hypertension.

The Sar^1 analog, [Sar^1,D-Phe^8]-angiotensin I, *322* was recently shown[60] to display potent antagonist activity against angiotensin II. The analog *322* displays potent *in vitro* antagonist activity in the rabbit aorta, ID_{50} 10 n*M* (30.2% rel. angiotensin II ED_{50}), inhibition of angiotensin II receptor binding (rat mesentery), IC_{50} 16 n*M,* and *in vivo* inhibition of angiotensin II pressor activity, ID_{50} 13.8 n*M.* The activities of *322* compare favorably with the known agonist activities of [Sar^1]-ANG I, *320* (9.0% rel. angiotensin II ED_{50}, rabbit aorta; ID_{50} 43 n*M,* vascular receptor binding, and ED_{50} 18.0 n*M in vivo* pressor activity). The receptor binding and *in vitro* antagonist activities of *322* are superior to [Sar^1,D-Phe^8]-angiotensin II-NH_2 *241* (IC_{50} 2000 n*M*), furthermore, which suggests that the -His-Leu extension in *322* compensates for the lack of the important C terminal carboxylate in *241.* Both sidechains in the -His-Leu extension are involved in receptor binding, as evidenced by the lack of affinity in the -Ala-Leu and -His-Ala analogs *325* and *326* of [Sar^1,D-Phe^8]-ANG I (IC_{50} 564 n*M* and $>10^5$ n*M,* respectively). In contrast, the sidechains in the -His-Leu extension of [Sar^1,Phe^8]-ANG I have only a minor role in receptor binding, as evidenced by the retention of affinity in the corresponding -Ala-Leu and -His-Ala analogs *323* and *324* of

[Sar^1,Phe^8]-ANG I (IC_{50} 8.98 and 2.23 n*M*, respectively). Although the *in vitro* antagonist activities of *322* are superior to *241,* the *in vivo* antagonist activity of *322* is comparable to *241.* Unlike other C-terminal carboxamides (e.g., angiotensin II-NH_2 *240* and [Sar^1,Ile^8]-angiotensin II-NH_2 *242*), the carboxamide *241* is equipotent *in vivo* with its corresponding C terminal acid analog. Surprisingly, both *322* and *241* are weak inhibitors of angiotensin-converting enzyme.[60] The fact that angiotensin-converting enzyme inhibition could be detected with the Sar^1-containing analog *322* but not the Asp^1-containing 321 suggests that the N terminus of angiotensin I may be important for interaction with the enzyme. Analog *241* is the first angiotensin II analog to demonstrate angiotensin-converting enzyme inhibitory activity. Analog *322* constitutes the most dramatic example of binding potentiation (200-fold) by Sar^1 substitution in any angiotensin analog.

G. NONADDITIVITY OF OPTIMIZED SIDECHAINS: MULTIPLE BINDING MODES?

One of the more astonishing aspects of the structure–activity studies is that modifications that are optimum individually in angiotensin II analogs lose their impact upon multiple combinations of such modifications (Table 9). For example, the modification (OMe)Thr^5 enhanced the activity of [Sar^1, Ile^8]-angiotensin II in analog *127* (Table 5.C) but had no impact upon the activity of [Sar^1, D-(2)Nal^8]-angiotensin II in analog *246* or of [Sar^1, Tpr^7, Ile^8]-angiotensin II in analog *245.* Similarly, the modification Tpr enhanced the activity of [Sar^1, Ile^8]-angiotensin II in analog *171* (Table 7.C) but had no impact upon the activity of [Sar^1, D-(2)Nal^8]-angiotensin II in analog *248.* Some other multiple substitutions are shown in Table 9. The lack of enhanced activity in these analogs casts doubt on the discovery of analogs bearing combinations of substituents with enhanced antagonist activity over [Sar^1, Ile^8]-angiotensin II. Some of these combinations were aimed at providing analogs with diminished residual agonist activity, e.g., [Sar^1,(pF)Phe^4,D-Phe^8]-angiotensin II *97* and [Sar^1,D-Arg^2, (OMe)Thr^5, Ile^8]-angiotensin II *243,* but the analogs displayed antagonist activities that were lower than [Sar^1, Ile^8]-angiotensin II.

If the final receptor-bound conformation for [Sar^1, Ile^8]-angiotensin II was a static conformation, it would seem logical that one or more of these multiple combinations might have shown enhanced activity. The fact that this was not observed suggests that another scenario might hold, e.g., that angiotensin II agonists and antagonists bind to the receptor in **several binding modes.** If the observed affinity of an angiotensin II agonist or antagonist is the sum of all binding modes, then analogs that bear multiple substitutions might not favor any particular binding mode and give rise to lower affinity. In other words, a particular modification that enhances one binding mode could be detrimental in another binding mode. This problem will be discussed further in Sections III and V.

H. MODIFICATIONS TO REDUCE ENZYMATIC DEGRADATION

A number of investigators have attempted to modify angiotensin II agonists and antagonists in ways that increase resistance to enzymatic degradation.[28,61–75]

A practical approach would be to replace the peptide bonds (ψ) with isosteric groups, such as olefinic groups or reduced amide groups (Figure 5).[76] As seen in Table 10, the single replacement of (ψ1,2) with $-CH_2NH-$ in [Sar[1],Leu[8]]-angiotensin II gave an analog *251,* which retained antagonist activity.[77] The double replacement of (ψ1,2) and (ψ3,4) with $-CH_2NH-$ in [Sar[1],Leu[8]]-angiotensin II gave an analog *252,* which also retained antagonist activity.[77] The triple replacement of (ψ1,2), (ψ3,4), and (ψ4,5) with $-CH_2NH-$ in [Sar[1],Leu[8]]-angiotensin II, however, gave an analog *253,* which lost two to three orders of magnitude in antagonist activity.[77] When all amide bonds were replaced with methylene amine groups, the resulting analog *254* was inactive.[78] Individual reduced amide replacements at (ψ5,6) and (ψ7,8) have not been reported. Olefin or other amide bond replacements into angiotensin II analogs have not been reported.

Another approach is to replace angiotensin II residues that are susceptible to peptide bond cleavage with amino acids that reduce recognition of the peptide as a substrate for the enzyme; e.g., replacement of Asp[1] with Sar[1] reduces aminopeptidase degradation,[22,79] or Tyr[4] with (αMe)Tyr[4] reduces chymotrypsin degradation.[74] These modifications have yielded analogs that resist enzymatic degradation.[22,79,74] Both modifications may be blocking enzymatic degradation by altering peptide backbone structure. Alteration of amino acid sidechain structure could lead to lack of recognition by a protease, e.g., replacement of Arg[2] with a noncationic amino acid or Tyr[4] with an alkyl amino acid, but such modifications in angiotensin II give analogs that are devoid of agonist or antagonist activity.[39,41] N-methyl and C-alpha methyl amino acids are also known to reduce the conformational flexibility of peptides.[80,81] The substitution of these amino acids into angiotensin II analogs to constrain conformation will be discussed in Section III.

The substitution of D-amino acids for L-amino acids in peptides can also diminish proteolytic degradation. With the exception of positions one and eight, single D-amino acid substitutions in [Sar[1],Ile[8]]-angiotensin II gave rise to angiotensin II antagonists with reduced potency (Tables 2 through 7). A number of multiple D-amino acid substitutions have been attempted (Table 12), none of which retained the potency of [Sar[1],Ile[8]]-angiotensin II. In this context, the retention of modest agonist activity in the so-called retroenantiomeric angiotensin II analogs *258* and *259*[82] (Table 10) is astonishing. In a retroenantiomeric peptide, the chirality of the amino acids is reversed, while the sequence is retained. Thus, a diaminoalkane is placed at the N terminus, and a malonate derivative is placed at the C terminus. The resulting peptide contains the same sequence, but the amide groups are reversed relative to the native L-amino acid-containing peptide. The angiotensin II analogs *258* and *259* contain D,L-α-benzylmalonate (D,L-Bml) in place of phenylalanine in position eight and either β-alanine (β-Ala) or D-alanine (D-Ala) in place of proline in position seven, to adjust for the fact that the pyrrolidine ring cannot be accommodated readily in the retroenantiomeric analogs. Although diaminoalkane replacements for the N terminal amino acid have been incorporated successfully into retroinverseo peptides, the authors chose β-alanine

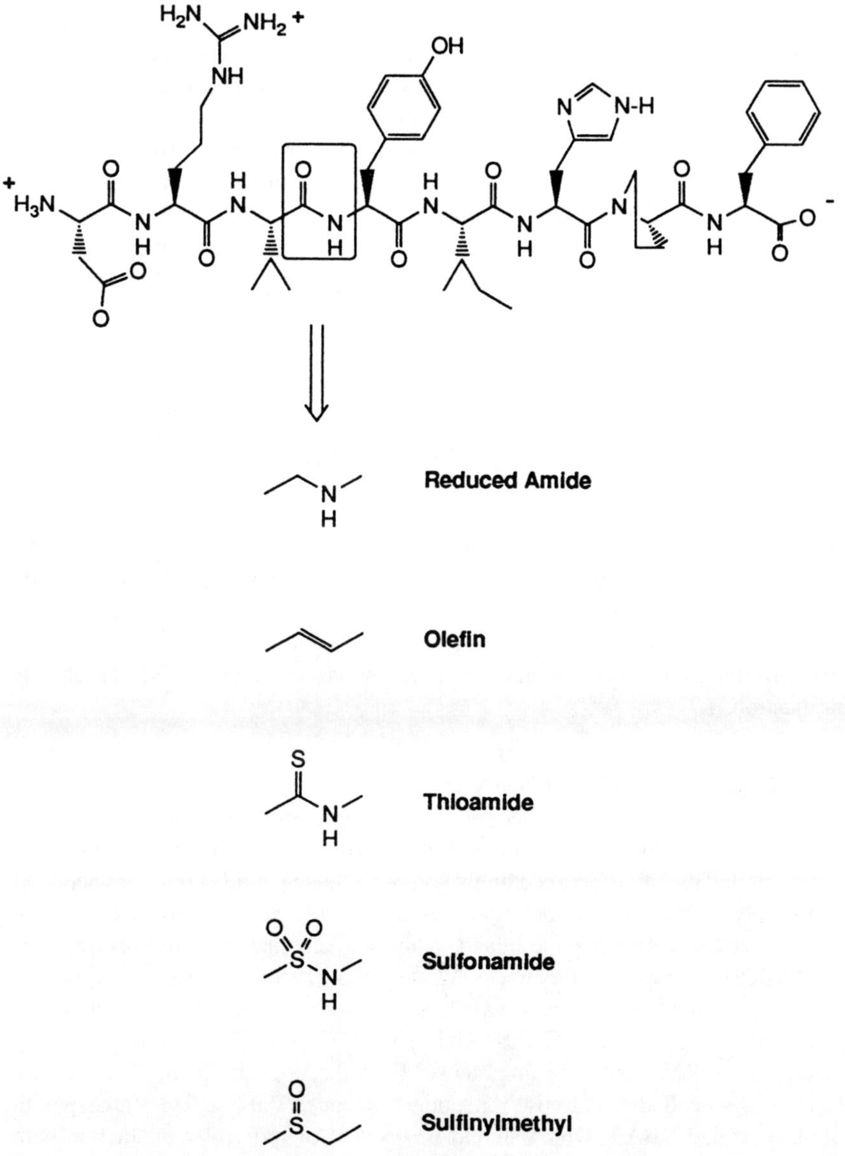

FIGURE 5. Peptide bond replacements.[76] Examples of functional groups that have been substituted for amide groups in peptides, shown here in relation to the amide group between Val^3-Tyr.[4]

as a des-amino aspartic acid version of the N terminal Asp. The ''retention'' of agonist activity in these analogs (see footnote i, Table 10) suggests that if any of the backbone amide groups are involved in receptor binding, their reversal does not eliminate receptor interaction. Surprisingly, these analogs were not evaluated as angiotensin II antagonists, even though the retroenantiomer of [D-Phe^8]-angiotensin II was contained apparently in the diastereomeric mixture of peptides resulting from D,L-Bml incorporation.

I. ANALOGS FOR RECEPTOR LABELING AND CROSSLINKING

Critical to the identification, isolation, and purification of angiotensin II receptors (Chapters 2 and 3 of Volume 1) has been the development of appropriate angiotensin II analogs that may be crosslinked to the receptor following receptor binding to tag the protein with radiolabel and/or to provide an affinity handle for affinity purification of the angiotensin II receptor. The addition of functionality to angiotensin II analogs to accomplish each task is not trivial, despite the abundance of success in the literature with such pursuits for other peptide receptors. For such uses, (1) angiotensin II analogs have been radioiodinated and tritiated[83] (the mono-iodo analog of angiotensin II retains high affinity and is most typically employed[84]); (2) a diverse set of analogs has been studied in a number of laboratories (this field was thoroughly reviewed recently by Escher[83]); and (3) a series of biotinylated angiotensin II analogs has been described for affinity purification.[85]

J. ANTISENSE ANGIOTENSIN II

On a molecular level, the recognition between complementary nucleotide sequences is highly defined. In contrast, the recognition between complementary hormones and receptors is exceedingly complex. Since peptide hormones and their receptors arise from nucleotide sequences, a molecular recognition code has been hypothesized in which peptide ligands and complementary binding peptides can be encoded by complementary nucleotide sequences.[86] Specifically, the hypothesis proposes that the complementary mRNA code for a peptide ligand, such as angiotensin II (CUG-GCG-CAU-AUG-UAG-GUG-GGG-AAA, obtained by reading the mRNA code for angiotensin II in the opposite 3′ to 5′ direction), encodes a so-called ''antisense'' peptide sequence, Val-Ala-Tyr-Val-Asp-Val-Gly-Lys, called ''IIA,'' which should be capable of molecular interaction with angiotensin II.[87] In fact, IIA inhibits ^{125}I-angiotensin II binding to rat adrenal particles with a $K'_d \sim 50$ n*M*. Antibodies raised to IIA, which are thought to mimic angiotensin II, specifically inhibit ^{125}I-angiotensin II binding to rat adrenal particles with a $K'_d \sim 400$ n*M*. This antibody inhibits aldosterone secretion from rat adrenal glomerulosa cells and has been employed in immunoaffinity purification of an angiotensin II binding protein that appears to be identical to angiotensin II receptors purified by other techniques.[87] The relationship of antisense peptides to native peptide receptors remains to be elucidated.

III. THE SEARCH FOR CONFORMATIONALLY CONSTRAINED PEPTIDES THAT DEFINE THE RECEPTOR-BOUND CONFORMATION

The linear octapeptide angiotensin II is a flexible molecule. To begin to understand the conformation of angiotensin II agonist and antagonist analogs at the receptor, it has long been recognized[81] that modifications need to be incorporated that reduce the degrees of conformational freedom in the molecule without diminishing receptor affinity. Peptide structure is typically viewed in terms of the polypeptide backbone and the different sidechains that branch off the backbone. Modifications that reduce conformational freedom include the addition of functionality to the backbone to sterically restrict mobility[81] and through ''short-range cyclizations'' (between backbone atoms within the region of a dipeptide).[88] ''Long-range'' cyclizations, then, give rise to cyclic tripeptides or larger cyclic peptides. The macrocycle in cyclic peptides can involve the peptide backbone, so-called homodetic peptides. The macrocycle can also involve the sidechains, so-called heterodetic peptides, most typically through a disulfide bridge between two cysteines. Even though a cyclic peptide contains fewer degrees of conformational freedom than a linear peptide, cyclic peptides, especially cyclic octapeptides, are still highly flexible. Thus, chemists have been developing an increasing array of nonnative amino acids and dipeptides that are less flexible due to the presence of functional groups, called conformational constraints, with defined, restricted conformational mobility and which can be incorporated into peptides to determine which conformational constraints maintain or enhance receptor affinity. The highly defined constraints which mimic types of protein secondary structure are called peptide mimetics.[89]

A. LINEAR ANALOGS BEARING CONFORMATIONAL CONSTRAINTS

To probe the receptor-bound conformational requirements of angiotensin II (angiotensin II) octapeptide agonists and antagonists, the synthesis and biological activities of [Sar^1]-angiotensin II agonist and [Sar^1,X^8]-angiotensin II antagonist analogs (X^8 = Ile, D-Phe, or Aib), bearing conformational constraints in different positions, have been investigated.[12,74,90–92] The conformational constraints that have been employed include Pro,[93] Dtc (5,5-Dimethylthiazolidine-4-carboxylic acid),[94] Aib, Cle, and other alphamethylamino acids,[81,95] (NMe)Ala, and other N-methylamino acids,[80] and the lactam modification of Freidinger[96,97]. Figure 6 displays these types of constraints within the context of modifications to Ac-L-Ala-NHMe. To compare the manner in which these constraints modify the conformation of Ac-L-Ala-NHMe, it is important to consider first the conformations that are available to Ac-L-Ala-NHMe and other L-amino acids in peptides. The allowed conformations have been determined from protein X-ray crystal structures[98] and by empirical conformational energy calculations.[99] In general, the

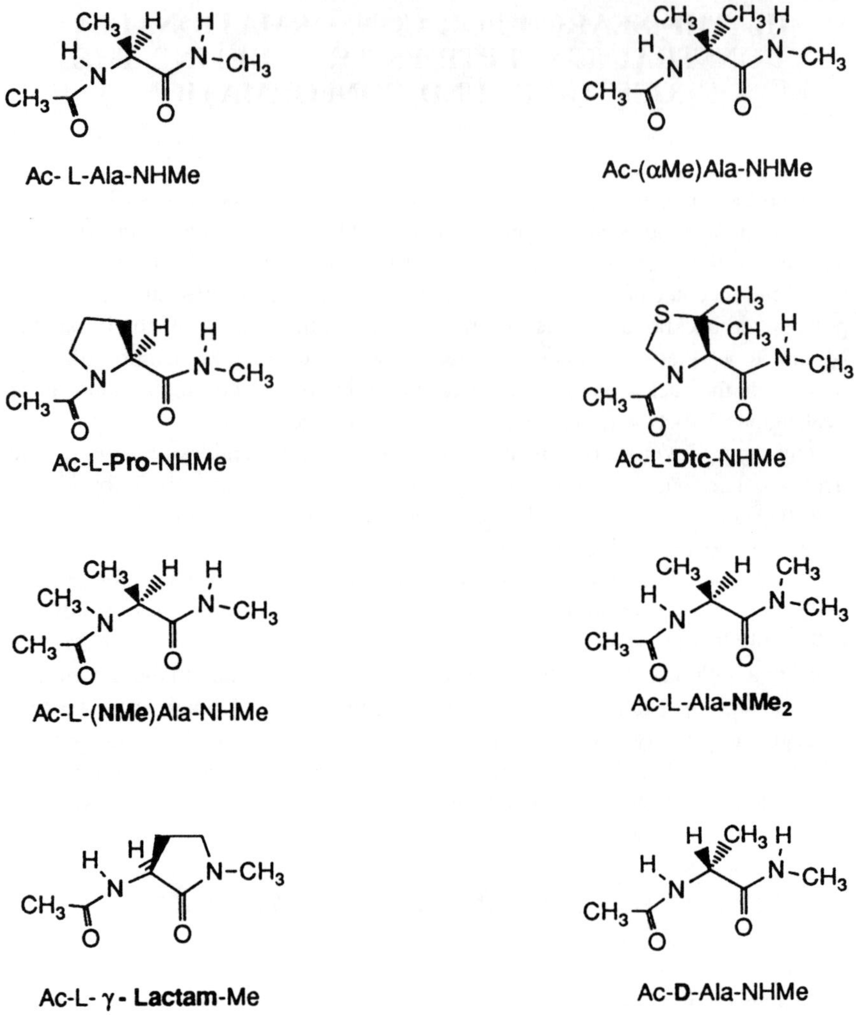

FIGURE 6. Structures of nonnative amino acids and dipeptides which display preferred conformations that differ from native L-amino acids.

overall topology of a peptide or protein is defined by the peptide backbone torsion angles ϕ and ψ (Figure 7).[98] The amide torsion angle θ is constrained to approximately 0 or 180°. The sidechains in acyclic amino acids have little influence upon the backbone conformations beyond the β-carbon; thus, the sidechain torsion angles χ-need not be considered. The regions of conformation accessible to an amino acid residue in a protein sequence have been delineated as shown in Figure 7, and include the β sheet, the α helical, and the γ turn conformations. Figure 8 displays the approximate conformational space (ϕ,ψ) accessible to Ac-Ala-NHMe

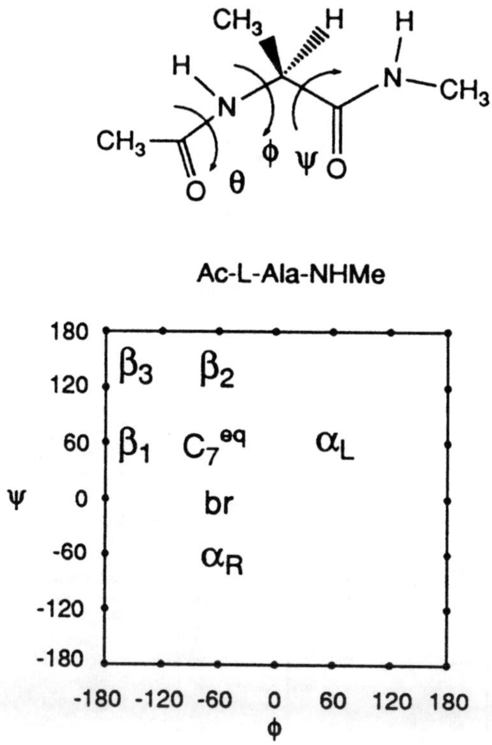

β_1–β_3 = Beta sheet conformation.
C_7^{eq} = C_7 Equitorial or gamma turn conformation.
br = Bridged conformation.
α_R = Right-handed helical conformation.
α_L = Left-handed helical conformation.

FIGURE 7. Conformational space accessed by native L-amino acids in peptides.

as well as to Ac-XX-Me, in which XX is one of the aforementioned conformationally constrained amino acids. For ease of comparison, the allowed regions of conformation of the constrained amino acids are roughly delineated with pluses (+) in Figure 8. (The reader should consult the preceding references for more detailed descriptions of the allowed conformations.) This analysis suggests that the backbone conformation of a peptide in the region of one of these modified amino acids may be found in one of those regions. (Factors that limit this hypothesis will be discussed in Chapter 6.) Note that the allowed conformations about a D-amino acid residue (shown for Ac-D-Ala-NHMe in Figure 8) are the mirror image of those allowed for an L-amino acid. Thus, it is incorrect to describe a D-amino acid as a conformational constraint, since the total range of conformational space is identical to an L-amino acid. From Figure 8, it is readily apparent

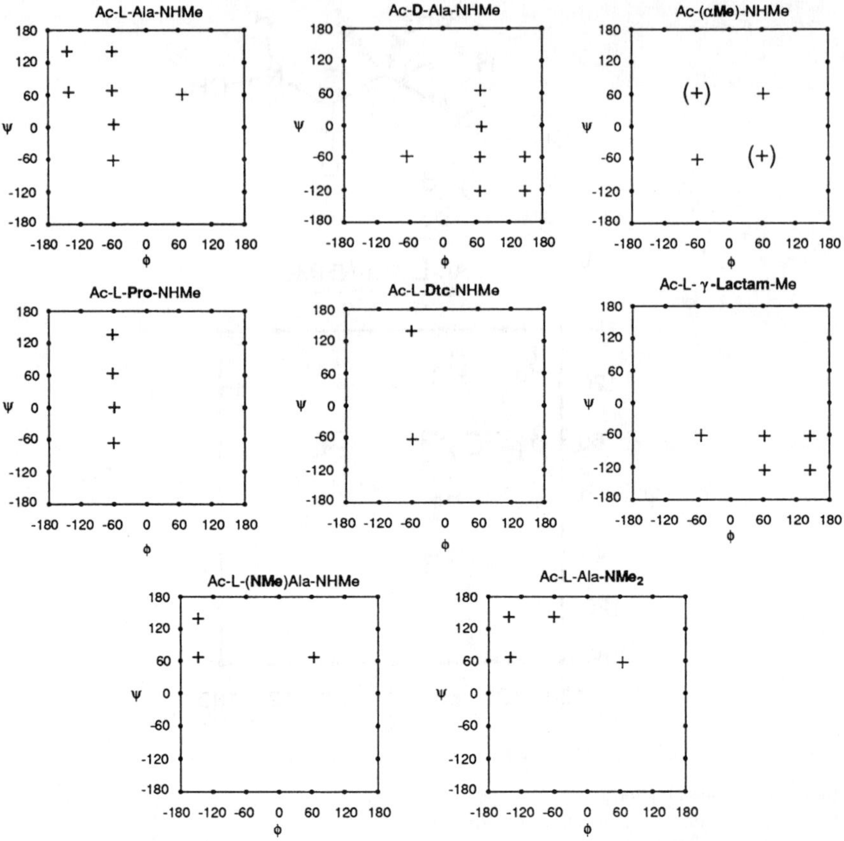

FIGURE 8. Conformational space accessed by dipeptides bearing different conformational constraints.

that the conformational space accessible to Ac-XX-Me is reduced to defined regions relative to Ac-Ala-NHMe. The Figure also shows how each different type of conformational constraint impacts upon the available conformational space in different ways. For example proline restricts conformation to ones in which φ is approximately −60°, whereas a (NMe)amino acid prevents such conformations. Notice that an N-alkyl substituent also constrains the preceding amino acid residue (shown as Ac-L-Ala-NMe_2 in Figure 8).[80] It has been suggested that multiple substitutions of these constrained amino acids at each position of a peptide would give some indication of the ''preferred'' conformation for each residue in the receptor-bound conformation.[81] This study has been conducted in a number of laboratories and is summarized below.

1. Position One Aspartic Acid

Formally, one cannot define the conformation about either the N terminal or C terminal amino acid since they lack neighboring amino acids that would serve to define either the missing φ or ψ backbone torsion angle.

2. Position Two Arginine

The appropriate NMe, αMe, or propyl analogs of arginine have not been incorporated into either [Sar^1]-angiotensin II or [Sar^1,Ile^8]-angiotensin II. N$^\alpha$methylarginine[100] has been described recently, as well as (4-guanidinoalkyl)proline.[48] Analogs bearing these amino acid analogs have not been described to date. In [des-Asp^1]-angiotensin II analogs, arginine becomes the N terminal amino acid and is undefinable, consequently, in angiotensin III.

3. Position Three Valine

As seen in Figure 8, each conformational constraint accesses rather different regions of conformational space.* As seen in Table 3.A and Table 3.B, both [Sar^1,(NMe)Ala^3]-angiotensin II *56* and [Pro^3]-angiotensin II *52* retain agonist activity. In contrast, among the antagonist analogs bearing constraining amino acids in Table 3.C, only [Sar^1,(NMe)Ala^3,Ile^8]-angiotensin II *61* has retained antagonist activity. Due to the different regions of conformational space accessed by (NMe)Ala[80] and Pro,[93] these results suggest that the conformational requirements may be more stringent in position three for angiotensin II antagonists than for angiotensin II agonists.

4. Position Four Tyrosine

The potent agonist activity of [Sar^1, (αMe)Tyr^4]-angiotensin II *98* (Table 4.E), suggests that similarly substituted angiotensin II antagonists would display high affinity, although other conformations about Tyr^4 might be expected in angiotensin II antagonists as well. We were unable to prepare [Sar^1, (αMe)Tyr^4,Ile^8]-angiotensin II by solid phase synthesis, obtaining a fragment terminated at (αMe)Tyr^4 instead. The (NMe)Tyr^4 and D-Tyr^4 analogs *100* and *101* were low in antagonist activity.

5. Position Five Isoleucine

[Sar^1,Dtc^5]-angiotensin II *114* displayed superior agonist activity over other [Sar^1,X^5]-angiotensin II analogs (Table 5.A and 5.B), while both [Sar^1,Cle^5 and Dtc^5,Ile^8]-angiotensin II *134* and *135* displayed superior antagonist activity over other [Sar^1,X^5,Ile^8]-angiotensin II analogs (Table 5.C). Due to the different regions of conformational space accessed by Dtc[94] and Cle,[95] these results suggest that the conformational requirements for angiotensin II agonists may be more stringent in position five than for angiotensin II antagonists.

6. Position Six Histidine

Until recently, the conformationally constrained histidine analogs, such as (αMe)His and (NMe)His, have not been available. Thus, the conformational preferences of His^6 in angiotensin II agonists and antagonists have not been examined. The study by Manavalan and Momany on the influence of N-alkyl amino acids

*Such conformational constraints may also influence the conformational freedom of the preceding and following amino acids, as well as noncontiguous amino acids in the peptide.

on peptide conformation shows an influence upon the conformation of the amino acid that precedes an N-alkyl amino acid in a peptide.[80] Thus, the proline in position seven imposes a certain degree of constraint on both position seven and position six.

7. Position Seven Proline

In contrast to position five, Dtc^7 substitution for Pro^7 of either [Sar^1]-angiotensin II or [Sar^1,Ile^8]-angiotensin II gave analogs *166* and *173* with reduced activities (Table 7). These results are consistent with the hypothesis that conformations of [Sar^1]-angiotensin II and [Sar^1,Ile^8]-angiotensin II containing a C_7 or gamma turn conformation in position seven are preferred for both angiotensin II agonist and antagonist activity, whereas conformations of [Sar^1]-angiotensin II and [Sar^1,Ile^8]-angiotensin II containing a C_7 conformation in position five are not preferred.

Incorporation of the L,L-lactam-Phe modification of Freidinger et al.[96] into [Sar^1]-angiotensin II gives a pure angiotensin II antagonist *174* (pA_2 8.3), comparable to saralasin (pA_2 8.1) (Table 7.C). Modeling studies indicate that the L,L-lactam-Phe group can essentially adopt the agonist conformation of Pro-Phe.[75] The high affinity antagonist activity, however, implies that the lactam ring either sterically prevents the receptor from adopting a conformation that results in a cellular response, or the receptor forces the L,L-lactam-Phe peptide into an antagonist conformation.

8. Position Eight Phenylalanine

Although the conformation about this C-terminal amino acid cannot be defined in a formal sense, the importance of the conformation about the C terminal tail of angiotensin II for agonist activity and high receptor affinity has led chemists to study a variety of the conformationally constrained amino acids (Table 8). As seen in the Table, [(αMe)Phe^8]-angiotensin II *186* is equipotent with angiotensin II. [Sar^1,(NMe)Phe^8]-angiotensin II *217,* however, appears to retain high receptor affinity despite its loss of agonist activity, since *217* is a potent angiotensin II antagonist. [Sar^1,(NMe)Ile^8]-angiotensin II *218* is also a potent angiotensin II antagonist. Interestingly, [Sar^1,D-(αMe)Phe^8]-angiotensin II *225* and [Sar^1,(αMe)Val^8]-angiotensin II *228* are both potent angiotensin II antagonists. [Sar^1,Pro^8]-angiotensin II *234* and [Sar^1,Dtc^8]-angiotensin II *235,* however, are poor angiotensin II antagonists.

9. Multiple Conformational Constraints

The aforementioned differences in conformational constraints in positions three, five, and seven that retain angiotensin II agonist activity in [Sar^1]-angiotensin II, compared to those that retain angiotensin II antagonist activity in [Sar^1,Ile^8]-angiotensin II, suggest that the conformational requirements for the angiotensin II agonist [Sar^1]-angiotensin II and the angiotensin II antagonist [Sar^1,Ile^8]-angiotensin II may be different in these three positions.

Individual substitution of (NMe)Ala^3, Dtc^5, D-Phe^8, and Aib^8 into [Sar^1,Ile^8]-angiotensin II gives analogs which retain antagonist activity. Multiple substitutions of these types of residues into [Sar^1,Ile^8]-angiotensin II give analog *268* [Sar^1,(NMe)Ala^3,Dtc^5,Aib^8]-angiotensin II, *261* [Sar^1(NMe)Ala^3,D-Phe^8]-angiotensin II, and *262* [Sar^1,Dtc^5,D-Phe^8]-angiotensin II, which display considerably reduced antagonist activity (Table 11). Consequently, in angiotensin II antagonists, the construction of highly constrained analogs may not be possible by the additive substitution of ''preferred'' constrained amino acids into a single analog.

One possible interpretation of the inactivity of these analogs may be that the constraints collectively lock the peptide into the final ligand conformation that lacks sufficient flexibility for a dynamic receptor binding process or ''zipper'' mechanism[101] that has been previously postulated for flexible peptide ligand and receptor interaction. If the zipper mechanism applies to angiotensin II receptor interaction, then a certain degree of flexibility will need to be retained in angiotensin II octapeptide antagonists to maintain high receptor affinity. A second possible interpretation was suggested previously, that of multiple binding modes for angiotensin II analogs.*

10. Multiple D-Amino Acid Modifications

Individual D-amino acid modifications into [Sar^1,Ile^8]-angiotensin II gave analogs with reduced antagonist activities (see the appropriate data tables). Some multiple D-amino acid modifications have been attempted (Table 12), which also gave analogs with considerably diminished antagonist activities. Curiously, D-amino acid substitutions into positions six and eight gave analogs *273* and *274,* which displayed weak antagonist activities.

B. CYCLIC ANALOGS

Few cyclic angiotensin II analogs have been reported. This is probably the result of the inactivity of most of such analogs (Table 13). cyclo-Lys-N$^\varepsilon$-C-terminal-CO-(Lys^1]-angiotensin II *318* was reported previously to display agonist activity in the rat uterus. In our hands this analog lacked affinity for the angiotensin II receptor.

The only analogs with noteworthy activity are the Cys^{3,5} and hCys^{3,5} analogs *281, 282, 292,* and *293,* described initially by Sugg et al.[90] and later by Spear et al.[91] (Table 13). More recently, this series has been extended by Marshall et al.[92,102] analogs *294* to *298.** The relevance of these results to the receptor-bound conformation will be discussed in Chapter 2.

*The possibility for multiple binding modes has been supported in the recently published work of Plucinska et al. with cyclic AII analogs: K. Plucinska, T. Kataoka, M. Yodo, W. L. Cody, J. X. He, C. Humblet, G. H. Lu, E. Lunney, T. C. Major, R. L. Panek, P. Schelkun, R. Skeean, and G. R. Marshall, Multiple binding modes for the receptor-bound conformations of cyclic AII agonists, *J. Med. Chem.,* 36, 1902, 1993.

IV. APPROACHES TOWARD THE DEVELOPMENT OF ORAL PEPTIDE ANGIOTENSIN II ANTAGONISTS

A number of investigators have sought modifications of angiotensin II that would reduce partial agonist activity and add resistance to enzymatic degradation. Others have sought fragments of angiotensin II that retained affinity for the angiotensin II receptor, while still others have attempted to develop nonpeptides, either by rational design from the peptide or from leads discovered through the screening of compound databases and biological samples. As has been indicated above, analogs have yet to be discovered that retain high antagonist activity and also display reduced partial agonist activity. The incorporation of multiple modifications to reduce enzymatic degradation have given analogs with diminished affinity. The other approaches, which attempt to get away from the octapeptide framework, are discussed below.

A. FRAGMENTS OF ANGIOTENSIN II

Fragments of angiotensin II that retain some or all of the critical binding groups (the guanidinium group of Arg, the phenolic group of Tyr, the imidazole group of His and a C-terminal carboxylate) have been evaluated as angiotensin II agonists and antagonists (Table 15). Deletion of one or more ''spacer'' amino acids, Val^3, Ile^5, or Pro^7, in either [Des-Asp^1]-angiotensin II or [Des-Asp^1,Ile^8]-angiotensin II give rise to shorter sequences, which are either inactive or low in affinity. Analogs of these shorter sequences bearing D-amino acids are also inactive. Sequences lacking the spacer amino acids, Val^3, Ile^5, or Pro^7, must not be capable of placing the important guanidinium, phenolic, imidazole, and carboxylate groups into proper orientation for high affinity receptor binding. The minimum chain length for high affinity peptide-based angiotensin II antagonists continues to be either the des-Asp^1 heptapeptide or the Sar^1,des-Phe^8 heptapeptide sequence. These results highlight the importance of recently described potent nonpeptide angiotensin II antagonists.

B. NONPEPTIDES AND SEMIPEPTIDES DESIGNED FROM ANGIOTENSIN II

1. Constrained Semipeptide Fragments

Fragments of angiotensin II may prefer to adopt conformations that do not place the binding groups in a proper orientation for receptor binding. An alternative approach would be to study fragments from potent angiotensin II analogs bearing conformationally constrained amino acids (Figure 9). Such fragments may be more likely to adopt the proper orientation of binding groups for receptor binding. These constrained fragments were also inactive (Table 15).

2. Nonpeptides from Angiotensin II Models

To date, the only high affinity nonpeptide ligands for peptide receptors have derived from leads discovered through screening of biological samples or compound databases.[103] The rational design of a nonpeptide ligand from a model of

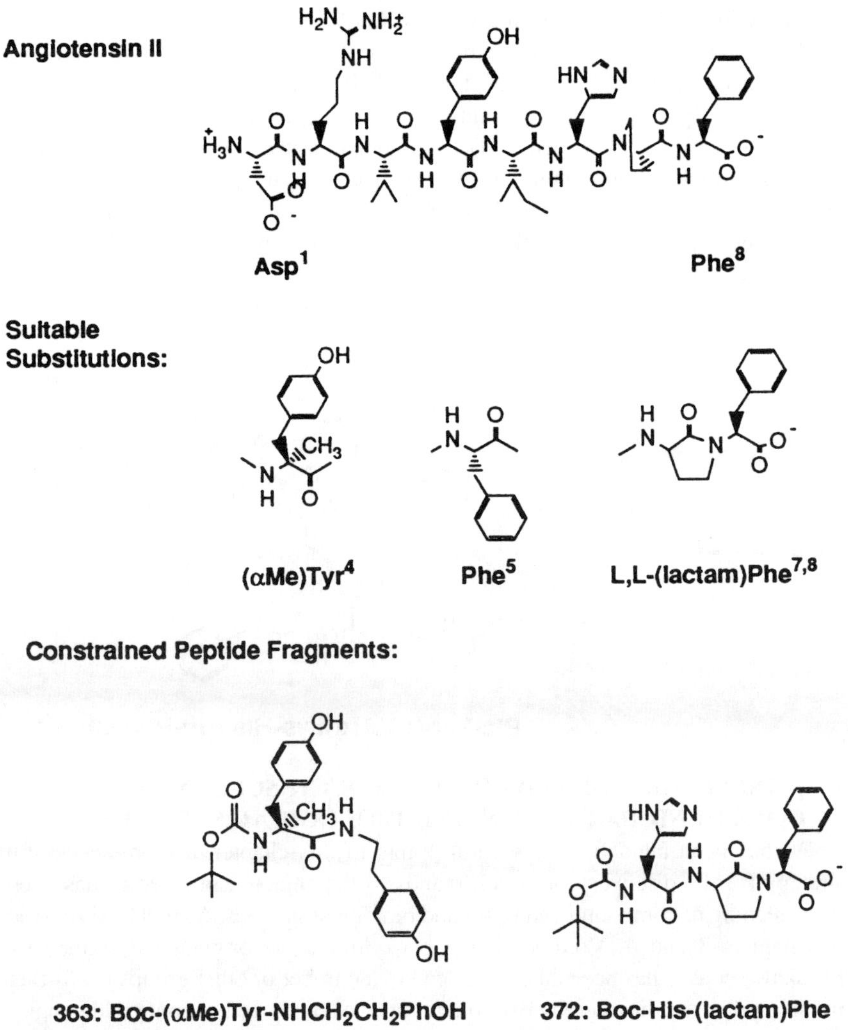

FIGURE 9. Derivation of some angiotensin II fragments bearing conformational constraints.

a peptide remains an elusive goal in medicinal chemistry. Hirschman has described monosaccharide-based nonpeptide ligands for the somatostatin receptor, designed from constrained cyclic somatostatin analogs. The low affinity of these compounds, however, leaves open the question of how closely they mimic the parent peptide.[104] A few forays into the design of nonpeptides from early models of angiotensin II were attempted at SmithKline Beecham, but the inactivity of these compounds, along with success with the benzylimidazoles, caused us to abandon this line of inquiry. As a method of generating nonpeptide leads, however, this approach clearly may be useful with other peptidergic systems.

3. The Ciba-Geigy Semipeptide CGP 42112A

A group at Ciba-Geigy[105] has reported a peptide analog that is of interest both for its selectivity for the AT_2 receptor (see below) and for its unusual structure, below. This ''pentapeptide'' contains all of the essential binding groups, strung together in an unusual sequence. The peptide is basically an angiotensin II(4–8) fragment, in which the arginine residue is attached to the position five sidechain. Although the method of design has not been published, it would seem that the Ciba-Geigy chemists were working from some model of angiotensin. Any model of angiotensin II should be able to overlay low energy conformations of CGP 42112A. This will be discussed in Chapter 3, Volume II.

CGP 42112A

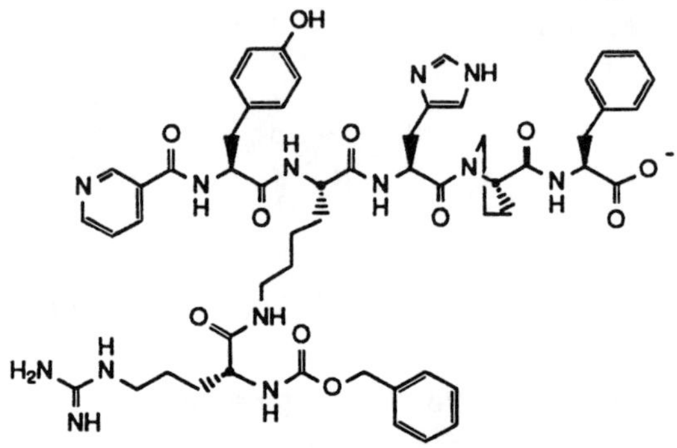

Pic-Tyr-(ε-ZArg)Lys-His-Pro-Phe-OH

C. NONPEPTIDES DISCOVERED THROUGH SCREENING COMPOUND DATABASES AND BIOLOGICAL SAMPLES

As has been alluded to at several points in this chapter, a more successful strategy for the discovery of novel ligands to the angiotensin receptor has been via screening of compound databases and biological samples. As will be discussed in Chapters 3 and 4, Volume II, the benzylimidazole originally described by Furukawa et al.[7,8] has been taken as a lead in a number of other groups for further modifications that have given rise to a number of orally active nonpeptide angiotensin II antagonists. Figure 10 displays some other low affinity angiotensin II antagonists that have been discovered through screening. Imai et al. reported that diaminoalkanes inhibit angiotensin II receptor binding, albeit at micromolar concentrations.[106] The optimal chain length for inhibition was found in $H_2N\text{-}(CH_2)_{6,7}\text{-}NH_2$, which they proposed simulated the distance between Arg guanidine and His imidazole. The corresponding guanidinylalkylimidazole, however, has not been described.

V. PEPTIDES THAT DISTINGUISH ANGIOTENSIN II RECEPTOR SUBTYPES

Most of the peptide and semipeptide structure–activity studies described in this chapter predate the discovery of receptor subtypes and isoforms AT_{1A}, AT_{1B}, AT_2,

Furukawa et al.:[7,8]

1-o-chlorobenzyl-2-butyl-3-chloroimidazole-5-acetic acid

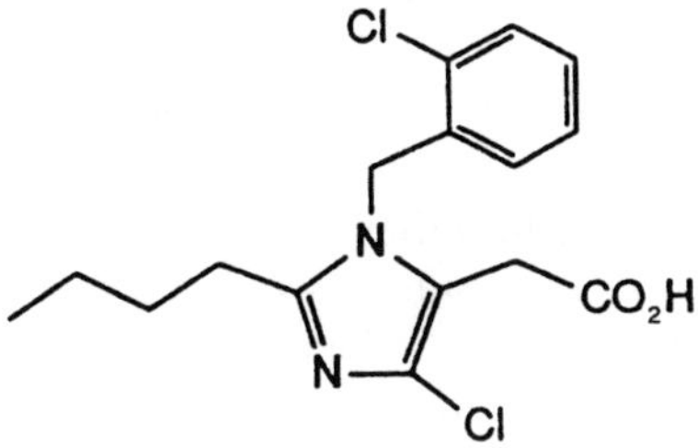

Ball et al.:[161]

Veratridine

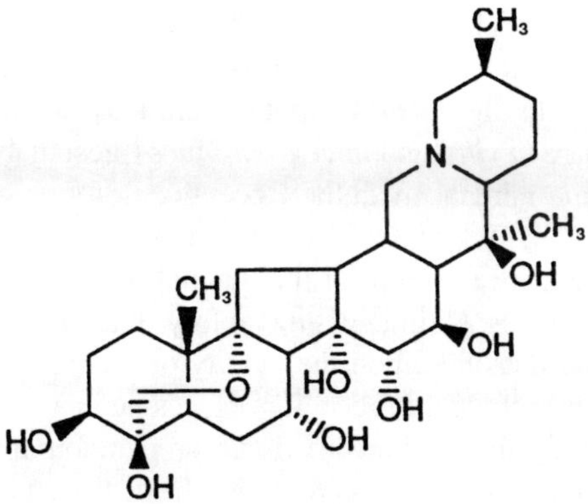

Imai et al.:[106]

Diaminoalkanes

FIGURE 10. Nonpeptides discovered through screening.

etc. Prior to the recent discovery of high affinity, selective nonpeptide angiotensin II antagonists, which led to the characterization of angiotensin II receptor subtypes,[23,107,108] it was commonly believed that all tissues contained the same receptor.[109] Thus, binding studies of angiotensin II analogs were focused upon demonstrating the existence of a functional angiotensin II receptor in the tissue under investigation.[110–117]

The task of characterizing the differences in the peptide pharmacophore of the different angiotensin II receptor subtypes is just beginning.[108] The first AT_2 selective peptide to be reported in the literature was [$(pNH_2)Phe^6$]-angiotensin II *147*.[118,119] Since then, the aforementioned novel semipeptide CGP 42112A has also been described as AT_2 selective (Section IV.B). Table 16 attempts to correlate previously described *in vitro* activities (see references in appropriate tables) with receptor affinities. The correlation of receptor affinity with either *in vitro* (myotropic) or *in vivo* (hypertensive) activity is not straightforward, since rabbit aorta were employed *in vitro* and rats were employed *in vivo,* while bovine adrenal cortex was employed regularly as a routine AT_1 receptor assay, and bovine ovary and cerebellum were employed routinely as AT_2 receptor beds. Early affinity measurements were conducted with the rat mesentery. The rat mesentery was initially employed as an AT_1 system but later was displaced by the bovine adrenal medulla to avoid species variation in comparison with the bovine AT_2 receptor-containing ovary and brain. Species variation still may be an important factor in attempting to compare *in vitro* and *in vivo* activities (presumably AT_1 driven) with affinity in the bovine adrenal medulla. Recently, isoforms of the AT_1 receptor were discovered.[120,121] AT_{1A} and AT_{1B} are widely expressed in rat tissues, with AT_{1A} in greater abundance, except in the adrenal gland and the pituitary gland, where AT_{1B} predominates. Ultimately, the ability of an analog to displace angiotensin II will depend not only upon the avidity of the ligands but also upon the levels of isoforms and subtypes of receptors expressed in a particular tissue, which could be influenced by the methods of tissue preparation and handling. Despite these problems, some interesting points arise from the study of angiotensin II analogs in Table 16.

Analogs displaying the highest affinities include [Sar^1,**Tal^8**]-angiotensin II *196* in the rat mesentery, [Sar^1,**Dtc^5**]-angiotensin II *114* in the bovine ovary, and [Sar^1,**Trp^6**,Ile^8]-angiotensin II *155* in the bovine cerebellum.

The most interesting point arises from comparison of the AT_1 system (adrenal medulla) with the AT_2 systems. For most of the analogs in Table 16, number of modifications to [Sar^1]-angiotensin II and [Sar^1,Ile^8]-angiotensin II that reduced *in vitro* and *in vivo* activities also reduce affinity in the adrenal medulla. One exception is [Sar^1,**Phe^4**,Ile^8]-angiotensin II *89,* which displays an IC_{50} of 1.2 nM in the adrenal medulla. Surprisingly, many of the modifications that reduce *in vitro* and *in vivo* activities as well as affinity to the adrenal medulla do not reduce affinity to one or both of the AT_2 systems. The most important of the analogs bearing these types of modifications include [Sar^1,**Cit^2**,Ile^8]-angiotensin II *41,* [Sar^1,**$(pF)Phe^4$**,Ile^8]-angiotensin II *91,* [Sar^1,**$(pOMe)Phe^4$**,Ile^8]-angiotensin

II *90,* [Sar[1],**Phe**$^{\mathbf{6}}$,Ile[8]]-angiotensin II *153,* [Sar[1],**Trp**$^{\mathbf{6}}$,Ile[8]]-angiotensin II *155,* [Sar[1],**Tal**$^{\mathbf{6}}$,Ile[8]]-angiotensin II *154,* [Sar[1],**(OMe)Thr**$^{\mathbf{5}}$,**Tal**$^{\mathbf{6}}$,Ile[8]]-angiotensin II *244,* [Sar[1],**(N**$^{\mathbf{3}}$**Me)His**$^{\mathbf{6}}$,Ile[8]]-angiotensin II *156,* [Sar[1],**Dtc**$^{\mathbf{5}}$,**D-Phe**$^{\mathbf{8}}$]-angiotensin II *262,* and [Sar[1],**(pCl)Phe**$^{\mathbf{7}}$,Ile[8]]-angiotensin II *185.* These peptides swell the ranks of octapeptides that prefer AT_2 (previously, [**(pNH**$_{\mathbf{2}}$**)Phe**$^{\mathbf{6}}$]-angiotensin II[118] and the Ciba-Geigy peptide CGP 42112A[105]). Notice that, with the exception of the Ciba-Geigy peptide, these analogs bear modifications in positions two, four, six, and eight, all critical residues for peripheral activities. These analogs suggest that the AT_2 receptor pharmacophore is quite distinct from the AT_1 pharmacophore.

The dramatic differences between AT_1 and AT_2 raise the issue of AT_2 function. AT_2 function can be best addressed with AT_2 selective peptide agonists. Until recently [(pNH$_2$)Phe^6]-angiotensin II was the only known AT_2 selective peptide agonist. This peptide displays only residual pressor activity (0.05% rel. angiotensin II, Table 6) but potent dipsogenic activity.[122] The fact that the affinity of the peptide for liver is decreased in the presence of GTPγs has been taken as evidence that it is an agonist in the liver as well.[118] *In vitro* and *in vivo* activities for CPG 42112A have not been described beyond its activities as an inhibitor of angiotensin II receptor binding in AT_2-containing tissue beds.[105,122–124] Most of the AT_2 selective peptides described in Table 16 are antagonists. [Sar[1],(NMe)Ala[3]]-angiotensin II *56,* on the other hand, is a potent agonist in the rabbit aorta and displays ~180-fold selectivity for AT_2 over AT_1 (Table 16). Such selective peptide agonists should be employed to further characterize the physiological roles of AT_2 receptor stimulation.

Note that some of these analogs prefer the bovine ovary over the cerebellum, i.e., [Sar[1],**Tal**$^{\mathbf{6}}$,Ile[8]]-angiotensin II *154* and [Sar[1],**(N**$^{\mathbf{3}}$**Me)His**$^{\mathbf{6}}$,Ile[8]]-angiotensin II *156,* while others prefer the cerebellum over the ovary, i.e., [Sar[1],**Phe**$^{\mathbf{6}}$,Ile[8]]-angiotensin II *153* and [Sar[1],**(OMe)Thr**$^{\mathbf{5}}$,**Tal**$^{\mathbf{6}}$,Ile[8]]-angiotensin II *244.* The substituent in position five, furthermore, determines the preference of the analog: [Sar[1],**(Ile**$^{\mathbf{5}}$**)**,**Tal**$^{\mathbf{6}}$,Ile[8]]-angiotensin II *154* prefers the ovary, while [Sar[1],**(OMe)Thr**$^{\mathbf{5}}$,**Tal**$^{\mathbf{6}}$,Ile[8]]-angiotensin II *244* prefers the cerebellum. Thus, the AT_2 receptor in the ovary may be an isoform of that in the cerebellum,[125] which is not without precedent, since isoforms of the AT_1 receptor have been described recently.[120,121] The angiotensin II receptor expressed in *Xenopus laevis* oocytes appears to be different from the AT_2 receptor as well, since CGP 42112A displays substantially greater affinity (IC_{50} 400 n*M*) for the *Xenopus* oocyte receptor than for either Dup 753 or PD 123177 ($IC_{50} > 10$ μ*M* each).[122] In the mammalian oocyte, Dup 753 displays high affinity (IC_{50} 400 p*M*) in contrast to CGP 42112A and PD 123177 ($IC_{50} > 10$ μ*M* each).

The implication of these results, which are summarized in Table 16, is that a considerable amount of original structure–affinity studies will need to be performed to determine the extent of the differences between the AT_1 and AT_2 receptors. In contrast to the low affinity for AT_1 in the adrenal cortex, the high affinity of [Sar[1],Dtc[5],D-Phe[8]]-angiotensin II *262* for AT_2 in the ovary suggests that conformational differences between AT_1 and AT_2 selective analogs may be uncovered as well.

Given the apparently different pharmacophore requirements for AT_1 and AT_2 receptors, the previously discussed issues concerning lack of substituent additivity may be understandable. If a tissue bed contains more than one type of angiotensin II receptor, then the degree to which that tissue responds to an analog will be related to the affinity and agonist efficacy of the analog at each type of receptor. The optimization of one position in angiotensin II for *in vitro* activity may result in optimization at one position of angiotensin II for a particular receptor. The combination of optimized modifications in a single analog may not succeed if the modifications favor different receptors. To test this hypothesis, however, the different AT subtypes of an *in vitro* tissue, such as rabbit aorta, will need to be quantitated and correlated with radioreceptor assays for each subtype, utilizing receptor beds from the same species or preferably from the same tissue. Only then can this hypothesis be tested in a series of analogs bearing single and multiple modifications.

Finally, we had hoped to be able to compare the structures and affinities of peptides with those of the nonpeptides discussed in Chapters 3 and 4, Volume II. This appears to be unlikely for the following reason. The compounds in Table 16 are ordered by position modified, agonists vs. antagonists, and then rank-ordered by decreasing antagonist potency in the rabbit aorta. As seen in the table, the rank ordering of affinities in the AT_1 systems is roughly parallel within the major groupings (position modified and agonist vs. antagonist) but in certain key positions, the parallelism breaks down, i.e., positions four and six. Species differences (rabbit vs. bovine) undoubtedly account for these differences.

VI. ISSUES IN THE DEVELOPMENT OF ORAL PEPTIDE ANGIOTENSIN II ANTAGONISTS

From the preceding studies of peptide structure and activity, a number of important issues have surfaced concerning the development of orally active peptide angiotensin II antagonists.

A. THE NECESSITY FOR THE HYDROPHILIC ARG GUANIDINIUM GROUP AND THE C TERMINAL CARBOXYLATE GROUP

All replacements of the position two arginine guanidinium, the position six histidine imidazole, and the C terminal carboxylate groups give angiotensin II analogs with lowered activities. Thus, a high affinity peptidergic angiotensin II antagonist appears to require these three hydrophilic groups. Assuming that oral absorption of an angiotensin II antagonist would be via passive diffusion, the apparent requirement of three hydrophilic groups **may** need to be compensated for by incorporation of lipophilic functionality in other parts of the molecule.

B. THE LARGE SIZE OF THE PEPTIDE PHARMACOPHORE

The peptide structure–activity relationship has defined a large number of binding groups that are critical for high receptor affinity. The peptide frameworks that have been examined to date, the native octapeptide framework and the Ciba-Geigy semi-pentapeptide, appear to be the only peptide-like frameworks that are capable of positioning binding groups into a high affinity array.

C. DIFFICULTIES IN STABILIZING AGAINST PROTEOLYTIC DEGRADATION

As was indicated in an earlier section, multiple modifications that would protect against proteolytic degradation have failed to date to give high affinity semipeptides.

D. PARTIAL AGONIST ACTIVITY

At the beginning of this chapter, the presence of residual or partial agonist activity in saralasin and [Sar^1,Ile^8]-angiotensin II was described as a critical problem with these antagonists in man. The partial agonist activity of the peptide antagonists can be readily observed in the rat, as shown in Figure 11. The doses required to elicit an agonist response are three orders of magnitude higher than the doses required to elicit an antagonist response. This transient increase in blood pressure that is observed upon injection of peptide prior to blood pressure decrease could be envisioned as presenting a serious threat to a patient experiencing a hypertensive crisis. The goal of eliminating partial agonist activity while retaining antagonist affinity has not been met to date. Some progress has been made in reducing partial agonist activity, however. Figure 11 and Table 17 display analogs that display high *in vitro* antagonist activity and yet display diminished agonist activity. There do not appear to be any particular amino acid residues in the octapeptides that are responsible for partial agonist activity, since modifications to different positions in [Sar^1,Ile^8]-angiotensin II that diminish antagonist activity do not reduce partial agonist activity (Table 18). Even analogs bearing multiple sidechain substitutions, which displayed reduced antagonist activity, displayed nearly the same level of partial agonist activity as [Sar^1,Ile^8]-angiotensin II, e.g., analogs *97, 243, 244, 247,* and *250* in Table 9 and analogs *261* and *262* in Table XI. These results suggest that the mechanism for partial agonist activity may not be mediated by an angiotensin II receptor. The extent to which the partial agonist activity has been diminished in the analogs of Figure 11 actually may be sufficient. The potent nonpeptide antagonists that are totally devoid of agonist activity represent, nonetheless, an improvement over these relatively expensive peptidic agents, which would not be expected to display oral activity.

This rather daunting set of issues presents some distinct hurdles in the development of oral peptide or semipeptide angiotensin II antagonists. Fortunately, the benzylimidazole-based nonpeptides have been able to circumvent these issues.

VII. CONCLUSION

Although the angiotensin II agonist and antagonist peptides have failed directly to provide oral antihypertensive agents, their study has provided a context in which

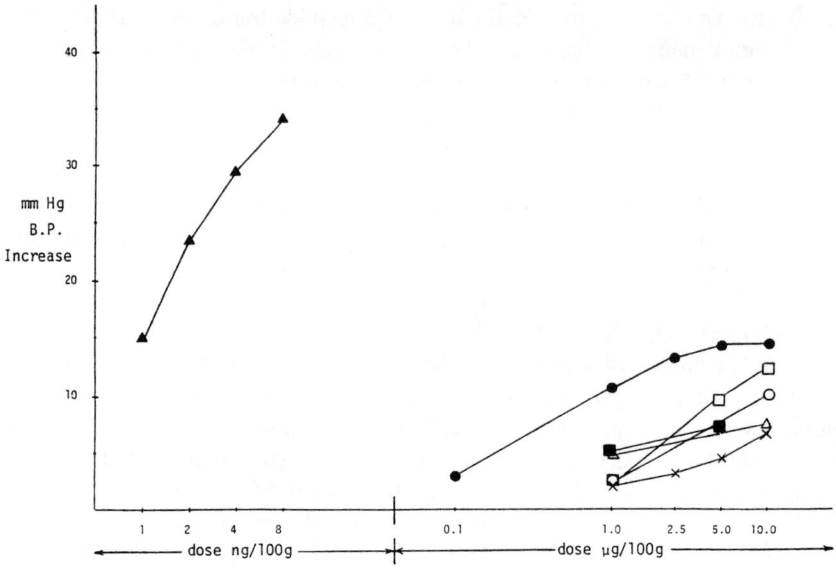

		ID_{50}
▲	Asp^1—Arg^2—Val^3—Tyr^4—Ile^5—His^6—Pro^7—Phe^8 = Angiotensin II (AII)	
	Antagonists:	
●	Sar^1 —— Ala^8 (Saralasin)	15
○	Sar^1 —— (OMe)Thr^5 —— Ile^8	2.5
×	Sar^1 —— Nle^7 – Ile^8	25
■	Sar^1 —— Ceg^7 – Ile^8	10
□	Sar^1 —— D-Phe^8-$CONH_2$	12.5
△	Sar^1 —— -Phg^5 —— Ile^8	5.0

FIGURE 11. The reduced partial agonist activity of some peptide angiotensin II antagonists.

to understand the novel nonpeptides. Since the nonpeptide antagonists bind to a small portion of the receptor and may not even bind to the receptor in the manner of the native peptide, in general, peptide agonists and antagonists will continue to be important tools for drug development. Perhaps the most important area in which peptide analogs will continue play a prominent role in the field of angiotensin is in the characterization of the angiotensin receptor subtypes, as reversible, selective competitive antagonists. Selective agonists will be important for investigating the molecular pharmacology of the AT_2 receptor. With the advent of receptor mutants, researchers may now be able to identify the receptor amino acid residues that compose the binding site for angiotensin II and to characterize inhibitory nonpeptides as mimetic, steric, or allosteric inhibitors. For the peptide aficionado, the field of angiotensin has yielded many intriguing discoveries, including the agonist hypothesis, the partial agonist activity of almost any octapeptide, and the suggestions that the activity of angiotensin analogs is the sum of multiple receptor interactions. The fact that position eight modifications of angiotensin I give high affinity receptor antagonists may inspire further investigations into the role of angiotensin I as more than a precursor to angiotensin II. The correlation between angiotensin II and at least one class of nonpeptide antagonist needs to be interrogated with complimentary peptide and nonpeptide structure–activity studies. With the advent of angiotensin receptor subtypes and isoforms, the study of angiotensin II agonist and antagonist peptides could continue well into the next decade.

ACKNOWLEDGMENTS

The original work cited in this chapter occurred over the period of a decade, beginning in the peptide research laboratories of the Beckman Bioproducts Group of Beckman Instruments Inc., in Palo Alto, California, proceeding through the relocation of those laboratories to the Department of Peptidomimetic Research, SmithKline and French Laboratories, a division of SmithKline Beckman (now known as SmithKline Beecham Pharmaceuticals) in Upper Merion, Pennsylvania. The authors wish to acknowledge the previous contributions of the following colleagues in peptide synthesis: William Adams, Jr., Fadia Ali, Ester Brandeis, Raul Calvo, David Martin, Theresa Cash, Dalgit Narindray, Joanne Silvestri, and Heidi Weideman; bioassay: Nambi Aiyar, Elliot Ohlstein, Elwood Stack; structural analysis by amino acid analysis: David Chang and Robert Sanchez; structural analysis by mass spectrometry: Steven Carr and Walter Johnson. The support of William Huffman, Sidney Hecht, Brian Metcalf, and Robert Ruffolo, Jr. is gratefully acknowledged.

TABLE 1
Activities of Position One Angiotensin II Analogs

No.	X-Arg-Val-Tyr-Ile-His-Pro-Z 1 2 3 4 5 6 7 8 X^1	Z^8	Agonist *In vitro*[a] AII-like[c]	Agonist *In vivo*[b] AII-like[d]	Antagonist *In vitro*[a] pA_2	Antagonist *In vivo*[b] ID_{50}[e]
1[g]	Asp	Phe	100 (pD_2 8.52)[f]	100		
2[h]	β-Asp	Phe	28	150		
3[h]	β-D-Asp	Phe	87	200		
4[h]	Asn	Phe	110	93		
5[h]	Ala	Phe		37		
6[h]	D-Ala	Phe	240	100		
7[h]	Gly	Phe	27	50		
8[h]	Sar	Phe	175 100[f]	70 190[f]		
9[h]	Me_2Gly	Phe		88		
10[h]	Me_3Gly	Phe	3.5	10		
11[h]	Pro	Phe	21	50		
12[i]	Suc	Phe	12	101		
13[h]	Asp	Leu			8.21	170
14[h]	Sar	Leu			8.71	40
15[h]	β-Asp	Leu			8.21	110
16[g]	Sar	Ile	1.0	(10.2)	9.1	10
17[g]	D-(NMe)Ala	Ile	0	(12.5)	9.0	25
18[g]	(NEt)Gly	Ile	0	(7.5)	9.0	50

[a] Agonist, ''AII-like'' activity, and antagonist activity, pA_2, were measured in the *in vitro* rabbit aorta strip assay, e.g., according to the method of Rioux et al.[35] unless noted otherwise.
[b] Residual ''AII-like'' activity and antagonist activity, ID_{50}, were measured *in vivo,* in the rat blood pressure assay, e.g., described by Regoli et al.[22] unless noted otherwise.
[c] ''AII-like'' activity *in vitro* is expressed as percent activity relative to AII, unless noted otherwise.
[d] ''AII-like'' activity *in vivo* is expressed either as the percent activity relative to AII, (or by the millimeters Hg of blood pressure increase produced by a 1 μg bolus intravenous injection of compound), unless noted otherwise.
[e] ID_{50} is in ng/rat/min (using 250 g rats), unless noted otherwise.
[f] The concentration of agonist that is able to reduce the effect of a double dose of agonist to that of a single dose according to Schild.[126]
[g] Samanen et al.[11]
[h] Regoli et al.[34]
[i] Paiva et al.[127]

TABLE 2
Activities of Position Two Angiotensin II Analogs

	X-Y-Val-Tyr-Ile-His-Pro-Z 1 2 3 4 5 6 7 8			Agonist		Antagonist	
No.	X^1	Y^2	Z^8	*In vitro*[a] AII-like[c]	*In vivo*[b] AII-like[d]	*In vitro*[a] pA_2	*In vivo*[b] ID_{50}[e]
1[f]	Asp	Arg	Phe	100	100		
19[g]	Asn	D-Arg	Phe	4.0			
20[h]	des-Asp	Arg	Phe	47.4 (RI)	15.0		
				2.4[j]	(75.0)[j]		
21[h]	des-Asp	D-Arg	Phe	43.3 (RI)	33.3		
				4.5[j]	(65.0)[j]		
22[h]	des-Asp	Lys	Phe	10.9 (RI)	3.0		
23[h]	des-Asp	D-Lys	Phe	20.6 (RI)	23.0		
24[h]	des-Asp	hArg	Phe	25.2 (RI)	2.07		
				4.5[j]	(65.0)[j]		
25[h]	des-Asp	D-hArg	Phe	26.8 (RI)	11.0		
26[h]	des-Asp	Orn	Phe	6.0 (RI)	3.0		
27[h]	des-Asp	D-Orn	Phe	20.0 (RI)	23.0		
28[i]	des-Asp	Abu	Phe	2.3 (RI)	0.92		
29[i]	des-Asp	D-Abu	Phe	20.6 (RI)	22.0		
30[i]	des-Asp	Ala	Phe	2.5 (RI)	3.0		
				0[j]	(40.0)[j]		
31[i]	des-Asp	D-Ala	Phe	20.6 (RI)	48.0		
				5.0[j]	(75.0)[j]		
32[j]	des-Asp	(αMe)Ala	Phe	0	(60.0)		
33[j]	des-Asp	D-(NMe)Ala	Phe	1.4	(85.0)		
34[k]	des-Asp	γ-Abu	Phe	0		7.2	
35[k]	des-Asp	ε-Ape	Phe	0		7.4	
36[k]	des-Asp	ε-Ahx	Phe	0		7.0	
37[k]	des-Asp	γ-Gbu	Phe	0		6.15	
38[k]	des-Asp	ε-Ghx	Phe	0		<5	
16[j]	Sar	Arg	Ile	0	(10.2)	9.1	10
39[j]	Sar	D-Arg	Ile	0	10	7.5	100
40[j]	Sar	Lys	Ile	0	(12.5)	7.75	50
41[j]	Sar	Cit	Ile	0	(25.0)	6.2	
42[j]	Sar	hArg	Ile	0	(7.5)	8.25	50.0
43[j]	des-Asp	Arg	Ile	0	(5.0)	7.75	200
44[j]	des-Asp	D-Arg	Ile	0	(5.0)	7.0	100
45[j]	des-Asp	Ala	Ile	0		6.0	
46[j]	des-Asp	D-Ala	Ile	0	(5.0)	7.0	100

TABLE 2 (continued)
Activities of Position Two Angiotensin II Analogs

No.	X-Y-Val-Tyr-Ile-His-Pro-Z (1 2 3 4 5 6 7 8) X[1]	Y[2]	Z[8]	Agonist: *In vitro*[a] AII-like[c]	Agonist: *In vivo*[b] AII-like[d]	Antagonist: *In vitro*[a] pA_2	Antagonist: *In vivo*[b] ID_{50}[e]
47[j]	des-Asp	(αMe)Ala	Ile	0		6.0	
48[j]	des-Asp	D-(NMe)Ala	Ile	0	(10.0)	7.1	25
49[l]	[Lys[1],Gly[2],Ala[7]]-AII			8 n*M* IC_{50} (Receptor binding)			
50[l]	[Lys[1],Gly[2],Met[5],Ser[7]]-AII			6 n*M* IC_{50} (Receptor binding)			
1[l]	AII			12 n*M* IC_{50} (Receptor binding)			

a Agonist, "AII-like" activity and antagonist activity, pA_2, were measured in the *in vitro* rabbit aorta strip assay, e.g., according to the method of Rioux et al.[35] unless noted otherwise.
b Residual "AII-like" activity and antagonist activity, ID_{50}, were measured *in vivo,* in the rat blood pressure assay, e.g., described by Regoli et al.[22] unless noted otherwise.
c "AII-like" activity *in vitro* is expressed as percent activity relative to AII, unless noted otherwise.
d "AII-like" activity *in vivo* is expressed either as the percent activity relative to AII (or by the millimeters Hg of blood pressure increase produced by a 1 μg bolus intravenous injection of compound), unless noted otherwise.
e ID_{50} is in ng/rat/min (using 250 g rats), unless noted otherwise.
f Samanen et al.[11]
g Visser et al.[128]
h de Graaf et al.[44]
i van Battenburg et al., data originally reported relative to **72.**[129]
j Samanen et al.[41]
k Bovy et al.[45]
l Soffer et al.[46]

TABLE 3
Activities of Position Three Angiotensin II Analogs

A. [**Asp**[1], X[3]]-Angiotensin II Agonist Analogs

No.	*Asp*-Arg-*X*-Tyr-Ile-His-Pro-*Phe*	Agonist: *In vitro*[a] Rabbit aorta	Agonist: *In vivo*[b] Rat BP
1[c]	Val	100	100
51	Ala	30(RU)[c]	80.5[d],68[d],35[e]
52	Pro	80(RU)[f]	53[g],40[e]
53	Aib	—	1.0[e]
54	Cle	—	1–2[h]

a Agonist activity *in vitro,* expressed as percent activity relative to AII, was measured in the rabbit aorta strip assay unless shown otherwise (RU = rat uterus, GPI = guinea pig ileum), according to methods described in the references.
b Agonist "AII-like" activity *in vivo,* expressed as percent activity relative to AII, was measured in the rat blood pressure assay described in references listed in the table.
c Samanen et al.[11]
c Peach et al.[61]
d Khosla et al.[63]
e Marshall et al.[66]
f Khosla et al.[64]
g Khosla et al.[65]
h Park et al.[67]

TABLE 3 (continued)
Activities of Position Three Angiotensin II Analogs

B. [**Sar**[1], X[3]]-Angiotensin II Agonist Analogs

		Agonist	
No.	*Sar*-Arg-*X*-Tyr-Ile-His-Pro-*Phe*	*In vitro*[a] Rabbit aorta	*In vivo*[b] Rat BP
8[c]	Val	180	125
55[c]	Ala	38	—
56[c]	(NMe)Ala	120	—

[a] Agonist "AII-like" activity *in vitro,* expressed as percent activity relative to AII, pD_2 8.5, was measured in the rabbit aorta strip assay according to the method of Rioux et al.,[35] unless noted otherwise.

[b] Agonist "AII-like" activity *in vivo,* expressed as percent activity, was measured in the rat blood pressure assay (n = 5) according to the method of Regoli et al.,[22] unless noted otherwise.

[c] Samanen et al.[12]

C. [Sar[1],X[3],**Ile**[8]]-Angiotensin II Antagonist Analogs

		Agonist		Antagonist	
No.	*Sar*-Arg-*X*-Tyr-Ile-His-Pro-*Ile*	*In vitro*[a] AII-like[c]	*In vivo*[b] AII-like[d]	*In vitro*[a] Rabbit aorta	*In vivo*[b] Rat BP[e]
16[f]	Val	0	(10.2)	9.1	10.0
57[g]	D-Val	0		6.0	
58[f]	Ala	0	(7.5)	8.3	25.0
59[h]	Pro	1.5		7.17	
60[f]	Aib	0	(3.0)	7.0	50
61[f]	(NMe)Ala	0	(18.0)	8.9	25.0
62[g]	Phe	0	(12.5)	8.9	50
63[g]	D-Phe	0		<6.0	
64[g]	(NMe)Phe	0	(10)	7.85	15–25
65[g]	(αMe)Phe	0		<6.0	
66[g]	Tyr	0	(10)	8.5	20
67[g]	Phg	0	(12.5)	8.5	20
68[g]	Trp	0	(3)	7.25	250
69[g]	Glu	0		<6.0	
70[g]	Lys	0	(10)	6.7	25

[a] Agonist, "AII-like" activity, and antagonist activity, pA_2, were measured in the *in vitro* rabbit aorta strip assay, e.g., according to the method of Rioux et al.,[35] unless noted otherwise.

[b] Residual "AII-like" activity and antagonist activity, ID_{50}, were measured *in vivo,* in the rat blood pressure assay, e.g., described by Regoli et al.,[22] unless noted otherwise.

[c] "AII-like" activity *in vitro* is expressed as percent activity relative to AII, unless noted otherwise.

[d] "AII-like" activity *in vivo* is expressed either as the percent activity relative to AII (or by the millimeters Hg of blood pressure increase produced by a 1 μg bolus intravenous injection of compound), unless noted otherwise.

[e] ID_{50} is in ng/rat/min (using 250 g rats), unless noted otherwise.

[f] Samanen et al.[12]

[g] Samanen et al.[11]

[h] Khosla et al.[63]

TABLE 4
Activities of Position Four Angiotensin II Analogs

A. [Asp,[1] Y[4]]-AII Analogs

No.	Asp-Arg-Val-Y-Ile-His-Pro-Z (1 2 3 4 5 6 7 8)	Y^4	Z^8	Agonist *In vitro*[a] AII-like[c]	Agonist *In vivo*[b] AII-like[d]	Antagonist *In vitro*[a] pA_2	Antagonist *In vivo*[b] ID_{50}[e]
71[f]	Asp	Phe	Phe	11.0	(20.0)		(+)[f]
72[h]	Asp	Phe	Tyr			6.62(RU)[g]	
73[h]	Asp	(pF)Phe	Phe			6.30(RU)[g]	

[a] Agonist, "AII-like" activity, and antagonist activity, pA_2, were measured in the *in vitro* rabbit aorta strip assay (n=5) according to the method of Rioux et al.,[35] unless noted otherwise.
[b] Residual "AII-like" activity and antagonist activity, ID_{50}, were measured *in vivo,* in the rat blood pressure assay (n = 5) described by Regoli et al.,[22] unless noted otherwise.
[c] "AII-like" activity *in vitro* is expressed as percent activity relative to AII, pD_2 8.5 (n = 5), unless noted otherwise.
[d] "AII-like" activity *in vivo* is expressed by the millimeters Hg of blood pressure increase produced by 1 μg bolus intravenous injection of compound (n=5), unless noted otherwise.
[e] ID_{50} is in ng/rat/min (using 250 g rats), unless noted otherwise.
[f] Park et al.[16], slight antagonist effect was observed.
[g] For comparative purposes, these values were derived from the negative log of the K_I reported in Needleman et al.,[15] (RU) = rat uterus.
[h] Marshall et al.[14] and Needleman et al.[15]

B. [Sar[1](pX)Phe[4],Phe[8]]-AII Analogs

No.	X^4	Agonist *In vitro*[a] ANG-II -like[c]	Agonist *In vivo*[b] ANG-II -like[d]	Antagonist *In vitro*[a] pA_2	Antagonist *In vivo*[b] ID_{50}[e]	(σ_p)[f]	π	Hydrogen[h] bonding
8[o]	$-OH$	100	(50)	0		−0.37	−0.67	H^+
74	$-NH_3^+$	15[i]				−0.66	−1.23	H^+
75[o]	$-B(OH)_2$	10		0		0.12	−0.55	H^+
76	$-SO_2NH_2$	0.9[j]		0[j]		0.57	−1.82	H^+
77	$-SH$	0.9[j]		≪6.0[j]		0.15	0.39	H^+
78	$-SAcm$	9.0[j]		8.09[j]		~0.21[k]	~0[k]	H^+
79	$-SO_3^-$	0[j]		≪6.0[j]		0.09	−4.76(W)	H^-
80	$-NO_2$	(+)[l]		(+)[l]		0.78	−0.28	H^-
81[o]	$-OCH_3$	0	(20)	7.1	250	−0.27	−0.02	H^-
		<0.1(RU)[m]		7.5(RU)[m]				
82	$-H$	22[m]		7.5(RU)[m]		0	0	H^-
83[o]	$-F$	0	(22)	7.5	150	0.06	0.14	H^-

TABLE 4 (continued)
Activities of Position Four Angiotensin II Analogs

B. [Sar[1](pX)Phe[4],Phe[8]]-AII Analogs

No.	X^4	Agonist: *In vitro*[a] ANG-II -like[c]	Agonist: *In vivo*[b] ANG-II -like[d]	Antagonist: *In vitro*[a] pA_2	Antagonist: *In vivo*[b] ID_{50}[e]	(σ_p)[f]	π	Hydrogen[h] bonding
84[o]	−OEt	0.1 <0.1(RU)[m]	(2.5)	pa[n] <6.0(RU)[m]	150	−0.24	0.38	H^-
85[o]	−Cl	0	(5.5)	7.5	30	0.23	0.71	H^-
86[o]	−1	0	(80.0)	6.6	pa[n]	0.18	1.12	H^-

[a] Agonist, "AII-like" activity, and antagonist activity, pA_2, were measured in the *in vitro* rabbit aorta strip assay, e.g., according to the method of Rioux et al.,[35] unless noted otherwise.
[b] Residual "AII-like" activity and antagonist activity, ID_{50}, were measured *in vivo,* in the rat blood pressure assay, e.g., described by Regoli et al.,[22] unless noted otherwise.
[c] "AII-like" activity *in vitro* is expressed as percent activity relative to AII, unless noted otherwise.
[d] "AII-like" activity *in vivo* is expressed either as the percent activity relative to AII (or by the millimeters Hg of blood pressure increase produced by a 1 μg bolus intravenous injection of compound), unless noted otherwise.
[e] ID_{50} in ng/rat/min (using 250 g rats), unless noted otherwise.
[f] (σ_p) is a measure of inductive effects[40] on phenyl using $\sigma p^- = e^-$, $\sigma p^+ = e^+$.
[g] π is a measure of lipophilicity (π = +).[40]
[h] Hydrogen bonding: H^+ = capable of donating H bond; H^- = incapable of donating H bond.
[i] Escher et al.[38]
[j] Escher et al.[17]
[k] Estimated from values in Hansch et al.[40]
[l] Mixed agonist-antagonist effects seen above 100ng/ml.
[m] Goghari et al.[18] (RU) = rat uterus.
[n] pa = partial agonist, quantitation not possible.
[o] Samanen et al.[39]

C. [Sar[1](pX)Phe[4],Ile[8]]-AII Analogs

No.	X^4	Agonist: *In vitro*[a] AII-like[c]	Agonist: *In vivo*[b] AII-like[d]	Antagonist: *In vitro*[a] pA_2	Antagonist: *In vivo*[b] ID_{50}[e]	(σ_p)[f]	π	Hydrogen[h] bonding
16[j]	−OH	0	(10.0)	9.1	10	−0.37	−0.67	H^+
87[j]	$-B(OH)_2$	0	(0)	5.7	500	0.12	−0.55	H^+
88[j]	$-NO_2$	0	(5.0)	7.1	1000	0.78	−0.28	H^-
89[j]	−H	0 <0.1(RU)[i]	(2.7)	8.5 6.7(RU)[i]	80	0	0	H^-

TABLE 4 (continued)
Activities of Position Four Angiotensin II Analogs

C. [Sar[1](pX)Phe[4],Ile[8]]-AII Analogs

		Agonist		Antagonist				
No.	X^4	*In vitro*[a] AII-like[c]	*In vivo*[b] AII-like[d]	*In vitro*[a] pA_2	*In vivo*[b] ID_{50}[e]	(σ_p)[f]	π	Hydrogen[h] bonding
90[j]	$-OCH_3$	0 <0.1(RU)[i]	(3.0)	7.1 6.6(RU)[i]	100	−0.27	−0.02	H^-
91[j]	−F	0	(0)	6.8	750	0.06	0.14	H^-
92[j]	−Cl	0	(2.0)	<6.0	500	0.23	0.71	H^-
93[j]	−OEt	0	(5.0)	<6.0	>1000	−0.24	0.38	H^-

[a] Agonist, "AII-like" activity, and antagonist activity, pA_2, were measured in the *in vitro* rabbit aorta strip assay, e.g., according to the method of Rioux et al.,[35] unless noted otherwise.
[b] Residual "AII-like" activity and antagonist activity, ID_{50}, were measured *in vivo,* in the rat blood pressure assay, e.g., described by Regoli et al.,[22] unless noted otherwise.
[c] "AII-like" activity *in vitro* is expressed as percent activity relative to AII, unless noted otherwise.
[d] "AII-like" activity *in vivo* is expressed either as the percent activity relative to AII (or by the millimeters Hg of blood pressure increase produced by a 1 μg bolus intravenous injection of compound), unless noted otherwise.
[e] ID_{50} is in ng/rat/min (using 250 g rats), unless noted otherwise.
[f] (σ_p) is a measure of inductive effects[40] on phenyl using $\sigma_p^- = e^-$, $\sigma_p^+ = e^+$.
[g] π is a measure of lipophilicity ($\pi = +$).[40]
[h] Hydrogen bonding: H^+ = capable of donating H bond, H^- = incapable of donating H bond.
[i] (RU) = rat uterus.[18]
[j] Samanen et al.[39]

D. [Sar[1],Y[4],Z[8]]-AII Analogs

	Sar-Arg-Val-Y-Ile-His-Pro-Z (1 2 3 4 5 6 7 8)		Agonist		Antagonist	
No.	Y^4	Z^8	*In vitro*[a] AII-like[c]	*In vivo*[b] AII-like[d]	*In vitro*[a] pA_2	*In vivo*[b] ID_{50}[e]
94[f]	Tyr	D-Phe	0	(10.0)	9.0	12.5
95[f]	Phe	Tyr	0	(5.5)	7.5	50.0
96[f]	Phe	D-Tyr	0	(5.0)	7.0	30.0
97[f]	(pF)Phe	D-Phe	0	(10.0)	6.85	50.0

[a] Agonist, "AII-like" activity, and antagonist activity, pA_2, were measured in the *in vitro* rabbit aorta strip assay, e.g., according to the method of Rioux et al.,[35] unless noted otherwise.
[b] Residual "AII-like" activity and antagonist activity, ID_{50}, were measured *in vivo,* in the rat blood pressure assay, e.g., described by Regoli et al.,[22] unless noted otherwise.
[c] "AII-like" activity *in vitro* is expressed as percent activity relative to AII, unless noted otherwise.
[d] "AII-like" activity *in vivo* is expressed either as the percent activity relative to AII (or by the millimeters Hg of blood pressure increase produced by a 1 μg bolus intravenous injection of compound), unless noted otherwise.
[e] ID_{50} is in ng/rat/min (using 250 g rats), unless noted otherwise.
[f] Samanen et al.[39]

TABLE 4 (continued)
Activities of Position Four Angiotensin II Analogs

E. Bearing Conformational Constraints

	X-Arg-Val-Y-Ile-His-Pro-Z 1 2 3 4 5 6 7 8			Agonist		Antagonist	
No.	***X*[1]**	***Y*[4]**	***Z*[8]**	***In vitro*[a]** **AII-like[c]**	***In vivo*[b]** **AII-like[d]**	***In vitro*[a]** **pA_2**	***In vivo*[b]** **ID_{50}[e]**
98[f]	Asp	(αMe) Tyr			92.6[f]		
99[g]	Asp	(αMe) Phe		19 (RU)[g]	9[g]		
100[h]	Sar	(NMe) Tyr	Ile	(0.11 CA)[h]	0.34[h]	7.44[h]	2.14 (DR)[h]
101[i]	Sar	D-Tyr	Ile	0		<6.0	

[a] Agonist, ''AII-like'' activity, and antagonist activity, pA_2, were measured in the *in vitro* rabbit aorta strip assay, e.g., according to the method of Rioux et al.,[35] unless noted otherwise.

[b] Residual ''AII-like'' activity and antagonist activity, ID_{50}, were measured *in vivo,* in the rat blood pressure assay, e.g., described by Regoli et al.,[22] unless noted otherwise.

[c] ''AII-like'' activity *in vitro* is expressed as percent activity relative to AII, unless noted otherwise.

[d] ''AII-like'' activity *in vivo* is expressed either as the percent activity relative to AII (or by the millimeters Hg of blood pressure increase produced by a 1 μg bolus intravenous injection of compound), unless noted otherwise.

[e] ID_{50} in ng/rat/min (using 250 g rats), unless noted otherwise.

[f] Khosla et al.[74]

[g] Turk et al.[130]

[h] Khosla et al.,[73] (CA) = *in vitro* ''AII-like'' activity in the cat adrenal medulla, (DR) = dose ratio, *in vivo* ''AII-like'' activity with 250 ng/kg/min for 3 min.

[i] Samanen et al.[11]

TABLE 5
Activities of Position Five Angiotensin II Analogs

A. [Asp[1], Y[5]]-Angiotensin II Agonist Analogs

	Asp-Arg-Val-Tyr-Y-His-Pro-Phe 1 2 3 4 5 6 7 8	Agonist	
No.	***Y*[5]**	***In vitro*[a]** **Rabbit aorta**	***In vivo*[b]** **Rat BP**
102	Ala	5.0(RU)[c]	4.9[d], 3[e]
103	Abu		16[e]
104	Nle		21[e]
105	Arg		0[f]
106	Leu		25[g]
107	Chg		114[h]
108	(OMe)Thr		118[h]

TABLE 5 (continued)
Activities of Position Five Angiotensin II Analogs

A. [Asp[1], Y[5]]-Angiotensin II Agonist Analogs

	Asp-Arg-Val-Tyr-Y-His-Pro-Phe 1 2 3 4 5 6 7 8	Agonist	
No.	Y^5	*In vitro*[a] Rabbit aorta	*In vivo*[b] Rat BP
109	Pro		10[i]
110	Aib		1.0[j],0.9[i]
111	Cle		24[i]

[a] Agonist activity *in vitro,* expressed as percent activity relative to AII, was measured in the rabbit aorta strip assay unless shown otherwise (RU = rat uterus, GPI = guinea pig ileum) according to methods described in the references.
[b] Agonist ''AII-like'' activity *in vivo,* expressed as percent activity relative to AII, was measured in the rat blood pressure assay described in references listed in the table.
[c] Peach.[61]
[d] Park et al.[62]
[e] Schwyzer and Turrian.[54]
[f] Paruszewski.[131]
[g] Jorgensen et al.[132]
[h] Jorgensen et al.[55]
[i] Jorgensen et al.[69]
[j] Marshall.[66]

B. [Sar[1], Y[5]]-Angiotensin II Agonist Analogs

	Sar-Arg-Val-Tyr-Y-His-Pro-Phe 1 2 3 4 5 6 7 8	Agonist	
No.	Y^5	*In vitro*[a] Rabbit aorta	*In vivo*[b] Rat BP
8[c]	Ile	180	125
112[c]	(NMe)Ile	20.0	
113[c]	Cle	70	
114[c]	Dtc	250	100
115[c]	(SMe)Pen	80	

[a] Agonist ''AII-like'' activity *in vitro,* expressed as percent activity relative to AII, pD_2 8.5, was measured in the rabbit aorta strip assay (n=5) according to the method of Rioux et al.[35]
[b] Agonist ''AII-like'' activity *in vivo,* expressed as percent activity, was measured in the rat blood pressure assay (n=5) according to the method of Regoli et al.[22]
[c] Samanen, et al.[12]

TABLE 5 (continued)
Activities of Position Five Angiotensin II Analogs

C. [Sar[1],Y[5],Ile[8]]-Angiotensin II Antagonist Analogs

No.	Sar-Arg-Val-Tyr-Y-His-Pro-Ile Y^5	Agonist *In vitro*[a] AII-like[c]	Agonist *In vivo*[b] AII-like	Antagonist *In vitro*[a] Rabbit aorta[d]	Antagonist *In vivo*[b] Rat BP[e]
16[f]	Ile	0	(10.2)	9.1	10.0
116[f]	Ala	0	(5.0)	7.5	50.0
117[f]	Phe	0	(7.5)	9.15	100
118[f]	Tyr	0	(2.5)	9.6	50.0
119[f]	(pNO_2)Phe	0	(7.5)	8.7	100
120[f]	His	0	(10)	9.0	5
121[f]	Lys	0	(10)	8.7	50
122[f]	Arg	0	(10)	9.0	15
123[f]	Glu	0	(2.0)	9.0	100
124[f]	Nle	0	(20)	8.85	20
125[f]	Peg	0	(10)	7.75	20
126[f]	Trp	0	(7.5)	7.25	50
127[f]	(OMe)Thr	0	(2.5)	9.7	2.5
128[f]	(SMe)Pen	0	(12.5)	9.0	20.0
129[f]	Phg	0	(5)	9.0	5.5
130[f]	(βMe)Phe	0	(10)	9.3	7.5
131[g]	Pro	0	(7.0)	8.7	20.0
132[g]	Aib	0	(12.5)	7.7	25.0
133[h]	(NMe)Ile	0[h]		7.64[h]	—
134[g]	Cle	0	(10.0)	8.3	10.0
135[g]	Dtc	0	(11.4)	9.0	25.0
136[g]	(αMe)Tyr	0	(0[i])	8.25	100
137[j]	D-Val	0		<6.0	

[a] Agonist, ''AII-like'' activity, and antagonist activity, pA_2, were measured in the *in vitro* rabbit aorta strip assay, e.g., according to the method of Rioux et al.,[35] unless noted otherwise.

[b] Residual ''AII-like'' activity and antagonist activity, ID_{50}, were measured *in vivo,* in the rat blood pressure assay, e.g., described by Regoli et al.,[22] unless noted otherwise.

[c] ''AII-like'' activity *in vitro* is expressed as percent activity relative to AII, unless noted otherwise.

[d] ''AII-like'' activity *in vivo* is expressed either as the percent activity relative to AII (or by the millimeters Hg of blood pressure increase produced by a 1μg bolus intravenous injection of compound), unless noted otherwise.

[e] ID_{50} is in ng/rat/min (using 250 g rats), unless noted otherwise.

[f] Samanen et al.[56]

[g] Samanen et al.[12]

[h] Khosla et al.[73]

[i] Analog 123 displays residual ''All-like'' activity at higher doses, e.g., 2.5 mm at 10 μg.

[j] Samanen et al.[11]

TABLE 6
Activities of Position Six Angiotensin II Analogs

	X-Arg-Val-Tyr-Ile-Y-Pro-Z 1 2 3 4 5 6 7 8			Agonist		Antagonist	
No.	X^1	Y^6	Z^8	*In vitro*[a] AII-like[c]	*In vivo*[b] AII-like[d]	*In vitro*[a] pA_2	*In vivo*[b] ID_{50}[e]
138[f]	Asp	D-His	Phe		4.0		
139[g]	Asp	(γ-Im)Abu	Phe		0.05		
140[h]	Asp	Phe	Phe	1.9 (RU)	5.1	0 (RU)	
141[h]	Asp	Tal	Phe	4.2 (RU)	19.3		
142[h]	Asp	Pza	Phe	20.7 (RU)	49.6	0 (RU)	
143[i]	Asp	(N-τMe)His	Phe	10.0 (RU)	5.0		
144[i]	Asp	(N-πMe)His	Phe	0.1 (RU)	0.05		
145[j]	Asp	Ima	Phe		0.40		
146[j]	Asp	Pya	Phe		3.5		
147[j]	Asp	(4-NH_2)Phe	Phe		0.05		
148[k]	Asp	Lys	Phe		0.10		
149[l]	Asp	Ala	Phe	0.1 (GI)	0.8		
150[m]	Asp	Asn	Phe	0			
				2200 n*M* Bind[l]			
151[m]	Asp	Gln	Phe	+[m]			
				67 n*M* Bind			
(compared to:)							
1[m]	Asp	His	Phe	4.2 n*M* Bind[m]			
152[h]	Asp	Arg	Phe	0 (RU)	0.012	0 (RU)	
16[n]	Sar	His	Ile	0	(10)	9.0	10
153[n]	Sar	Phe	Ile	0	(10)	6.75	100
154[n]	Sar	Tal	Ile	0	(12.5)	7.5	100
155[n]	Sar	Trp	Ile	0	(7.5)	7.4	100
156[n]	Sar	(N-τMe)His	Ile	0	(2.5)	6.75	100
157[n]	Sar	Lys	Ile	0		≪6.0	
158[n]	Sar	Asp	Ile	0		≪6.0	
159[n]	Sar	D-His	Ile	0		≪6.0	

[a] Agonist, ''AII-like'' activity, and antagonist activity, pA_2, were measured in the *in vitro* rabbit aorta strip assay, e.g., according to the method of Rioux et al.,[35] unless noted otherwise.

[b] Residual ''AII-like'' activity and antagonist activity, ID_{50}, were measured *in vivo,* in the rat blood pressure assay, e.g., described by Regoli et al.,[22] unless noted otherwise.

[c] ''AII-like'' activity *in vitro* is expressed as percent activity relative to AII, unless noted otherwise.

[d] ''AII-like'' activity *in vivo* is expressed either as the percent activity relative to AII (or by the millimeters Hg of blood pressure increase produced by a 1μg bolus intravenous injection of compound), unless noted otherwise.

[e] ID_{50} is in ng/rat/min (using 250 g rats), unless noted otherwise.

[f] Khosla et al.[133]

[g] Bloemhoff et al.[134]

[h] Freer and Stewart.[50]

[i] Needleman et al.[49]

[j] Hsieh et al.[135]

[k] Schroder and Hampel.[136]

[l] Bumpus et al.[137]

[m] Pottorf et al.[52]

[n] Samanen et al.[41]

TABLE 7
Activities of Position Seven Angiotensin II Analogs

A. [**Asp**[1], Z[7]]-Angiotensin II Agonist Analogs

No.	Asp-Arg-Val-Tyr-Ile-His-Z-Phe Z^7	Agonist *In vitro*[a] Rabbit aorta	Agonist *In vivo*[b] Rat BP
160	Ala	0.1(RU)[c]	1.5[d],7.5[e]
161	Val[5] (NMe)Ala	4.3(GPI)[f]	22[f]
162	Cle	—	1–1.5[g]
163	Aib	—	1.0[e]

[a] Agonist activity *in vitro,* expressed as percent activity relative to AII, was measured in the rabbit aorta strip assay unless shown otherwise (RU = rat uterus, GPI = guinea pig ileum) according to methods described in the references.

[b] Agonist "AII-like" activity *in vivo,* expressed as percent activity relative to AII, was measured in the rat blood pressure assay described in references listed in the table.

[c] Peach.[61]

[d] Park et al.[62]

[e] Marshall.[66]

[f] Andreatta and Scheraga.[70]

[g] Park et al.[67]

B. [Sar[1], Z[7]]-Angiotensin II Agonist Analogs

No.	Sar-Arg-Val-Tyr-Ile-His-Z-Phe Z^7	Agonist *In vitro*[a] Rabbit aorta	Agonist *In vivo*[b] Rat BP
8	Pro	180[c]	125[c] 160(RU)[d]
164[d]	D-Pro	0.1 (RU)[d]	
165[d]	Tpr	238(RU)[d]	
166	Dtc	4.0[c]	
167	Aib	18.0[c]	

[a] Agonist "AII-like" activity *in vitro,* expressed as percent activity relative to AII, pD_2 8.5, was measured in the rabbit aorta strip assay (n=5) according to the method of Rioux et al.[35]

[b] Agonist "AII-like" activity *in vivo,* expressed as percent activity, was measured in the rat blood pressure assay (n=5) according to the method of Regoli et al.[22]

[c] Samanen et al.[12]

[d] Moore.[138]

TABLE 7 (continued)
Activities of Position Seven Angiotensin II Analogs

C. [Sar[1], Z[7], Ile[8]]-Angiotensin II Antagonist Analogs

		Agonist		Antagonist	
No.	Sar-Arg-Val-Tyr-Ile-His-Z-Ile Z[7]	*In vitro*[a] AII-like[c]	*In vivo*[b] AII-like[d]	*In vitro*[a] Rabbit aorta	*In vivo*[b] Rat BP[e]
16[f]	Pro	0	(10.2)	9.1	10.0
168[g]	D-Pro	0.1 (RU)[g]		<5.0 (RU)[g]	
169[f]	(NMe)Ala	0	(10.0)	8.4	20.0
170[f]	Aib	0	(20.0)	8.0	60
171[f]	Tpr	0	(15.0)	9.7 7.2 (RU)[g]	12.5
172[f]	(O)Tpr	0		6.7	
173[f]	Dtc	4.0	(40)	7.5	250
174[f]	L,L-lactam-Phe	0	(0[h])	8.3	100

[a] Agonist, ''AII-like'' activity, and antagonist activity, pA_2, were measured in the *in vitro* rabbit aorta strip assay, e.g., according to the method of Rioux et al.,[35] unless noted otherwise.

[b] Residual ''AII-like'' activity and antagonist activity, ID_{50}, were measured *in vivo,* in the rat blood pressure assay, e.g., described by Regoli et al.,[22] unless noted otherwise.

[c] ''AII-like'' activity *in vitro* is expressed as percent activity relative to AII, unless noted otherwise.

[d] ''AII-like'' activity *in vivo* is expressed either as the percent activity relative to AII (or by the millimeters Hg of blood pressure increase produced by a 1μg bolus intravenous injection of compound), unless noted otherwise.

[e] ID_{50} is in ng/rat/min (using 250 g rats), unless noted otherwise.

[f] Samanen et al.[12]

[g] (RU) = rat uterus, Moore.[138]

[h] Agonist activity was not observed at 1 μg/rat, but at higher doses, e.g., 5.0 mm Hg at 5 μg.

D. [Sar[1], Z[7], Ile[8]]-Angiotensin II Antagonist Analogs

			Agonist		Antagonist	
No.	Sar-Arg-Val-Tyr-Ile-His-Z-Ile Z[7]		*In vitro*[a] AII-like[c]	*In vivo*[b] AII-like[d]	*In vitro*[a] Rabbit aorta	*In vivo*[b] Rat BP[e]
175[f]	Arg		0		<6.0	
176[f]	Glu		0	(0)	6.0	>1000
177[f]	Nle		0	(2.0)	8.85	25
178[f]	Leu	Thr[8]	0	(2.0)	8.5	25
179[f]	Ile	Ala[8]	0	(0)	7.1	1000
180[f]	Phe		1–2	(25)	<6.0	35
181[f]	(NMe)Phe		0	(10)	7.5	15–25
182[f]	Phe	Ala[8]	0		<6.0	

TABLE 7 (continued)
Activities of Position Seven Angiotensin II Analogs

D. [Sar[1], Z[7], Ile[8]]-Angiotensin II Antagonist Analogs

		Agonist		Antagonist	
No.	Sar-Arg-Val-Tyr-Ile-His-Z-Ile *Z*[7]	*In vitro*[a] AII-like[c]	*In vivo*[b] AII-like[d]	*In vitro*[a] Rabbit aorta	*In vivo*[b] Rat BP[e]
183[f]	Peg	0		<6.0	
184[f]	Ceg	0	(5)	9.0	10
185[f]	(pCl)Phe	(+)[g]			

[a] Agonist, "AII-like" activity, and antagonist activity, pA_2, were measured in the *in vitro* rabbit aorta strip assay, e.g., according to the method of Rioux et al.,[35] unless noted otherwise.
[b] Residual "AII-like" activity and antagonist activity, ID_{50}, were measured *in vivo,* in the rat blood pressure assay, e.g., described by Regoli et al.,[22] unless noted otherwise.
[c] "AII-like" activity *in vitro* is expressed as percent activity relative to AII, unless noted otherwise.
[d] "AII-like" activity *in vivo* is expressed either as the percent activity relative to AII (or by the millimeters Hg of blood pressure increase produced by a 1μg bolus intravenous injection of compound), unless noted otherwise.
[e] ID_{50} is in ng/rat/min (using 250 g rats), unless noted otherwise.
[f] Samanen et al.[11]
[g] Binding data only, refer to Table 16.

TABLE 8
Activities of Position Eight Angiotensin II Analogs

A. [Z[8]]-AII Analogs

	X-Arg-Val-Tyr-Ile-His-Pro-Z 1 2 3 4 5 6 7 8		Agonist		Antagonist	
No.	*X*[1]	*Z*[8]	*In vitro*[a] AII-like[c]	*In vivo*[b] AII-like[d]	*In vitro*[a] pA_2	*In vivo*[b] ID_{50}[e]
Agonists:						
1	Asp	L-Phe	100	100		
186[f]	Asp	L-(αMe)Phe	100 (RU)	80		
187[g]	Sar	ΔPhe	100 (RG)			
188[h]	Asp	L-Thiq*	~1.0 (RU)			
189[i]	Asp	L-Apb	45 (RM)			
190[j]	Sar	(βMe)Phe	80–100			
191[k]	Sar	L-Dip	284 (RU)			
192[k]	Sar	D-Dip	48 (RU)			
193[l]	Asn	Ahp		11.5		
194[l]	Asn	Nle		7.3		
Partial agonists:						
195[m]	Asp	Tal		26.6	9.36	
196[n]	Sar	Tal	pa	(10)	8.6	2.5

TABLE 8 (continued)
Activities of Position Eight Angiotensin II Analogs

A. [Z^8]-AII Analogs

	X-Arg-Val-Tyr-Ile-His-Pro-Z 1 2 3 4 5 6 7 8		Agonist		Antagonist	
No.	***X*^{1}**	***Z*^{8}**	***In vitro*[a]** **AII-like[c]**	***In vivo*[b]** **AII-like[d]**	***In vitro*[a]** **pA$_2$**	***In vivo*[b]** **ID$_{50}$[e]**
Partial agonists:						
197[m]	Asp	Trp		22.2	8.36	
198[i]	Asp	Cha	20			
199[g]	Sar	ΔOal	55(RA)		7.0(RG)	
200[g]	Sar	(SBzl)Cys	44(RA)		7.1 (RG)	
201[j]	Sar	Peg	+ ← 10 ng/ml →		+	
202[o]	Sar	Car	15		6.12	
203[p]	Sar	Phe(Br^5)	14	<0.1	7.83	
204[i]	Asp	Apib	7.0 (RM)		5.57 (RM)	
205	Asp	D-Phe	5.0[q]		5.4[q]	
Antagonists:						
206[r]	Asp	Ala			8.32[r]	
207[s]	Asp	Ile			8.32 (RU)	
208[t]	Asp	Leu			8.21	170
209[s]	Asp	Cys			8.22 (RU)	
210	Sar	Ala	0[n]	(10)[n]	8.61,[u] 8.6[n] 6.5 (RG)[g]	15[n]
16	Sar	Ile	0[p]	(10.2)[n]	9.1[n] 8.0 (RG)[g] 9.33[v]	10[n]
14[t]	Sar	Leu			8.71	40
211	Sar	Thr	0[n] 0.5[w]	(9.8)[n] 0.6[w]	8.9[n] 8.79[w]	15[n]
212	Sar	His	0[n]	(0)[n]	7.2[n]	750[n]
213[k]	Sar	L-Bip	0		8.87 (RU)	
214[k]	Sar	D-Bip	0		8.7 (RU)	
94[j]	Sar	D-Phe	0	(7.5)	9.0	12.5
215[j]	Sar	D-Val	0	(10)	7.2	75
216[j]	Sar	D-(2)Nal	0	(5)	9.3	25
217[x]	Sar	(NMe)Phe		4.8	8.0 (RU)	
218[y]	Sar	(NMe)Ile		1.1	8.0	
219[l]	Asn	Ind		0.1		@[l]
220[z]	Sar	Pgl		(17% 2 10 μg)	8.41 (GI)	1800
221[o]	Sar	D-Phe(Br$_5$)	0	0	7.78	
222[aa]	Sar	Hfv	1.1	—	9.08	
223[aa]	Sar	D-Hfv	0	—	8.10	
224[h]	Asp	D-Thiq*			~7.0 (RU)	
225[j]	Sar	D-(αMe)Phe	0	(20.0)	9.7	10.0
226[g]	Sar	ΔAla			7.41 (RG)	
227[g]	Sar	ΔVal			6.57 (RG)	
16[j]	Sar	Ile	0	(10.2)	9.1	10.0

TABLE 8 (continued)
Activities of Position Eight Angiotensin II Analogs

A. [Z⁸]-AII Analogs

	X-Arg-Val-Tyr-Ile-His-Pro-Z 1 2 3 4 5 6 7 8		Agonist		Antagonist	
No.	***X*[1]**	***Z*[8]**	***In vitro*[a]** **AII-like[c]**	***In vivo*[b]** **AII-like[d]**	***In vitro*[a]** **pA_2**	***In vivo*[b]** **ID_{50}[e]**
Agonists:						
228[bb]	Sar	(αMeVal)		0.1	8.75 (RU)	
229[w]	Sar	Aib	0	11.0	8.7	19.76 (DR)
230[cc]	Sar	(SMe)Cys			9.2	
					8.5 (GI)	
					7.9 (RU)	
231[n]	Sar	(SMe)Pen	0		6.75	
232[w]	Sar	(OMe)Ser		0.49	9.66	
233[w]	Sar	(OMe)Thr		0.60	8.76	
234[n]	Sar	Pro	0		7.75	
235[n]	Sar	Dtc	0		<6.0	

* Stereochemistry assumed.

[a] Agonist, "AII-like" activity, and antagonist activity, pA_2, were measured in the *in vitro* rabbit aorta strip assay, e.g., according to the method of Rioux et al.,[35] unless noted otherwise.

[b] Residual "AII-like" activity and antagonist activity, ID_{50}, were measured *in vivo*, in the rat blood pressure assay, e.g., described by Regoli et al.,[22] unless noted otherwise.

[c] "AII-like" activity *in vitro* is expressed as percent activity relative to AII, unless noted otherwise.

[d] "AII-like" activity *in vivo* is expressed either as the percent activity relative to AII (or by the millimeters Hg of blood pressure increase produced by a 1 μg bolus intravenous injection of compound), unless noted otherwise.

[e] ID_{50} is in ng/rat/min (using 250 g rats), unless noted otherwise.

[f] Turk et al.[130]

[g] RA = rat adrenal glomerulosa cell aldosterone production, ΔOal = dehydro(β-octanyl)alanine, Hallinan et al.[139]

[h] Balasubramanian and Marshall.[140]

[i] Peach and Ober.[141]

[j] Samanen et al.[28]

[k] Hsieh et al.[29]

[l] @ = more potent *in vivo* than [Leu⁸]-AII, but less potent than [Sar¹,Leu⁸]-AII, Hsieh et al.[25]

[m] Khosla et al.[142]

[n] Samanen et al.[11]

[o] Escher et al.[27]

[p] Leduc et al.[26]

[q] Regoli et al.[20]

[r] Turker et al.[19]

[s] Needleman et al. (3)[15]

[t] Regoli et al.[34]

[u] Pals et al.[2]

[v] Turker et al.[21]

[w] Khosla et al.[143]

[x] Pena and Stewart.[144]

[y] Khosla et al.[73]

[z] Schoelkens et al.[145]

[aa] Hsieh et al.[146]

[bb] Turk et al.[72]

[cc] Freer et al.[58]

TABLE 8 (continued)
Activities of Position Eight Angiotensin II Analogs

B. [Z^8]-Angiotensin II C-Terminal Analogs

	X-Arg-Val-Tyr-Ile-His-Pro-Z (1 2 3 4 5 6 7 8)		Agonist		Antagonist	
No.	X^1	Z^8	*In vitro*[a] AII-like[c]	*In vivo*[b] AII-like[d]	*In vitro*[a] pA_2	*In vivo*[b] ID_{50}[e]
4[f]	Asn	Phe *L*—$NHCH(CH_2Ph)CO_2H$	100	100		
236[f]	Asn	Phe-ol *L*—$NHCH(CH_2Ph)$**CH_2OH**	0.09	0.001		
237[f]	Asn	*L*—$NHCH(CH_2Ph)$**$COCH_3$**	0.4	0.03		
238[f]	Asn	(αMe)Phe-OMe *L*—N(**CH_3**)(CH_2Ph)**CO_2CH_3**	0.002	0.001		
239[f]	Asn	Phe-OMe *L*—$NHCH(CH_2Ph)$**CO_2CH_3**	20	13		
240[g]	Asp	Phe-NH_2 *L*—$NHCH(CH_2Ph)$**$CONH_2$**		3.0		
94[h]	Sar	D-Phe *D*—$NHCH(CH_2Ph)CO_2H$	0	(7.5)	9.0	12.5
241[h]	Sar	D-Phe-NH_2 *D*—$NHCH(CH_2Ph)$**$CONH_2$**	0	(2.5)	7.5	12.5
16[h]	Sar	Ile *L*—$NHCH(iBu)CO_2H$	0	(10.2)	9.0	10.0
242[i]	Sar	Ile-NH_2 *L*—$NHCH(iBu)$**$CONH_2$**	0	(5.0)	7.1	75

[a] Agonist, "AII-like" activity, and antagonist activity, pA_2, were measured in the *in vitro* rabbit aorta strip assay, e.g., according to the method of Rioux et al.,[35] unless noted otherwise.

[b] Residual "AII-like" activity and antagonist activity, ID_{50}, were measured *in vivo,* in the rat blood pressure assay, e.g., described by Regoli et al.,[22] unless noted otherwise.

[c] "AII-like" activity *in vitro* is expressed as percent activity relative to AII, unless noted otherwise.

[d] "AII-like" activity *in vivo* is expressed either as the percent activity relative to AII (or by the millimeters Hg of blood pressure increase produced by a 1μg bolus intravenous injection of compound), unless noted otherwise.

[e] ID_{50} is in ng/rat/min (using 250 g rats), unless noted otherwise.

[f] Hsieh and Marshall.[30]

[g] Riniker and Schwyzer.[147]

[h] Samanen et al.[28]

[i] Samanen et al.[60]

TABLE 9
Activities of Angiotensin II Analogs Bearing Multiple Sidechain Modifications

									Agonist		Antagonist	
No.	**Sar-1**	**Arg-2**	**Val-3**	**Tyr-4**	**Ile-5**	**His-6**	**Pro-7**	**Phe 8**	***In vitro*[a] AII-like[c]**	***In vivo*[b] AII-like[d]**	***In vitro*[a] pA_2**	***In vivo*[b] ID_{50}[e]**
16[f]								Ile	0	(10.2)	9.0	50
97[f]				(pF) Phe				D-Phe	0	(10)	6.85	50
243[f]		D-Arg			(OMe) Thr			Ile	0	(10)	7.8	250
244[f]					(OMe) Thr	Tal		Ile	0	(5)	7.3	30
245[f]					(OMe) Thr		Tpr	Ile	0	(20)	8.9	25
246[f]					(OMe) Thr			D-(2Nal)	0	(5.5)	9.25	100
247[f]					His		Tpr	Ile	0	(5)	7.7	200
248[f]					His		Tpr	D-Phe-NH_2	0		<7.0	
249[f]							Tpr	D-(2)Nal	0	(7.5)	7.9	25
250[f]							Nle	D-Phe	0	(8)	7.0	25

[a] Agonist, ''AII-like'' activity, and antagonist activity, pA_2, were measured in the *in vitro* rabbit aorta strip assay, e.g., according to the method of Rioux et al.,[35] unless noted otherwise.

[b] Residual ''AII-like'' activity and antagonist activity, ID_{50}, were measured *in vivo,* in the rat blood pressure assay, e.g., described by Regoli et al.,[22] unless noted otherwise.

[c] ''AII-like'' activity *in vitro* is expressed as percent activity relative to AII, unless noted otherwise.

[d] ''AII-like'' activity *in vivo* is expressed either as the percent activity relative to AII (or by the millimeters Hg of blood pressure increase produced by a 1μg bolus intravenous injection of compound), unless noted otherwise.

[e] ID_{50} is in ng/rat/min (using 250 g rats), unless noted otherwise.

[f] Samanen et al.[11]

TABLE 10
Activities of Angiotensin II Analogs Bearing Amide Group Modifications

No.	Sar- 1	Arg- 2	Val- 3	Tyr- 4	Ile- 5	His- 6	Pro- 7	Phe 8	Agonist: *In vitro*[a] AII-like[c]	Agonist: *In vivo*[b] AII-like[d]	Antagonist: *In vitro*[a] pA_2	Antagonist: *In vivo*[b] ID_{50}[e]
16								Ile	0	(10.2)	9.0	50
14[f]								Leu			9.45 (AT)[f]	
251[f]		(ψ)[g]									9.43 (AT)	
252[f]		(ψ)	(ψ)								9.28 (AT)	
253[f]		(ψ)	(ψ)	(ψ)							7.92 (AT)	
254[h]		(ψ)	(ψ)	(ψ)	(ψ)	(ψ)	(ψ)	(ψ)			"ia" (AT)	
210[i]								Ala	0.7 (GI)			
									0.4 (RU)			
255[i]								β-Ala	0.08 (GI)			
									0.02 (RU)			
Retro-inverseo analogs												
256[i]	D-Asp D-Arg D-Val D-Tyr D-Val D-His D-Pro-D,L-Bml								ia (GI)			
									ia (RU)			
257[i]	β-Ala D-Arg D-Val D-Tyr D-Val D-His D-Pro-D,L-Bml								ia (GI)			
									ia (RU)			
258[i]	β-Ala D-Arg D-Val D-Tyr D-Val D-Hisβ-Ala-D,L-Bml								0.06 (GI)			
									0.03 (RU)			
259[i]	β-Ala D-Arg D-Val D-Tyr D-Val D-His D-Ala-D,L-Bml								0.019 (GI)			
									0.017 (RU)			

[a] Agonist, "AII-like" activity, and antagonist activity, pA_2, were measured in the *in vitro* rabbit aorta strip assay, e.g., according to the method of Rioux et al.,[35] unless noted otherwise.
[b] Residual "AII-like" activity and antagonist activity, ID_{50}, were measured *in vivo,* in the rat blood pressure assay, e.g., described by Regoli et al.,[22] unless noted otherwise.
[c] "AII-like" activity *in vitro* is expressed as percent activity relative to AII, unless noted otherwise.
[d] "AII-like" activity *in vivo* is expressed either as the percent activity relative to AII (or by the millimeters Hg of blood pressure increase produced by a 1 μg bolus intravenous injection of compound), unless noted otherwise.
[e] ID_{50} is in ng/rat/min (using 250 g rats), unless noted otherwise.
[f] AT = Negative logarithm of IC_{50} ^{125}I-AII binding to rat adrenal tissue, Roarke et al.[77]
[g] ψ = $-CH_2NH-$ amide bond replacement.
[h] Kaltenbronn et al.[78]
[i] Goissis et al.[82] The authors note that *258* retained 8.5% (GI) and 7.7% (RU) of the activities of [Ala[7]]-AII, while *259* retained 24.2% (GI) and 8.3% (RU) of the activities of [β-Ala[7]]-AII.

TABLE 11
Activities of Angiotensin II Analogs Bearing Multiple Conformational Constraints

							Agonist		Antagonist	
							In vitro[a]	*In vivo*[b]	*In vitro*[a] Rabbit	*In vivo*[b]
No.	Sar-Arg-	*W*-	Tyr-	*X*-	His-	*Y*- *Z*	AII-like[c]	AII-like[d]	aorta	Rat BP[e]
61[f]		(NMe) Ala		Ile		Pro-Ile	0	(18.0)	8.9	25.0
135[f]		Val		Dtc		Pro-Ile	0	(11.4)	9.0	25.0
229		Val		Ile		Pro-Aib	0[g]	11.0[g]	8.7[g]	19.76(DR)[g]
93		Val		Ile		Pro-D-Phe	0[g]	(7.5)[h]	9.0[h]	12.5[h]
260[f]		(NMe) Ala		Dtc		Pro-Aib	0		≪6.0	
261[f]		(NMe) Ala		Ile		Pro-D-Phe	0	(10)	<6.0	100
262[f]		Val		Dtc		Pro-D-Phe	0	(5)	7.5	1000
Compared to										
16		Val		Ile		Pro-Ile	0	(10.2)	9.1	10.0
210		Val		Ile		Pro-Ala	0	(10.0)	8.6	15.0
							0.50[g]	15.11[g]	8.61[g]	7.92 (DR)[g]

[a] Agonist, ''AII-like'' activity, and antagonist activity, pA_2, were measured in the *in vitro* rabbit aorta strip assay, e.g., according to the method of Rioux et al.,[35] unless noted otherwise.

[b] Residual ''AII-like'' activity and antagonist activity, ID_{50}, were measured *in vivo,* in the rat blood pressure assay, e.g., described by Regoli et al.,[22] unless noted otherwise.

[c] ''AII-like'' activity *in vitro* is expressed as percent activity relative to AII, unless noted otherwise.

[d] ''AII-like'' activity *in vivo* is expressed either as the percent activity relative to AII (or by the millimeters Hg of blood pressure increase produced by a 1 μg bolus intravenous injection of compound), unless noted otherwise.

[e] ID_{50} is in ng/rat/min (using 250 g rats), unless noted otherwise.

[f] Samanen et al.[12]

[g] Described by Khosla et al.[143] (DR) = dose ratio, *in vivo* AII-like activity with 250ng/kg/min for 3 min.

[h] Samanen, et al.[28]

TABLE 12
Activities of Angiotensin II Analogs Bearing Multiple D-Amino Acid Modifications

								Agonist		Antagonist	
										In vitro[a]	
No.	Sar-Arg-	Val-	Tyr-	Ile-	His-	Pro-	Phe	*In vitro*[a] AII-like[c]	*In vivo*[b] AII-like[d]	Rabbit aorta	*In vivo*[b] Rat BP[e]
263[f]	D-Arg		D-Tyr				Ile	0		<6.0	
264[g]		D-Val		D-Val			Phe	0		<6.0	
265[f]		D-Val		D-Val			Ile	0		<6.0	
266[g]			D-Tyr		D-His		Phe	0		<6.0	
267[f]			D-Tyr		D-His		Ile	0		<6.0	
268[g]				D-Val		D-Pro	Phe	0		<6.0	
269[g]				D-Val		D-Leu	Phe	0		<6.0	
270[g]				D-Val		D-Pro	D-Phe	0		<6.0	
271[f]					D-His		D-Phe	0		5.7	
272[f]					D-His		D-Val	0		6.0	
273[g]	D-Arg			D-Val		D-Pro		0		<6.0	
274[g]		D-Val		D-Val		D-Pro		0		<6.0	
Compared to											
16		Val		Ile		Pro-Ile		0	(10.2)	9.1	10.0
210		Val		Ile		Pro-Ala		0	(10.0)	8.6	15.0
								0.50[g]	15.11[g]	8.61[g]	7.92(DR)[g]

[a] Agonist, ''AII-like'' activity, and antagonist activity, pA_2, were measured in the *in vitro* rabbit aorta strip assay, e.g., according to the method of Rioux et al.,[35] unless noted otherwise.

[b] Residual ''AII-like'' activity and antagonist activity, ID_{50}, were measured *in vivo,* in the rat blood pressure assay, e.g., described by Regoli et al.,[22] unless noted otherwise.

[c] ''AII-like'' activity *in vitro* is expressed as percent activity relative to AII, unless noted otherwise.

[d] ''AII-like'' activity *in vivo* is expressed either as the percent activity relative to AII (or by the millimeters Hg of blood pressure increase produced by a 1 μg bolus intravenous injection of compound), unless noted otherwise.

[e] ID_{50} in ng/rat/min (using 250 g rats), unless noted otherwise.

[f] Samanen et al.[28]

[g] Samanen et al.[11]

TABLE 13
Activities of Cyclic Angiotensin II Analogs (Macrocycle Formed via Underlined Residues)

									Agonist		Antagonist	
No.	**Asp-**	**Arg-**	**Val-**	**Tyr-**	**Ile-**	**His-**	**Pro-**	**Phe**	***In vitro*[a] AII-like[c]**	***In vivo*[b] AII-like[d]**	***In vitro*[a] pA_2**	***In vivo*[b] Rat BP[e]**
Cyclic disulfides												
275[f]	Cys				Cys			Ile	(aff. 8000 n*M*)			
276[f]	D-Cys				Cys			Ile	(aff. 9000 n*M*)			
277[f]	Sar		Cys					Cys	(aff. >1 m*M*)			
278[f]	Sar		Cys					D-Cys	(aff. 18 μ*M*)			
279[f]	Sar		hCys					D-Cys	(aff. 0.8 μ*M*)			
280[f]	Sar		Cys					Cys	(aff. >1 m*M*)			
281[f]	Sar		Cys		Cys			Ile	(aff. 140 n*M*)		7.1	
282[f]	Sar		hCys		hCys			Ile	(aff. 4 n*M*)		9.16	
									ia 10^{-6}M[g]	ia 10^{-6} M[g]	9.0[g]	
Compared to the linear analogs												
283[f]	Sar		Ser		Ser			Ile	(aff. 7900 n*M*)		5.3	
284[f]	Sar		Met		Met			Ile	(aff. 26 n*M*)			

285[g]	(Asp)		Cys		Cys				ia 10^{-6} *M*	—		
286[g]			D-Cys		Cys				ia 10^{-6} *M*	—		
287[g]			Cys		D-Cys				ia 10^{-6} *M*	—		
288[g]			Pen		Cys				0.01			

TABLE 13 (continued)
Activities of Cyclic Angiotensin II Analogs
(Macrocycle Formed via Underlined Residues)

No.	Asp-	Arg-	Val-	Tyr-	Ile-	His-	Pro-	Phe	Agonist: *In vitro*[a] AII-like[c]	Agonist: *In vivo*[b] AII-like[d]	Antagonist: *In vitro*[a] pA_2	Antagonist: *In vivo*[b] Rat BP[e]
289[g]			Cys		Pen				0.1			
290[g]			Pen		Pen				0.01			
291[g]			hCys		Cys				0.1			
292[g]			Cys		hCys				1.2			
293[g]			hCys		hCys				49.0			
294[h]	Sar		Cys		Cys				(aff. 2.7 n*M*)[h]			
295[h]	Sar		hCys		hCys				(aff. 8.0 n*M*)[h]			
296[h]	Sar		t-Mpr		Cys				(aff. 0.5 n*M*)[h]			
297[h]	Sar		Cys		t-Mpr				(aff. 0.8 n*M*)[h]			
298[h]	Sar		hCys		t-Mpr				(aff. 0.65 n*M*)[h]			
Compared to the linear AII[h]									(aff. 2.5 n*M*)[h]			

299[i]	Cys							Cys			3.3 MEC (RU)[i]	
									0[j]	0[j]	<6.0[j]	
300[j]	D-Cys							Cys	0	0	<6	
301[j]	Cys-Sar							Cys	0	0	<6.0	
302[j]	D-Cys-Sar							Cys	0	0	<6.0	
303[j]	Cys							D-Cys	0	0	6.0–6.5	
304[j]	Asp		Cys				Cys	Phe	0	0	<6.0	
305[j]	Sar		Cys				Cys	Ile	0	0	<6.0	

TABLE 13 (continued)
Activities of Cyclic Angiotensin II Analogs
(Macrocycle Formed via Underlined Residues)

No.	Asp- 1	Arg- 2	Val- 3	Tyr- 4	Ile- 5	His- 6	Pro- 7	Phe 8	Agonist: *In vitro*[a] AII-like[c]	Agonist: *In vivo*[b] AII-Like[d]	Antagonist: *In vitro*[a] pA_2	Antagonist: *In vivo*[b] Rat BP[e]
Compared to the linear analog												
16[j]	Sar							Ile	0	(10.2)	9.1 2.1 MEC (RU)[i]	10
306[k]	Cys-Asp					Cys-Ile		Phe	0.02 (GI)	0.03		
307[k]	Cys-Asp					Cys		Phe	0.02 (GI)	0.04		
308[k] Cys	—					Cys		Phe	0.01 (GI)	0.02		
309[k]			Cys-Val				Cys	Phe		0.17		
310[k]			Cys				Cys	Phe	0.003 (GI)	0.01		
311[k]		Cys					Cys	Phe		0.01		
Cyclic amide peptides												
312[h]	Sar		Asp-CO$^{\beta}$		LysN$^{\varepsilon}$			Phe	(aff. 53 n*M*)[h]			
Compared to AII[e]									(aff. 2.5)[h]			

313[l]	Asp							Phe	ia 0.1 μg/kg			
314[m]	—		N$^{\alpha}$Val					Phe-CO$^{\alpha}$	0.1 (GPI)[m]	0.03		
315[m]	—		N$^{\alpha}$Gly-	Tyr	Gly	Gly	Gly	His-CO$^{\alpha}$	1.5 (GPI)[m]	0		
316[n]	N$^{\alpha}$-Arg-Lys							Phe-CO$^{\alpha}$				+
317[o]	des-Asp-N$^{\varepsilon}$Lys				Val			Phe-CO$^{\alpha}$			pD 6.87 (RU)[o]	

TABLE 13 (continued)
Activities of Cyclic Angiotensin II Analogs
(Macrocycle Formed via Underlined Residues)

No.	Asp-	Arg-	Val-	Tyr-	Ile-	His-	Pro-	Phe	Agonist: *In vitro*[a] AII-like[c]	Agonist: *In vivo*[b] AII-like[d]	Antagonist: *In vitro*[a] pA_2	Antagonist: *In vivo*[b] Rat BP[e]
Compared to AII[o]											pD 9.7 (RU)[o]	

318[p]	NH$^{\varepsilon}$-Lys							Phe-CO$^{\alpha}$	0.4[l] 0[q]	 0[q]	8.8 (RC)[p] <6.0[q]	

[a] Agonist, ''AII-like'' activity, and antagonist activity, pA_2, were measured in the *in vitro* rabbit aorta strip assay, e.g., according to the method of Rioux et al.,[35] unless noted otherwise.
[b] Residual ''AII-like'' activity and antagonist activity, ID_{50}, were measured *in vivo,* in the rat blood pressure assay, e.g., described by Regoli et al.,[22] unless noted otherwise.
[c] ''AII-like'' activity *in vitro* is expressed as percent activity relative to AII, unless noted otherwise.
[d] ''AII-like'' activity *in vivo* is expressed either as the percent activity relative to AII (or by the millimeters Hg of blood pressure increase produced by a 1μ bolus intravenous injection of compound), unless noted otherwise.
[e] ID_{50} is in ng/rat/min (using 250 g rats), unless noted otherwise.
[f] Sugg et al.[90]
[g] Spear et al.[91]
[h] Aff. = Receptor affinity measured in the rabbit uterus, Marshall et al.[102]
[i] MEC (RU) = Minimum effective concentration of peptide required to abolish the response to an ED75 dose of AII (10 n*M*) in the rat uterus assay, Matsoukas et al.[148]
[j] Samanen et al. unpublished results.
[k] Miranda and Juliano.[149]
[l] De Coen et al.[150]
[m] Jorgensen and Patton.[151]
[n] Ancans et al.[152]
[o] Chipens et al.[153]
[p] Chipens et al.[154]
[q] Samanen et al.[155]

TABLE 14
Comparison of Angiotensin II and Angiotensin I Analogs

		Agonist		Antagonist		Binding
				In vitro[a]		rat[c]
No.	Structure	*In vitro*[a] AII-like[d]	*In vivo*[b] AII-like[e]	Rabbit aorta	*In vivo*[b] Rat BP[f]	mesentery IC_{50}[g]
1	AII	100 (ED_{50} 3.02 n*M*)[h]		100		3.2
						6.4[i]
8	[Sar^{1}]-AII	190	100			1.3
240	AII-NH_2		3.0[j]			
319	AI	14.0	55			67.0
		5.0[k]				
320	[Sar^{1}]-AI	(ED_{50} 18 n*M*)	N.C.[l]			43.0
		9.0				
205	[D-Phe^{8}]-AII	5.0[m]		4300[m]		
94[n]	[Sar^{1},D-Phe^{8}]-AII		(7.5)	1.0	12.5	1.3
241[n]	[Sar^{1},D-Phe^{8}]-AII-NH_2		(2.5)	31.6	12.5	2000
321	[D-Phe^{8}]-AI	<1.0[o]	$\ll 1.0$[o]	>600[o]	>400[o]	
				$\gg 6000$[n]	$\gg 4600$[n]	4900[n]
322[n]	[Sar^{1},D-Phe^{8}]-AI		(7.5)	10.0	13.8	16.0
206[n]	[Ala^{8}]-AII		(8.8)	63.1	47.0	27.0
				4.78[o]		
210[n]	[Sar^{1},Ala^{8}]-AII	2.5	(10)	2.5	15.0	3.4
207	[Ile^{8}]-AII		0.4[m]	4.3[m]	210[m]	
16[n]	[Sar^{1},Ile^{8}]-AII		(10.2)	0.8	10.0	0.75
242[n]	[Sar^{1},Ile^{8}]-AII-NH_2		(5.0)	79.4	75	148.7
323[n]	[Sar^{1},Ala^{9}]-ANG I					8.9
324[n]	[Sar^{1},Ala^{10}]-ANG I					2.2
325[n]	[Sar^{1},D-Phe^{8},Ala^{9}]-ANG I					564
326[n]	[Sar^{1},D-Phe^{8},Ala^{10}]-ANG I					$>10^{5}$
327[p]	[Ile^{8}]-AI				90 (2900*)	

TABLE 14 (continued)
Comparison of Angiotensin II and Angiotensin I Analogs

No.	Structure	Agonist: *In vitro*[a] AII-like[d]	Agonist: *In vivo*[b] AII-like[e]	Antagonist: *In vitro*[a] Rabbit aorta	Antagonist: *In vivo*[b] Rat BP[f]	Binding rat[c] mesentery IC_{50}[g]
328[p]	[Cys^8]-AI				6.4 (50*)	
329[p]	[Val^8]-AI				7.0 (64*)	
330[p]	[Leu^8]-AI				7.8 (13*)	

[a] Agonist, "AII-like" activity, expressed as percent activity relative to AII and antagonist activity, pA_2, were measured in the *in vitro* rabbit aorta strip assay according to the method of Rioux et al.[35]

[b] Residual "AII-like" activity and antagonist activity, ID_{50}, were measured *in vivo,* in the rat blood pressure assay described by Regoli et al.[22]

[c] Binding to rat mesenteric vascular membranes was performed according to the method of Gunther et. al.,[156] competitive binding experiments were performed using 80 pM ^{125}I-AII and at least four concentrations of the analogs.

[d] For AII agonists, "AII-like activity" *in vivo* is expressed as the percent activity relative to AII.

[e] For AII antagonists, residual "AII-like" activity is expressed by the millimeters Hg of blood pressure increase produced by a 1 μg bolus intravenous injection of compound and displayed in parentheses.

[f] *In vivo* antagonist activity is expressed as the ID_{50} in ng/rat/min (using 250 g rats).

[g] IC_{50} is expressed in n*M* concentration of injected peptide.

[h] ED_{50} is the concentration, expressed in n*M,* which causes 50% maximal blood pressure increase.

[i] Gunther et al.[115]

[j] de Graaf, et al.[44]

[k] Riniker and Schwyzer.[147]

[l] N.C. = Noncompetitive. The maximal response to [Sar^1]-ANG I = 50% of the maximal response to AII.

[m] Devynck et al.[157]

[n] Samanen et al.[60]

[o] Oparil et al.[158]

[p] IC_{50} (n*M*) for inhibition of AII in nontreated or captopril-treated (*) rats, Jackson.[59]

TABLE 15
Angiotensin II Fragments

No.	Sequence	Agonist: *In vitro*[a] AII-like[c]	Agonist: *In vivo*[b] AII-like[e]	Antagonist: *In vitro*[a] pA_2	Antagonist: *In vivo*[b] ID_{50}[e]
331	Arg-Tyr-His-Phe	ia[f]			
332	Arg-Tyr-His-Ile	ia[f]			
333	D-Arg-Tyr-His-Phe	ia[f]			

TABLE 15
Angiotensin II Fragments (continued)

		Agonist		Antagonist	
No.	**Sequence**	***In vitro*[a] AII-like[c]**	***In vivo*[b] AII-like[e]**	***In vitro*[a] pA$_2$**	***In vivo*[b] ID$_{50}$[e]**
334	Arg-D-Tyr-His-Phe	ia[f]			
335	Arg-Tyr-D-His-Phe	ia[f]			
336	Arg-Tyr-His-D-Phe	ia[f]			
337	Arg-D-Tyr-D-His-Phe	ia[f]			
338	Arg-Tyr-His Pro Phe	ia[f]			
339	Arg-Tyr-His Pro Ile	ia[f]			
340	D-Arg-Tyr-His Pro Phe	ia[f]			
341	Arg-D-Tyr-His Pro Phe	ia[f]			
342	Arg-Tyr-Ile-His Pro Ile	ia[f]			
343	D-Arg-Tyr-Ile-His Pro Ile	ia[f]			
344	Arg-D-Tyr-Ile-His Pro Ile	ia[f]			
345	Arg-Tyr-D-Val-His Pro Ile	ia[f]			
346	D-Arg-D-Tyr-Ile-His Pro Ile	ia[f]			
347	D-Arg-Tyr-D-Val-His Pro Ile	ia[f]			
348	γ-Abu-Tyr-Ile-His-Pro-Ile			5.2[g]	
349	δ-Ape-Tyr-Ile-His-Pro-Ile			6.6[g]	
350	ε-Ahx-Tyr-Ile-His-Pro-Ile			5.3[g]	
351	γ-Gbu-Tyr-Ile-His-Pro-Ile			6.1g[b]	
20	Arg-Val-Tyr-Ile-His Pro Phe	47.4 (RI)[h]	15[h]		
43	Arg-Val-Tyr-Ile-His Pro Ile	0	(5.0)	7.75	200
1	Asp-Arg-Val-Tyr-Ile-His-Pro-Phe	100			
207	Asp-Arg-Val-Tyr-Ile-His Pro Ile	—	0.4[i]	8.4[i]	210[i]
210	Sar-Arg-Val-Tyr-Ile-His-Pro-Ala	0	(10)	8.6	15
				8.3[j]	10,000 (ID$_{100}$)[j]
352	Asp-Arg-Val-Ile-His-Pro		<0.005[j] 0.028% (Hum)[l]		
353	Sar-Arg-Val-Tyr-Ile-His Pro			7.5	>10^6 (ID$_{100}$)[j]
354	Sar-Arg-Val-Tyr-Ile-His-Pro-NH$_2$			7.6[j]	60,000 (ID$_{100}$)[j]
355	Asp-Arg-Val-Tyr-Ile-His		0.024% (Hum)[l]		
356	Sar-Arg-Val-Tyr-Ile-His-NH$_2$			<6.0[j]	
357	Sar-Arg-Val-Tyr-Ile-His-NMe$_2$			7.05[j]	100,000 (ID$_{100}$)[j]
358	Asp-Arg-Val-Tyr		<0.005[k]		
359	Val-Tyr-Ile-His-Pro	ia[f] 0.10[m]			
360	Boc-Val-Tyr-Ile-OMe	ia[f]			
361	Boc-Tyr-Ile-OMe	ia[f]			
362	Boc-Tyr-NHCH$_2$CH$_2$OMe	ia[f]			

TABLE 15 (continued)
Angiotensin II Fragments

		Agonist		Antagonist	
No.	Sequence	*In vitro*[a] AII-like[c]	*In vivo*[b] AII-like[e]	*In vitro*[a] pA_2	*In vivo*[b] ID_{50}[e]
363	Boc-(αMe)Tyr-NHCH$_2$CH$_2$PhOH	ia[f]			
364	Val-Tyr-Ile-His-Pro-Phe	2.0[n]	0.1[k] 0.2%(Hum)[o]		
365	Ac-Val-Tyr-Ile-His-Pro-Ile			6.6[g]	
366	Tyr-Ile-His Pro Phe	ia[f,n] 0.5(AS)[n]	0.04[k]		
367	Ile-His-Pro-Phe		0.0007[k]		
368	Boc-Pro-Phe	ia[f]			
369	Boc-Pro-D-Phe	ia[f]			
370	Boc-His-Pro-Phe	ia[f]			
371	Boc-His-Pro-D-Phe	ia[f]			
372	Boc-His-L-lactam-Phe	ia[f]			

[a] Agonist, ''AII-like'' activity, and antagonist activity, pA_2, were measured in the *in vitro* rabbit aorta strip assay, e.g., according to the method of Rioux et al.[35] unless noted otherwise.
[b] Residual ''AII-like'' activity and antagonist activity, ID_{50}, were measured *in vivo,* in the rat blood pressure assay, e.g., described by Regoli et al.,[22] unless noted otherwise.
[c] ''AII-like'' activity *in vitro* is expressed as percent activity relative to AII, unless noted otherwise.
[d] ''AII-like'' activity *in vivo* is expressed either as the percent activity relative to AII (or by the millimeters Hg of blood pressure increase produced by a 1 μg bolus intravenous injection of compound), unless noted otherwise.
[e] ID_{50} is in ng/rat/min (using 250 g rats), unless noted otherwise.
[f] ia = inactive at 10^{-6} M, Samanen et al.[11]
[g] Bovy et al.[45]
[h] de Graaf et al.[44]
[i] Regoli et al.[20]
[j] ID_{100} = Smallest dose (μg/kg/min) to completely block AII at 30 ng/kg, Bovy et al.[33]
[k] Park et al.[31]
[l] Kono et al.[32]
[m] Paiva et al.[127]
[n] Tsai et al.,[159] AS = adrenal steroidogenic activity.
[o] Hum = percent pressor activity relative to AII in humans, Kono et al.[160]

TABLE 16
Comparison of *In Vitro* and *In Vivo* Activities with Receptor Affinities

	Agonist activity		Antagonist activity		Binding ^{125}I-AII to bovine tissue		
No.	***In vitro*[a] % Rel. AII[c]**	***In vivo*[b] % Rel. AII (mm BP Hg)[d]**	***In vitro*[a] Rab. aorta pA_2**	***In vivo*[b] Rat BP ID_{50} (ng/r/m)[e]**	**(AT_1) Adr. cortex[f] IC_{50} (n*M*)**	**(AT_2) Ovary[g] IC_{50} (n*M*)**	**(AT_2) Brain[h] IC_{50} (n*M*)**
8 [Sar[1]]-AII	190	100			3.0	1.2	1.02
---							
16 [Sar[1],Ile[8]]-AII	0	(10.2)	9.1	10	3.0	0.5	3.6
Position two							
48 [des-Asp[1],**D-(NMe)Ala[2]**,Ile[8]]-AII	0	(10)	7.1	25	537	11.2	
41 [Sar[1],**Cit[2]**,Ile[8]]-AII	0	(25)	6.2		1220	80	5.0
Position three							
56 [Sar[1],**(NMe)Ala[3]**]-AII	120				49.9 1.44 (RM)[i]	0.27	
---							
61 [Sar[1],**(NMe)Ala[3]**,Ile[8]]-AII	0	(18)	8.9	25	1.9 (RM)		
Position four							
80 [Sar[1],**(pNO$_2$)Phe[4]**]-AII	pa[j]		pa[j]		1350	200	58
84 [Sar[1],**(pOEt)Phe[4]**]-AII	0.1	(2.5)	pa[k]	150	2200	1600	

TABLE 16 (continued)
Comparison of *In Vitro* and *In Vivo* Activities with Receptor Affinities

No.	Agonist activity		Antagonist activity		Binding ^{125}I-AII to bovine tissue		
	In vitro[a] % Rel. AII[c]	*In vivo*[b] % Rel. AII (mm BP Hg)[d]	*In vitro*[a] Rab. aorta pA_2	*In vivo*[b] Rat BP ID_{50} (ng/r/m)[e]	(AT_1) Adr. cortex[f] IC_{50} (n*M*)	(AT_2) Ovary[g] IC_{50} (n*M*)	(AT_2) Brain[h] IC_{50} (n*M*)
83 [Sar[1],**(pF)Phe**[4]]-AII	0	(22)	7.5	150	6000	10	5.5
---							
89 [Sar[1],**Phe**[4],Ile[8]]-AII	0	(2.7)	8.5	80	1.2	2.2	1.2
88 [Sar[1],**(pNO$_2$)Phe**[4],Ile[8]]-AII	0	(5.0)	7.1	1000	3000	110	26
90 [Sar[1],**(pOMe)Phe**[4],Ile[8]]-AII	0	(3)	7.1	100	150	30	59
91 [Sar[1],**(pF)Phe**[4],Ile[8]]-AII	0	(0)	6.8	750	8000	8	43
93 [Sar[1],**(pOEt)Phe**[4],Ile[8]]-AII	0	(5)	<6.0	>1000	6000	3800	178
Position five							
114 [Sar[1],**Dtc**[5]]-AII	250	100			14 10 (RM)	0.136	
113 [Sar[1],**Cle**[5]]-AII	70				33 0.8 (RM)	0.11	
			---				
127 [Sar[1],**(OMe)Thr**[5],Ile[8]]-AII	0	(7.5)	9.7	2.5	13.7 (RM)		
120 [Sar[1],**His**[5],Ile[8]]-AII	0	(10)	9.0	5	1.0 (RM)		
134 [Sar[1],**Cle**[5],Ile[8]]-AII	0	(10)	8.3	10	1.3 (RM)		
262 [Sar[1],**Dtc**[5],**D-Phe**[8]]-AII	0		<6.0		683 246 (RM)	0.82	

TABLE 16 (continued)
Comparison of *In Vitro* and *In Vivo* Activities with Receptor Affinities

	Agonist activity		Antagonist activity		Binding ^{125}I-AII to bovine tissue		
No.	*In vitro*[a] % Rel. AII[c]	*In vivo*[b] % Rel. AII (mm BP Hg)[d]	*In vitro*[a] Rab. aorta pA_2	*In vivo*[b] Rat BP ID_{50} (ng/r/m)[e]	(AT_1) Adr. cortex[f] IC_{50} (n*M*)	(AT_2) Ovary[g] IC_{50} (n*M*)	(AT_2) Brain[h] IC_{50} (n*M*)
Position six							
244 [Sar^1,**(OMe)Thr,^{5}Tal^6**,Ile^8]-AII	0	(5)		50	300	400	2.0
147 [Sar^1,**(pNH$_2$)Phe^6**]-AII	0.05[l]				8637 (PC12)[m]	12.2 (RL)[m]	
154 [Sar^1,**Tal^6**,Ile^8]-AII	0	(12.5)	7.5	100	950	0.5	13
155 [Sar^1,**Trp^6**,Ile^8]-AII	0	(7.5)	7.4	100	700	1.0	0.11
156 [Sar^1,**(N^{3}Me)His^6**,Ile^8]-AII	0	(2.5)	6.75	100	30	3	9.5
153 [Sar^1,**Phe^6**,Ile^8]-AII	0	(10.0)	6.5	100	300	100	2.0
157 [Sar^1,**Lys^6**,Ile^8]-AII	0		<6.0		>10^5	300	263
Position seven							
185 [Sar^1,**(pC/I)Phe^7**,Ile^8]-AII					>10^5	15	1.7
Position eight							
225 [Sar^1,**D-(αMe)Phe^8**]-AII	0	(20)	9.7	10	32.8	1.65	
					1.3 (RM)		
94 [Sar^1,**D-Phe^8**]-AII	0	(7.5)	9.0	12.5	1.3 (RM)		
211 [Sar^1,**Thr^8**]-AII	0	(9.8)	8.9	15	24	2.5	0.8
					4.4 (RM)		
196 [Sar^1,**Tal^8**]-AII	pa[j]	(10)	8.6	2.5	10	38	18
					0.11 (RM)		

TABLE 16 (continued)
Comparison of *In Vitro* and *In Vivo* Activities with Receptor Affinities

	Agonist activity		Antagonist activity		Binding ^{125}I-AII to bovine tissue		
No.	***In vitro***[a] **% Rel. AII**[c]	***In vivo***[b] **% Rel. AII (mm BP Hg)**[d]	***In vitro***[a] **Rab. aorta pA_2**	***In vivo***[b] **Rat BP ID_{50} (ng/r/m)**[e]	**(AT_1) Adr. cortex**[f] **IC_{50} (n*M*)**	**(AT_2) Ovary**[g] **IC_{50} (n*M*)**	**(AT_2) Brain**[h] **IC_{50} (n*M*)**
210 [Sar[1],**Ala**[8]]-AII	0	(10)	8.6	15	3.4 (RM)		
212 [Sar[1],**His**[8]]-AII	0	(0)	7.2	750	120 26 (RM)	3.0	264
241 [Sar[1],**D-Phe**[8]]-AII-NH_2	0	(2.5)	7.5	12.5	300 2000 (RM)	500	509
300 [**D-Cys**[1],**Cys**[8]]-AII	0		<6.0		1387 (RM)		
Miscellaneous							
373 LHRH					1500 (CA)[n] 1300 (RU)[n]		

[a] Agonist, "AII-like" activity, and antagonist activity, pA_2, were measured in the *in vitro* rabbit aorta strip assay, e.g., according to the method of Rioux et al.,[35] unless noted otherwise.
[b] Residual "AII-like" activity and antagonist activity, ID_{50}, were measured *in vivo,* in the rat blood pressure assay, e.g., described by Regoli et al.,[22] unless noted otherwise.
[c] "AII-like" activity *in vitro* is expressed as percent activity relative to AII, unless noted otherwise.
[d] "AII-like" activity *in vivo* is expressed either as the percent activity relative to AII (or by the millimeters Hg of blood pressure increase produced by a 1 μg bolus intravenous injection of compound), unless noted otherwise.
[e] ID_{50} is in ng/rat/min (using 250 g rats), unless noted otherwise.
[f] Bovine adrenal cortex, Aiyar et al.[125]
[g] Bovine ovary, Aiyar et al.[125]
[h] Bovine cerebellum, Aiyar et al.[125]
[i] RM = rat mesentery, Samanen et al.[60]
[j] Partial agonist activity, difficult to quantitate.
[k] Partial agonist activity seen above 100 ng/ml l) Hsieh et al.[135]
[m] RL = rat liver, PC12 = PC12 cells, Speth and Kim.[118]
[n] CA = canine adrenal glomerulosa membrane; RU = rat uterine membranes.[112]

TABLE 17
Peptide Angiotensin II Antagonists with Diminished Partial Agonist Activity[f]

No.	Asp- 1	Arg- 2	Val- 3	Tyr- 4	Ile- 5	His- 6	Pro- 7	Phe 8	Agonist: *In vitro*[a] AII-like[c]	Agonist: *In vivo*[b] AII-like[d]	Antagonist: *In vitro*[a] pA_2	Antagonist: *In vivo*[b] ID_{50}[e]
210	Sar							Ala	0	(10.0)	7.5	50
16	Sar							Ile	0	(10.2)	9.1	10
127	Sar				(OMe) Thr			Ile	0	**(2.5)**	9.7	2.5
129	Sar				Phg			Ile	0	**(5)**	9.0	5.5
177	Sar				Nle			Ile	0	**(2.0)**	8.85	25
184	Sar				Ceg			Ile	0	**(5)**	9.0	10
178	Sar				Leu			Thr	0	**(2.0)**	8.5	25
241	Sar							D-Phe-NH_2	0	**(2.5)**	7.5	12.5
216	Sar							D-(2)Nal	0	**(5)**	9.3	25

[a] Agonist, ''AII-like'' activity, and antagonist activity, pA_2, were measured in the *in vitro* rabbit aorta strip assay, e.g., according to the method of Rioux et al.,[35] unless noted otherwise.

[b] Residual ''AII-like'' activity and antagonist activity, ID_{50}, were measured *in vivo,* in the rat blood pressure assay, e.g., described by Regoli et al.,[22] unless noted otherwise.

[c] ''AII-like'' activity *in vitro* is expressed as percent activity relative to angiotensin II, unless noted otherwise.

[d] ''AII-like'' activity *in vivo* is expressed either as the percent activity relative to angiotensin II (or by the millimeters Hg of blood pressure increase produced by a 1 μg bolus intravenous injection of compound), unless noted otherwise.

[e] ID_{50} is in ng/rat/min (using 250 g rats), unless noted otherwise.

[f] See earlier tables for appropriate references.

TABLE 18
Peptide Angiotensin II Antagonists with Diminished Antagonist Activity but Retained Partial Agonist Activity[f]

No.	Asp- 1	Arg- 2	Val- 3	Tyr- 4	Ile- 5	His- 6	Pro- 7	Phe 8	Agonist: *In vitro*[a] AII-like[c]	Agonist: *In vivo*[b] AII-like[d]	Antagonist: *In vitro*[a] pA_2	Antagonist: *In vivo*[b] ID_{50}[e]
210	Sar							Ala	0	(10.0)	7.5	50
16	Sar							Ile	0	(10.2)	9.1	10
40	Sar	Lys						Ile	0	(12.5)	7.75	75
119	Sar				(pNO_2) Phe			Ile	0	(7.5)	8.7	100
155	Sar					Trp		Ile	0	(7.5)	7.4	100
153	Sar					Phe		Ile	0	(10)	6.75	100

TABLE 18 (continued)
Peptide Angiotensin II Antagonists with Diminished Antagonist Activity but Retained Partial Agonist Activity[f]

	Asp-Arg-Val-Tyr-Ile-His-Pro-Phe								Agonist		Antagonist	
No.	1	2	3	4	5	6	7	8	*In vitro*[a] AII-like[c]	*In vivo*[b] AII-like[d]	*In vitro*[a] pA_2	*In vivo*[b] ID_{50}[e]
181	Sar						(NMe) Phe	Ile	0	(10)	7.5	20
173	Sar						Dtc	Ile	4.0	(40)	7.5	250
215	Sar							D-Val	0	(10)	7.25	75

Note: See also the analogs of Table 9 and 11.

[a] Agonist, "AII-like" activity, and antagonist activity, pA_2, were measured in the *in vitro* rabbit aorta strip assay according to the method of Rioux et al.[35]
Angiotensin II antagonists bearing sarcosine in position 1 are slowly reversible antagonists depressing both the slope and maximum of the angiotensin II dose-response curve *in vitro* at high doses but not at low doses.[35] The pA_2 values reported here for comparative purposes were determined at low doses in the range of competitive inhibition, and consequently may be overestimated.[20]

[b] Residual "AII-like" activity and antagonist activity, ID_{50}, were measured *in vivo,* in the rat blood pressure assay described by Regoli et al.[22]

[c] "AII-like" activity *in vitro* is expressed as percent activity relative to AII.

[d] "AII-like" activity *in vivo* is expressed by the millimeters Hg of blood pressure increase produced by a 1 μg bolus intravenous injection of compound.

[e] ID_{50} in ng/rat/min (using 250 g rats).

TABLE 19
Abbreviations

Abbreviations employed in Chapters 2 and 3 that have not been described by IUPAC convention and that are described in Table 17 include:

ε-Ahx, epsilon-aminohexanoic acid
γ-Abu, gamma-aminobutyric acid
Aib, α-aminoisobutyric acid
ε-Ape, epsilon-aminopentanoic acid
Apb, alpha-amino-gamma-phenylbutyric acid
Apib, beta-amino-gamma-phenylbutyric acid
Abu, α-aminobutyric acid
(αMe)Ala, also Aib, α-aminoisobutyric acid
(αMe)Phe, α-methylphenylalanine
(αMe)Tyr, α-methyltyrosine
(αMe)Val, α-methylvaline
Cha, β-cylcohexylalanine
Chg, β-cyclohexylglycine
Cit, citrulline
Cle, cycloleucine, α-amino-cyclopentane-α-carboxylic acid
ΔAla, dehydroalanine, α-aminoacrylic acid

TABLE 19 (continued)
Abbreviations

ΔPhe, Z-dehydrophenylalanine, Z-α-amino-β-phenylacrylic acid
Dtc, 5,5-dimethylthiazolidine-4-carboxylic acid
ε-Ghx, 6-guanidinohexanoic acid
γ-Gbu, 4-guanidinobutyric acid
(γ-Im)Abu, 1-amino-4-(4-imidazolyl)butyric acid
hArg, homoarginine, 1-amino-6-guanidinohexanoic acid
hCys, homocysteine, 1-amino-3-mercaptobutyric acid
Ima, β-(2-imidazolyl)alanine
Ind, 2-aminoindane-2-carboxylic acid
L,L-lactam-Phe, (S-3-amino-2-oxo-1-pyrrolidine)-2-S-phenylbutyric acid
Mpr, mercaptoproline, thiazolidine-4-carboxylic acid
(NMe)Ala, N-α-methylalanine
(N-πMe)His, pros-methyl^{im}-histidine (in the peptide literature, typically referred to as 1-Me-His, contrary to organic nomenclature)
(N-τMe)His, tele-methyl^{im}-histidine (in the peptide literature, typically referred to as 3-Me-His, contrary to organic nomenclature)
(NMe)Ile, N-α-methylisoleucine
(NMe)Phe, N-α-methylphenylalanine
(4-NH_2)Phe, para-aminophenylalanine
(OMe)Thr, O-methylthreonine
(OMe)Tyr, O-methyltyrosine, also para-methoxyphenylalanine
(OEt)Tyr, O-ethyltyrosine, also para-ethoxyphenylalanine
Peg, phenethylglycine
(pF)Phe, para-fluorophenylalanine
Pgl, phenylglycine, also α-aminophenylacetic acid
(βMe)Phe, beta-methylphenylalanine
(SAcm)Phe, para-(thioacetamidomethyl)phenylalanine
(SMe)Pen, S-methylpenicillamine
Pza, 2-(3-pyrazolyl)alanine
Tal, Thi 2-(2-thiazolyl)alanine
Thiq, 1,2,3,4-tetrahydroisoquinoline-3-carboxylic acid

REFERENCES

1. **IUPAC-IUB,** Nomenclature, joint commission on biochemical nomenclature, nomenclature and symbolism for amino acids and peptides, *J. Biol. Chem.,* 247, 977, 1971.
2. **Pals, D. T., Denning, G. S., Jr., and Kennan, R. E.,** Historical development of saralasin, *Kidney Int.,* 15 (Suppl. 9), 7, 1979.
3. **Hata, T., Ogihara, T., Mikami, M., Nakamura, M., Maruyama, A., Mandai, T., and Kumahara, Y.,** Comparison of the biological effects of two angiotensin II analogues in hypertensive patients with sodium depletion, *Life Sci.,* 22, 1955, 1978.
4. **Case, D. B., Wallace, J. M., Keim, H. J., Sealey, J. E., and Laragh, J. M.,** Usefulness and limitations of saralasin, a partial competitive agonist of angiotensin II, for evaluating the renin and sodium factors in hypertensive patients, *Am. J. Med.,* 60, 825, 1976.

5. **Kleeman, H. W., Heitsch, H., Henning, R., Kramer, W., Kocher, W., Lerch, U., Linz, W., Nickel, W. U., Ruppert, D., Urbach, H., Utz, R., Wagner, A., Weck, R., and Wiegand, F.,** Renin inhibitory pentols showing improved enteral bioavailability, *J. Med. Chem.,* 35, 559, 1992.
6. **Boyd, S. A., Fung, A. K. L., Baker, W. R., Mantei, R. A., Armiger, Y. L., Stein, H. H., Cohen, J., Egan, D. A., Barlow, J. L., Klinghøfer, V., Verburg, K. M., Martin, D. L., Young, G. A., Polakowski, J. S., Hoffman, D. J., Garren, K. W., Perun, T. J., and Kleinert, H. D.,** C-Terminal modifications of nonpeptide renin inhibitors: improved oral bioavailability via modification of physicochemical properties, *J. Med. Chem.,* 35, 1735, 1992.
7. **Furukawa, Y., Kishimoto, S., and Nishikawa, K.,** Hypotensive imidazole-5-Acetic Acid Derivatives. U.S. Patent 4, 355, 040, 1982.
8. **Furukawa, Y., Kishimoto, S., and Nishikawa, K.,** Hypotensive Imidazole Derivatives. U.S. Patent 4,340,598, 1982.
9. **Bumpus, F. M. and Khosla, M. C.,** Angiotensin analogs as determinants of the physiologic role of angiotensin and its metabolites, in *Hypertension: Physiopathology and Treatment,* Genest J. K. and Kuchel, O., Eds., McGraw-Hill, New York, 1977, 183.
10. **Khosla, M. C., Smeby, R. R., and Bumpus, F. M.,** Structure-activity relationship in angiotensin II analogs, in *Handbook of Experimental Pharmacology,* 37, 126, 1974.
11. **Samanen, J. et al.,** unpublished results.
12. **Samanen, J., Cash, T., Narindray, D., Brandeis, E., Adams, W., Weideman, H., Yellin, T., and Regoli, D.,** An investigation of angiotensin II agonist and antagonist analogues with 5,5-dimethylthiazolidine-4-carboxylic acid and other constrained amino acids, *J. Med. Chem.,* 34, 3036, 1991.
13. **Khairallah, P. A., Toth, A., and Bumpus, F. M.,** Analogs of angiotensin II. II. Mechanism of receptor interactions, *J. Med. Chem.,* 13, 181, 1970.
14. **Marshall, G. R., Vine, W., and Needleman, P.,** A specific competitive inhibitor of angiotensin II, *Proc. Natl. Acad. Sci. U.S.A.,* 67, 1624, 1970.
15. **Needleman, P., Johnson, E. M., Vine, W., Flannigan, E., and Marshall, G. R.,** Comparative pharmacology of angiotensin antagonists, in *Chem. and Biol. Peptides, Proc. 3rd Am. Peptide Symp.,* Meienhofer, J., Ed., Ann Arbor Science Publishers, Ann Arbor, MI, 1972, 501.
16. **Park, W. K., Regoli, D., and Rioux, F.,** Characterization of angiotensin receptors in vascular and intestinal smooth muscle, *Br. J. Pharmacol.,* 48, 288, 1973.
17. **Escher, E., Bernier, M., and Parent, P.,** Angiotensin II analogues. II. Synthesis and incorporation of the sulfur-containing aromatic amino acids: L-(4′-SH)Phe, L-(4′-SO_2NH_2)Phe, L-(4′-SO_3^-)Phe and L-(4′-S-CH_3)Phe, *Helv. Chim. Acta,* 66, 1355, 1983.
18. **Goghari, M. H., Franklin, K. J., and Moore, G. J.,** Structure–activity relationships for the competitive antagonist [sarcosine[1],O-methyltyrosine[4]]angiotensin II (sarmesin), *J. Med. Chem.,* 29, 1121, 1986.
19. **Turker, R. K., Yamamoto, P., Khairallah, P. A., and Bumpus, F. M.,** Competitive antagonism of 8-Ala angiotensin II to angiotensins I and II on isolated rabbit aorta and rat ascending colon, *Eur. J. Pharmacol.,* 15, 285, 1971.
20. **Regoli, D., Park, W. K., and Rioux, F.,** Pharmacology of angiotensin, *Pharmacol. Rev.,* 26, 69, 1974.
21. **Turker, R. K., Hall, M. M., Yamamoto, M., Sweet, C. S., and Bumpus, F. M.,** A new long lasting competitive inhibitor of angiotensin II, *Science,* 177, 1203, 1972.
22. **Regoli, D. and Park, W. K.,** The pressor and myotropic effects and the antagonistic properties of several analogs of angiotensin II, *Can. J. Physiol. Pharmacol.,* 50, 99, 1972.
23. **Timmermans, P. B. M. W. M., Wong, P. C., Chiu, A. T., and Herblin, W. F.,** Nonpeptide angiotensin II receptor antagonists, *Trends Pharmacol. Sci.,* 12, 55, 1991.
24. **Marshall, G. R.,** Determination of the receptor bound conformation of angiotensin, *Dtsch. Apoth. -Ztg.,* 85, 2783, 1986.
25. **Hsieh, K., Jorgenson, E. C., and Lee, T. C.,** Angiotensin II analogues. XII. Role of the aromatic ring of position 8 phenylalanine in pressor activity, *J. Med. Chem.,* 22, 1038, 1979.

26. **Leduc, R., Bernier, M., and Escher, E.,** Angiotensin II analogues. Synthesis and incorporation of halogenated amino acids 3-(4′-idophenyl)alanine, 3-(3′,5′-dibromo-4′-chlorophenyl)alanine, 3-(3′,4′,5′-tribromophenyl)alanine, and 3-(2′,3′,4′,5′,6′-penta-bromophenyl)alanine, *Helv. Chim. Acta,* 66, 960, 1983.
27. **Escher, E., Guillemette, G., Leukart, O., and Regoli, D.,** Pharmacological properties of two analogues of angiotensin II containing carboranylalanine (Car), *Eur. J. Pharmacol.,* 66, 267, 1980.
28. **Samanen, J., Narindray, D., Adams, W., Cash, T., Yellin, T., and Regoli, D.,** Effects of D-amino acid substitution on antagonist activities of angiotensin II analogues, *J. Med. Chem.,* 31, 510, 1988.
29. **Hsieh, K., LaHann, T. R., and Speth, R. C.,** Topographic probes of angiotensin and receptor: potent angiotensin II agonist containing diphenylalanine and long-acting antagonists containing biphenylalanine and 2-indian amino acid in position 8, *J. Med. Chem.,* 32, 898, 1989.
30. **Hsieh, K., Marshall and G. R.,** Role of C-terminal carboxylate in angiotensin II activity: alcohol, ketone, and ester analogues of angiotensin II, *J. Med. Chem.,* 29, 1968, 1986.
31. **Park, W. K. C., Choi, C., Rioux, F., and Regoli, D.,** Synthesis of peptides by the solid phase method. I. Tetradecapeptide renin substrate and intermediate peptides, nonpeptides, and fragments of angiotensin II, *Can. J. Biochem.,* 52, 106, 1974.
32. **Kono, T., Taniguchi, A., Imura, H., Oseko, F., and Khosla, M. C.,** Biological activities of angiotensin II-(1-6)-hexapeptide and angiotensin II-(1-7)-heptapeptide in man, *Life Sci.,* 38, 1515, 1986.
33. **Bovy, P. R., O'Neal, J. M., Olins, G. M., Patton, D. R., McMahon, E. G., Palomo, M., Koepke, J. P., Salles, K. S., Trapani, A. J., Smits, G. J., McGraw, D. E., and Hutton, W. C.,** Structure–activity relationships for the carboxyl-terminus truncated analogues of angiotensin II, a new class of angiotensin II antagonists, *J. Med. Chem.,* 33, 1477, 1990.
34. **Regoli, D., Rioux, F., Park, W. K., and Choi, C.,** Role of the N-terminal amino acid for the biological activities of angiotensin and inhibitory antagonists, *Can. J. Physiol. Pharmacol.,* 52, 39, 1974.
35. **Rioux, F., Park, W. K., and Regoli, D.,** Application of drug-receptor theories to angiotensin, *Can. J. Physiol. Pharmacol.,* 51, 665, 1973.
36. **Riniker, B., Brunner, M., and Schwyzer, R.,** β-Asp[1]-Val[5]-hypertensin II, a hypertensin analogue of extremely high biological activity, *Angew. Chem. Int. Ed. Engl.,* 1, 405, 1962.
37. **Scanlon, M. N., Matsoukas, J. M., Franklin, K. J., and Moore, G. J.,** A new approach to angiotensin antagonists: methylation of the tyrosine hydroxyl in angiotensin II, *Life Sci.,* 34, 317, 1984.
38. **Escher, E. and Bernier, M.,** Structure–activity studies on position 4 of angiotensin II, in *Peptides 1980, Proc. 16th Eur. Peptide Symp.,* Brunfeldt, K., Ed., Scriptor, Copenhagen, 1981, 457.
39. **Samanen, J., Cash, T., Narindray, D., Brandeis, E., Yellin, T., and Regoli, D.,** The role of position four in angiotensin II antagonism: a structure–activity study, *J. Med. Chem.,* 32, 1366, 1989.
40. **Hansch, C., Leo, A., Unger, S. H., Kim, K. H., Nikaritani, D., and Lien, E. J.,** "Aromatic" substituent constants for structure–activity correlations, *J. Med. Chem.,* 16, 1207, 1973.
41. **Samanen, J., Brandeis, E., Narindray, D., Adams, W., Cash, T., Yellin, T., and Regoli, D.,** The importance of residues two (arginine) and six (histidine) in high affinity angiotensin II antagonists, *J. Med. Chem.,* 31, 737, 1988.
42. **Schattenkerk, C. and Havinga, E.,** Studies on polypeptides. IV. Syntheses and activities of some angiotensin-like peptides, *Rec. Trav. Chim. Pays-Bas,* 84, 653, 1965.
43. **Schroder, E.,** Synthese von Val[5]-angiotensin II-Asp[1]-β-amide, *Justus Liebigs Ann. Chem.,* 680, 132, 1964.
44. **de Graaf, J. S., Jansen, A. C. A., Kerling, K. E. T., Schattenkerk, C., and Havinga, E.,** Studies on polypeptides. XII. Myotropic activities of some des-Asp[1]-[Ile[5]]-angiotensin analogues, *Rec. Trav. Chim. Pays-Bas,* 90, 301, 1971.

45. **Bovy, P. R., O'Neal, J. M., McMahon, E., Palomo, M., Smits, G. J., Trapani, A. J., and McGraw, D.,** Synthesis and biological activity of N-terminus modified [Ile_8] angiotensin II analogues, *Eur. J. Med. Chem.,* 25, 589, 1990.
46. **Soffer, R. L., Bandyopadhyay, S., Rosenberg, E., Hoeprich, P., Teitelbaum, A., Brunck, T., Colby, C. B., and Gloff, C.,** Unexpected binding of an octapeptide to the angiotensin II receptor, *Proc. Natl. Acad. Sci. U.S.A.,* 84, 9219, 1987.
47. **Hartman, G. D. and Halczenko, W.,** A convenient synthesis of 4-aminomethyl-L-phenylalanine, *Synth. Commun.,* 21, 2103, 1991.
48. **Webb, T. R. and Eigenbrot, C.,** Conformationally restricted arginine analogues, *J. Org. Chem.,* 56, 3009, 1991.
49. **Needleman, P., Marshall, G. R., and Rivier, J.,** Angiotensin II. Synthesis and biological activity of 6-(1-methylhistidine) and 6-(3-methylhistidine) analogs, *J. Med. Chem.,* 16, 968, 1973.
50. **Freer, R. J. and Stewart, J. M.,** Synthesis and pharmacology of position 6 analogs of angiotensin II, *J. Med. Chem.,* 16, 733, 1973.
51. **Juliano, L. and Paiva, A. C. M.,** Conformation of angiotensin II in aqueous solution. Titration of several peptide analogs and homologs, *Biochemistry,* 13, 2445, 1974.
52. **Pottorf, R. S., Ripka, W. C., Chiu, A. T., and Wong, P. C.,** The role of the histidine nitrogens of angiotensin II in receptor binding, in *Peptides, Chemistry, Structure and Biology, Proc. 11th Am. Peptide Symp.,* Rivier, J. E. and Marshall G. R., Eds., ESCOM, Leiden, Holland, 1990, 312.
53. **Smeby, R. R. and Fermandjian, S.,** Conformation of angiotensin II, in *Chemistry and Biochemistry of Amino Acids, Peptides and Proteins,* Vol. 5, Weinstein, B., Ed., Marcel Dekker, New York, 1978, 117.
54. **Schwyzer, R. and Turrian, H.,** Chemistry and pharmacology of angiotensin, in *Vitamins and Hormones,* Vol. 18, Harris, R. S. and Thimann, K. V. Eds., Academic Press, New York, 1960, 237.
55. **Jorgensen, E. C., Rapaka, S. R., Windridge, G. C., and Lee, T. C.,** Angiotensin II analogs. 6. Stereochemical factors in the 5 position influencing pressor activity, *J. Med. Chem.,* 14, 899, 1971.
56. **Samanen, J., Narindray, D., Cash, T., Brandeis, E., Adams, W., Yellin, T., Eggleston, D., DeBrosse, C., and Regoli, D.,** Potent angiotensin II antagonists with non-β-branched amino acids in position 5, *J. Med. Chem.,* 32, 466, 1989.
57. **Oparil, S., Koerner, T., Traegear, G. W., Barnes, B. A., and Haber, E.,** Mechanism of pulmonary conversion of AI to angiotensin II in the dog, *Circ. Res.,* 32, 415, 1973.
58. **Freer, R. J., Sutherland, J. C., and Day, A. R.,** Organ selective angiotensin antagonists: sarcosyl^1-cysteinyl(S-methyl)-angiotensin I, *Eur. J. Pharmacol.,* 65, 349, 1980.
59. **Jackson, E. K.,** In vivo evaluation of the antagonist activity of angiotensin I analogs, *J. Pharmacol. Exp. Ther.,* 260, 223, 1992.
60. **Samanen, J., Narindray, D., Weideman, H., Cash, T., Sylvestri, J., Edwards, J., Aiyar, N., Stack, E., and Regoli, E.,** [Sar^1,D-Phe^8]-Angiotensin I, an antagonist to angiotensin II, manuscript in preparation, 1994.
61. **Peach, M. J.,** Physiological roles of angiotensin, in *Chemical and Biological Peptides,* Meienhofer, J., Ed., Ann Arbor Science Publishers, Ann Arbor, MI, 1972, 471.
62. **Park, W. K., Choi, C., Rioux, R., and Regoli, D.,** Synthesis of peptides with the solid phase method. I. Octapeptide analogs of angiotensin II, *Can. J. Biochem.,* 52, 113, 1974.
63. **Khosla, M. C., Smeby, R. R., and Bumpus, F. M.,** Solid phase peptide synthesis of [L-alanine^3-L-isoleucine^5]-angiotensin II, *Biochemistry,* 6, 754, 1967.
64. **Khosla, M. C., Chaturverdi, N. C., Smeby, R. R., and Bumpus, F. M.,** Synthesis of [1-isoleucine,3-proline, and 5-alanine]-angiotensins, *Biochemistry,* 7, 3417, 1968.
65. **Khosla, M. C., Hall, M. M., Smeby, R. R., and Bumpus, F. M.,** Solid-phase synthesis of [3-proline,8-isoleucine]-, [1-sarcosine,3-proline,8-isoleucine]- and [4-phenylalanine,8-isoleucine] angiotensin II as the antagonists of the parent hormone, *J. Med. Chem.,* 16, 1184, 1973.
66. **Marshall, G. R.,** Angiotensin II: a competitive inhibitor and analogs with restricted conformation, in *Prog. Peptide Res., Proc. 2nd Am. Peptide Symp.,* Lande, S., Ed., Gordon and Breach, New York, 1972, 15.

67. **Park, W. K., Asselin, J., and Berlinguet, L.,** Solid-phase peptide synthesis of 1-aminocyclopentanecarboxylic acid (Acpc) analogs of angiotensin II, using a new apparatus, in *Chemical and Biological Peptides,* Meienhofer, J., Ed., Ann Arbor Science Publishers, Ann Arbor, MI, 1972, 49.
68. **Jorgenson, E. C., Rapaka, S. R., Windridge, G. C., and Lee, T. C.,** Angiotensin II analogs. VI. Stereochemical factors in the 5 position influencing pressor activity. 1., *J. Med. Chem.,* 14, 899, 1971.
69. **Jorgenson, E. C., Rapaka, S. R., Windridge, G. C., and Lee, T. C.,** Angiotensin II analogs. VII. Stereochemical factors in the 5 position influencing pressor activity. 2., *J. Med. Chem.,* 14, 904, 1971.
70. **Andreatta, R. H. and Scheraga, H. A.,** Synthesis of [5-valine, 7-N-methylalanine]-angiotensin II, a hypertensive peptide, *J. Med. Chem.,* 14, 489, 1971.
71. **Moore, G. J. and Ko, E. M.,** Antagonists of angiotensin II containing N-methyl-L-alanine and D,L-nipecotic acid in position seven, *Can. J. Physiol. Pharmacol.,* 57, 763, 1979.
72. **Turk, J., Needleman, P., and Marshall, G. R.,** Analogues of angiotensin II with restricted conformational freedom, including a new antagonist, *Mol. Pharmacol.,* 12, 217, 1976.
73. **Khosla, M. C., Munoz-Ramierez, H., Hall, M. M., Smeby, R. R., Khairallah, P. A., Bumpus, F. M., and Peach, M. J.,** Synthesis of angiotensin antagonists containing N- and O-methylated and other amino acid residues, *J. Med. Chem.,* 19, 244, 1976.
74. **Khosla, M. C., Stachowiak, K., Smeby, R. R., Bumpus, F. M., Piriou, F., Lintner, K., and Fermandjian, S.,** Synthesis of [α-methyltyrosine-4]-angiotensin II: studies of its conformation, pressor activity and mode of enzymatic degradation, *Proc. Natl. Acad. Sci. U.S.A.,* 78, 757, 1981.
75. **Samanen, J., Hempel, J. C., Narindray, D., and Regoli, D.,** *A Position Seven Analog of Angiotensin II with Potent Antagonist Activity,* ESCOM, Leiden, Holland, 1988, 137.
76. **Spatola, A. F.,** Amide bond isosteres, surrogates, in *Chemistry and Biochemistry of Amino Acids,* Weinstein, B. Ed., Marcel Dekker, New York, 1983, 267.
77. **Roark, W. H., Tinney, F. J., and Nicolaides, E. D.,** Synthesis and antagonist activities of backbone-modified angiotensin II analogs, in *Peptides, Chemistry and Biology, Proc. 10th Am. Peptide Symp.,* Marshall, G. R., Ed., ESCOM, Leiden, Holland, 1988, 134.
78. **Kaltenbronn, J. S., Dejohn, D. E., Hudspeth, J. P., Humblet, C.C., Lunney, E. A., Nicolaides, E. D., Repine, J. T., Roark, W. H., and Tinney, F. J.,** Synthesis of a saralasin derivative completely modified at every amide bond with a methyleneamino isostere, in *Peptides, Chemistry, Structure and Biology, Proc. 11th Am. Peptide Symp.,* Rivier, J. E. and Marshall, G. R., Eds., ESCOM, Leiden, Holland, 1990, 969.
79. **Stachowiak, K., Khosla, M. C., Plucinska, K., Khairallah, P. A., and Bumpus, F. M.,** Synthesis of angiotensin II analogues by incorporating β-Homotyrosine or β-homoisoleucine residues, *J. Med. Chem.,* 22, 1128, 1979.
80. **Manavalan, P. and Momany, F. A.,** Conformational energy studies on N-methylated analogs of thyrotropin releasing hormone, enkephalin, and luteinizing hormone-releasing hormone, *Biopolymers,* 19, 1943, 1980.
81. **Marshall, G. R. and Bosshard, H. E.,** Studies on the biologically active conformation, *Circ. Res.,* 30,31 (Suppl. 2), 143, 1972.
82. **Goissis, G., Nouailhetas, V. L. A., and Paiva, A. C. M.,** Synthesis of biologically active retroenantiomers of angiotensin peptides, *J. Med. Chem.,* 19, 1287, 1976.
83. **Escher, E.,** Photoaffinity labeling of angiotensin II and bradykinin receptors, *Pharmacol. Ther.,* 37, 37, 1988.
84. **Capponi, A. M. and Catt, K. J.,** Solubilization and characterization of adrenal and uterine angiotensin II receptors after photoaffinity labeling, *J. Biol. Chem.,* 255, 12081, 1980.
85. **Bonnafous, J. C., Seyer, R., Tence, M., Marie, J., Kabba, M., and Aumelas, A.,** Design of angiotensin II derivatives suitable for indirect affinity techniques: potential applications to receptor studies, *J. Receptor Res.,* 8, 295, 1988.
86. **Bost, K. L., Smith, E. M., and Blalock, J. E.,** Antisense sequences, *Proc. Natl. Acad. Sci. U.S.A.,* 82, 1372, 1985.

87. **Elton, T. S., Dion, L. D., Bost, K. L., Oparil, S., and Blalock, J. E.,** Purification of an angiotensin II binding protein by using antibodies to a peptide encoded by angiotensin II complementary RNA, *Proc. Natl. Acad. Sci. U.S.A.,* 85, 2518, 1988.
88. **Toniolo, C.,** Conformationally restricted peptides through short-range cyclizations, *Int. J. Peptide Protein Res.,* 35, 287, 1990.
89. **Holzemann, G.,** Peptide conformation mimetics, part 1, *Kontakte (Darmstadt),* 1, 3, 1991.
90. **Sugg, E. F., Dolan, C. A., Patchett, A. A., Chang, R. S. L., Faust, K. A., and Lotti, V. J.,** Cyclic disulfide analogues of [Sar^2, Ile^8]-angiotensin II, in *Peptides: Chemistry, Structure, and Biology,* Rivier, J. E. and Marshall, G. R., Eds., ESCOM, Leiden, Holland, 1990, 305.
91. **Spear, K. L., Brown, M. S., Reinhard, E. J., McMahon, E. G., Olins, G. M., Paloma, M. A., and Patton, D. R.,** Conformational restriction of angiotensin II: cyclic analogues having high potency, *J. Med. Chem.,* 33, 1935, 1990.
92. **Kataoka, T., Beusen, D. D., Clark, J. D., Yodo, M., and Marshall, G. R.,** The utility of side-chain cyclization in determining the receptor-bound conformation of peptides: cyclic tripeptides and angiotensin II, *Biopolymers,* 32, 1519, 1992.
93. **Madison, V. and Kopple, K. D.,** Solvent-dependent conformational distributions of some dipeptides, *J. Am. Chem. Soc.,* 102, 4855, 1980.
94. **Samanen, J., Zuber, G., Bean, J., Romoff, T., Kopple, K., and Saunders, M.,** 5,5-Dimethylthiazolidine-4-carboxylic acid (Dtc) as a proline analog with restricted conformation, *Int. J. Peptide Protein. Res.,* 35, 501, 1990.
95. **Toniolo, C.,** Structure of conformationally constrained peptides: from model compounds to bioactive peptides, *Biopolymers,* 28, 247, 1989.
96. **Freidinger, R. M., Perlow, D. S., and Veber, D. F.,** Protected lactam-bridged dipeptides for use as conformational constraints in peptides, *J. Org. Chem.,* 47, 104, 1982.
97. **Freidinger, R. M.,** Computer graphics and chemical synthesis in the study of conformation of biologically active peptides, in *Peptides: Synthesis, Structure and Function, Proc. 7th Am. Peptide Symp.,* Rich, D. H. and Gross, E., Eds., Pierce Chemical Co., Rockford, IL, 1981, 673.
98. **Burgess, A. W., Pennuswamy, P. K., and Scheraga, H. A.,** Analysis of conformations of amino acid residues and prediction of backbone topography in proteins, *Isr. J. Chem.,* 12, 239, 1974.
99. **Lewis, P. N., Momany, F. A., and Scheraga, H. A.,** Energy parameters in polypeptides. VI. Conformational energy analysis of the N-acetyl N′-methyl amides of the twenty naturally occurring amino acids, *Isr. J. Chem.,* 11, 121, 1973.
100. **Samanen, J., Ali, F. E., Romoff, T., Calvo, R., Sorenson, E., Vasko, J., Berry, D., Bennett, D., Strohsacker, M., Powers, D., J., S., and Nichols, A.,** Development of a small RGD-peptide fibrinogen receptor antagonist with potent antiaggregatory activity in vitro, *J. Med. Chem.,* 34, 3114, 1991.
101. **Burgen, A. S. V., Roberts, G. C. K., and Feeney, J.,** Binding of flexible ligands to macromolecules, *Nature,* 253, 753, 1975.
102. **Marshall, G. R., Kataoka, T., Plucinska, K., Cody, W., He, J., Humblet, C., Lu, G., Lunney, E., Panek, R., and Skeean, R.,** Optimization of constraints forcing receptor-bound turn conformations of angiotensin, in *Peptides, Chemistry and Biology, Proc. 12th Am. Peptide Symp.,* Smith, J. A. and Rivier, J. E. Eds., ESCOM, Leiden, Holland, 1992, 260.
103. **Freidinger, R. M.,** Non-peptide ligands for peptide receptors, *Trends Pharmacol. Sci.,* 10, 270, 1989.
104. **Hirschmann, R., Nicolaou, K. C., Pietranico, S., Salvino, J., Leahy, E. M., Sprengeler, P. A., Furst, G., Smith, A. B., Strader, C. D., Cascieri, M. A., Candelore, M. R., Donaldson, C., Vale, W., and Maechler, L.,** Nonpeptidal peptidomimetics with a β-D-glucose scaffolding. A parital somatostatin agonist bearing a close structural relationship to a potent, selective substance P antagonist, *J. Am. Chem. Soc.,* 114, 9217, 1992.
105. **Whitebread, S., Mele, M., Kamber, B., and de Gasparo, M.,** Preliminary biochemical characterization of two angiotensin II receptor subtypes, *Biochem. Biophys. Res. Commun.,* 163, 284, 1989.

106. **Imai, N., Masugi, F., Ogihara, T., and Kumahara, Y.,** Structure of angiotensin II-binding site probed with diaminoalkanes, *Biochem. Biophys. Res. Commun.,* 123, 452, 1984.
107. **Chiu, A. T., Herblin, W. F., McCall, D. E., Ardecky, R. J., Carini, D. J., Duncia, J. V., Pease, L. J., Wong, P. C., Wexler, R. R., Johnson, A. L., and Timmermans, P. B. M. W. M.,** Identification of angiotensin II receptor subtypes, *Biochem. Biophys. Res. Commun.,* 165, 196, 1989.
108. **Hodges, J. C., Hamby, J. M., and Blankley, C. J.,** Angiotensin II receptor binding inhibitors, *Drugs Future,* 17, 575, 1992.
109. **Capponi, A. M., Aguilera, G., Fakunding, J. L., and Catt, K. J.,** Angiotensin II: receptors and mechanisms of action, in *Biochemical Regulation of Blood Pressure,* Soffer, R. L., Ed., John Wiley & Sons, New York, 1981, 205.
110. **Brown, G., Douglas, J. G., and Krontiris-Litowitz, J.,** Properties of angiotensin II receptors of isolated rat glomeruli: factors influencing binding affinity and comparative binding of angiotensin analogs, *Endocrinology,* 106, 1923, 1980.
111. **Brown, G. P. and Douglas, J. G.,** Angiotensin II binding sites on isolated rat renal brush border membranes, *Endocrinology,* 111, 1830, 1982.
112. **Capponi, A. M. and Catt, K. J.,** Angiotensin II receptors in adrenal cortex and uterus, *J. Biol. Chem.,* 254, 5120, 1979.
113. **Cox, J. M., Munday, K. A., and Poat, J. A.,** Identification of selective, high affinity [^{125}I-]-angiotensin and [^{125}I]-bradykinin binding sites in rat intestinal epithelia, *Br. J. Pharmacol.,* 87, 201, 1986.
114. **Glossman, H., Baukal, A., and Catt, K. J.,** Properties of angiotensin II receptors in the bovine and rat adrenal cortex, *J. Biol. Chem.,* 249, 825, 1974.
115. **Gunther, S., Alexander, R. W., Atkinson, W. I., and Gimbrone, M. A., Jr.,** Functional angiotensin II receptors in cultured vascular smooth muscle cells, *J. Cell Biol.,* 92, 289, 1982.
116. **Moore, T. J. and Williams, G. H.,** Angiotensin II receptors on human platelets, *Circ. Res.,* 51, 314, 1982.
117. **Sraer, J., Baud, L., Cosyns, J. P., Verroust, P., Nivez, M. P., and Ardaillou, R.,** High affinity binding of ^{125}I-angiotensin II to rat glomerular basement membranes, *J. Clin. Invest.,* 59, 69, 1977.
118. **Speth, R. C. and Kim, K. H.,** Discrimination of two angiotensin II receptor subtypes with a selective agonist analogue of angiotensin II, p-aminophenylalanine-^{6}angiotensin II, *Biochem. Biophys. Res. Commun.,* 169, 997, 1990.
119. **Hsieh, K. and Speth, R. C.,** Angiotensin II analogues selectively active in the brain, in Eleventh American Peptide Symposium, The American Peptide Society, La Jolla, CA, 1989, Abst. P-132.
120. **Kakar, S. S., Riel, K. K., and Neill, J. D.,** Differential expression of angiotensin II receptor subtype mRNAs (AT-1A and AT-1B) in the brain, *Biochem. Biophys. Res. Commun.,* 185, 688, 1992.
121. **Kitami, Y., Okura, T., Marumoto, K., Wakamiya, R., and Hiwada, K.,** Differential gene expression and regulation of type-1 angiotensin II receptor subtypes in the rat, *Biochem. Biophys. Res. Commun.,* 188, 446, 1992.
122. **Ji, H., Sandberg, K., and Catt, K. J.,** Novel angiotensin II antagonists distinguish amphibian from mammalian angiotensin II receptors expressed in *Xenopus laevis* oocytes, *Mol. Pharmacol.,* 39, 120, 1991.
123. **Rogg, H., Schmid, A., and de Gasparo, M.,** Identification and characterization of angiotensin II receptor subtypes in rabbit ventricular myocardium, *Biochem. Biophys. Res. Commun.,* 173, 416, 1990.
124. **Sandberg, K., Ji, H., Millan, M. A., and Catt, K. J.,** Amphibian myocardial angiotensin II receptors are distinct from mammalian AT_1 and AT_2 receptor subtypes, *FEBS Lett.,* 284, 281, 1991.

125. **Aiyar, N., Griffin, E., Edwards, R., Weinstock, R., Samanen, J., and Nambi, P.,** Characterization of bovine ovary angiotensin II receptors using subtype-selective antagonists, *Pharmacology,* 46, 1, 1992.
126. **Schild, H. O.,** pA_2, a new scale for the measurement of drug antagonism, *Br. J. Pharmacol.,* 2, 189, 1947.
127. **Paiva, T. B., Goissis, G., Juliano, L., Miyamoto, M. E., and Paiva, A. C. M.,** Angiotensin-like and antagonistic activities of N-terminal modified [8-leucine] angiotensin II peptides, *J. Med. Chem.,* 17, 238, 1974.
128. **Visser, G. H., Schattenkerk, C., Kerling, K. E. T., and Havinga, E.,** *Rec. Trav. Chim. Pays-Bas,* 83, 684, 1964.
129. **van Battenburg, O. D., de Graaf, J. S., Kerling, K. E. T., and Havinga, E.,** *Rec. Trav. Chim. Pays-Bas,* 96, 37, 1977.
130. **Turk, J., Needleman, P., and Marshall, G. R.,** Sterically restricted analogs of angiotensin II and bradykinin containing L-a-methylphenylalanine, *Fed. Proc.,* 34, 769, 1975.
131. **Paruszewski, R.,** (Tyr-5-AII), *Roczniki Chemii Ann. Soc. Chim. Polonorum,* 455, 289, 1971.
132. **Jorgenson, E. C., Windridge, G. C., and Lee, T. C.,** Angiotensin II analogs. X. Stereochemical factors in the 5 position influencing pressor activity, *J. Med. Chem.,* 16, 467, 1973.
133. **Khosla, M. C., Hall, M. M., Smeby, R. R., and Bumpus, F. M.,** Factors that influence the antagonistic properties of angiotensin II antagonists, *J. Med. Chem.,* 16, 829, 1973.
134. **Bloemhoff, W., de Graaf, J. S., Havinga, E., and Kerling, K. E. T.,** Synthesis and myotropic activity of [5-isoleucine, 6-homohistidine]-angiotensin II and of des-(1-aspartic acid)-[5-isoleucine, 6-homohistidine]-angiotensin II, in *Peptides 1971, Proc. 11th Eur. Peptide Symp.,* Nesvadba, H., Ed., North Holland, Amsterdam, 1973, 351.
135. **Hsieh, K. H., Jorgensen, E. C., and Lee, T. C.,** Angiotensin II analogues. XIV. Role of the imidazole nitrogens of position-6 histidine in pressor activity, *J. Med. Chem.,* 22, 1199, 1979.
136. **Schroder, E. and Hampel, R.,** Synthese von Lys^6-Val^5-, von Tyr^3-Val^5-, von $Tyr(Me)^1$-Val^5 und von Tyr^8-Val^5-angiotensin II -Asp^1-β-amid, *Justus Liebigs Ann. Chem.,* 684, 243, 1965.
137. **Bumpus, F. M., Smeby, R. R., and Khairallah, P. A.,** Synthesis and biological properties of angiotensin II analogs, in *Peptides, Chem. Biochem., Proc. 1st Am. Peptide Symp.,* B. Weinstein, Ed., Marcel Dekker, New York, 1970, 127.
138. **Moore, G. J.,** Angiotensins II and III: The Functional Role of the Penultimate Proline Residue, in *Peptides: Synthesis, Structure, Function, Proc. 7th Am. Peptide Symp.,* D. H. Rich Gross, Ed., Pierce Chemical Co., Rockford, IL, 1981, 245.
139. **Hallinan, E. A. and Mazur, R. H.,** Angiotensin II Analogs Containing Unsaturated Amino Acids, in *Peptides: Structure and Function, Proc. 6th Am. Peptide Symp.,* E. Gross and J. Meienhofer, Ed., Pierce Chemical Company, Rockford, IL., 1979, 475.
140. **Balasubramanian, P. and Marshall, G. R.,** Angiotensin II. Synthesis and biological activity of 8-(R and S)-tetrahydroisoquinoline and 7-(R and S)-(proline)thiazole analogs, in *Peptides, Chemistry and Biology, Proc. 10th Am. Peptide Symp.,* Marshall, G. R. Ed., ESCOM, Leiden, Holland, 1988, 452.
141. **Peach, M. J. and Ober, M.,** Inhibition of angiotensin induced adrenal catecholamine release by 8-substituted analogs of angiotensin II, *J. Pharmacol. Exp. Ther.,* 190, 49, 1974.
142. **Khosla, M. C., Hall, M. M., Smeby, R. R., and Bumpus, F. M.,** Agonist and antagonist relationships in 1- and 8-substituted analogs of angiotensin II., *J. Med. Chem.,* 17, 1156, 1974.
143. **Khosla, M. C., Munoz-Ramirez, H., Hall, M. M., Khairallah, P. A., and Bumpus, F. M.,** Synthesis of angiotensin II antagonists by incorporating α-methylalanine or O-methylthreonine residues in angiotensin analogues, *J. Med. Chem.,* 20, 1051, 1977.
144. **Pena, C. and Stewart, J. M.,** A new class of angiotensin inhibitors: N-methylphenylalanine analogs, *Life Sci.,* 14, 1331, 1974.
145. **Schoelkens, B. A., Wissmann, H., Lindner, E., and Geiger, R.,** 1,8-Disubstituted analogues of $[Ile^5]$ and $[Val^5]$angiotensin II: difference in potency and specificity of angiotensin II–antagonistic activity, *Hoppe-Seyler's Z. Physiol. Chem.,* 357, 825, 1976.

146. **Hsieh, K., Needleman, P., and Marshall, G. R.,** Long-acting angiotensin II inhibitors containing hexafluorovaline in position 8, *J. Med. Chem.,* 30, 1097, 1987.
147. **Riniker, B. and Schwyzer, R.,** Synthetische analoge des hypertensins. III. Gly^1-Val^5-hypertensin II, Orn^2-Val^5-hypertensin II, Phe^4-Val^5-hypertensin II und derivat, *Helv. Chim. Acta,* 44, 674, 1961.
148. **Matsoukas, J. M., Scanlon, M. N., and Moore, G. J.,** A cyclic angiotensin agonist: [1,8-cysteine]-angiotensin II, *J. Med. Chem.,* 27, 404, 1984.
149. **Miranda, A. and Juliano, L.,** Conformationally restricted analogs of angiotensin-II: titration and biological activity, *Brazilian J. Med. Biol. Res.,* 21, 903, 1988.
150. **De Coen, J. L., Ralston, E., Durieux, J. P., and Loffet, A.,** Cyclic Angiotensin, in *Peptides: Chem. Struct. Biol., Proc. 4th Am. Peptide Symp.,* R. Walter and J. Meienhofer, Eds., Ann Arbor Science, Ann Arbor, MI, 1975, 553.
151. **Jorgensen, E. C. and Patton, W.,** Angiotensin II analogs. II. Cyclo-(Val-Tyr-Ile-His-Pro-Phe), *J. Med. Chem.,* 12, 935, 1969.
152. **Ancans, J. E., Biseniece, D. A., Myshlyakova, N. V., Porunkevich, E. A., Ratkevich, M. P., Afanas'eva, G. A., Kublis, G., Skujins, A., Zalitis, G., and Cipens, G.,** Angiotensin cyclic analogs showing depressor and histamine releasing activity, *Biorg. Khim.,* 12, 1118, 1986.
153. **Chipens, G., Nikiforovich, G., Mululis, F., Veretennikova, N., Vosekalna, I., Sosnov, A., Polevaya, I., Ancans, J., Mishlyakova, N., Liepinsh, E., Sekacis, I., and Berslav, M.,** *Cyclic Analogs of Linear Peptides,* Pierce Chemical Co., Rockford, IL. 1979, 567.
154. **Chipens, G., Ancans, J., Berga, D., Vosekalna, I., Mishlyakova, N., and Krikis, A.,** *Analogues of Angiotensin with Vasodepressor Activity,* Vol. 2, Walter de Gruyter, New York, 1984, 261.
155. **Samanen, J.,** Cyclic peptide prepared by Peninsula Laboratories, unpublished, 1990.
156. **Gunther, S., Gimbrone, M. A., and Alexander, R. W.,** Identification and characterization of the high affinity vascular angiotensin II receptor in rat mesenteric artery, *Circ. Res.,* 47, 278, 1980.
157. **Devynck, M. A., Pernollet, M., Meyer, P., Fermandjian, S., and Fromageot, P.,** Angiotensin receptors in smooth muscle cell membranes, *Nat. N. Biol.,* 245, 55, 1973.
158. **Oparil, S., Koerner, T., Traegear, G. W., Barnes, B. A., and Haber, E.,** Mechanism of pulmonary conversion of angiotensin I to angiotensin II in the dog, *Circ. Res.,* 32, 415, 1973.
159. **Tsai, B. S., Peach, M. J., Khosla, M. C., and Bumpus, F. M.,** *J. Med. Chem.,* Synthesis and evaluation of [Des-Asp^1]-angiotensin I as a precursor for [Des-Asp^1]-angiotensin II (''Angiotensin III''), *J. Med. Chem.,* 18, 1180, 1975.
160. **Kono, T., Ikeda, F., Oseko, F., Ohmori, Y., Nakano, R., Muranaka, H., Taniguchi, A., Imura, H., Knosla, M. C., and Bumpus, F. M.,** Biological activity of des-asp^1-des-arg^2-angiotensin II in man, *Acta Endocrinol.,* 99, 577, 1982.
161. **Ball, D., Elliiot, M. E., Hadjokas, N., Goodfriend, T. L., and Green, M. J.,** Veratridine, angiotensin receptors and alderosteronogenesis in bovine adrenal glomerulosa cells, *Clin. Exp. Theor. Pract.,* A8, 323, 1986.

Chapter 3

CONFORMATIONAL MODELS OF ANGIOTENSIN II FOR NONPEPTIDE ANTAGONIST DESIGN

James Samanen, Judith Hempel, Richard Keenan, Catherine Peishoff, and Joseph Weinstock

TABLE OF CONTENTS

0-8493-8545-8/94/$0.00+$.50
© 1994 by CRC Press, Inc.

I. INTRODUCTION: MODELING GOALS

As described in the previous chapter, medicinal chemists have followed two approaches in the pursuit of orally active antagonists to angiotensin II: (1) modification and removal of peptidic elements of structure, and (2) discovery and empirical enhancement of nonnative ligands to the angiotensin II receptor. An important tool for either approach is the determination of the bioactive or receptor-bound conformation of angiotensin II. As is described in this and the following chapter, the discovery of the benzylimidazole angiotensin II receptor antagonists[1] led a number of groups to modify the benzylimidazole structure to yield high affinity antagonists. In the design of benzylimidazole analogs, models of angiotensin II served as a template to suggest those elements of angiotensin II structure that could be appended into the benzylimidazole structure to enhance affinity for angiotensin II receptors.

A. CONFORMATIONAL MODELS OF THE RECEPTOR-BOUND LIGAND

Decades of research have been devoted to theoretical and physicochemical studies of the conformations of angiotensin II and its analogs. This work has involved the study of highly flexible linear peptides, because constrained analogs of angiotensin II that retained high affinity for angiotensin II receptors were not known. Thus, the conformational models of angiotensin II that have been derived from these studies reflect solution conformations that do not likely relate to receptor-bound conformations. Models that attempt to incorporate the structure–activity relationships of linear angiotensin II analogs may come closer in defining a bioactive conformation, but these models can only be treated as working hypotheses. Because the literature on these models serves as the foundation for several of the conformational models that have been employed in the design of nonpeptide angiotensin II receptor antagonists, it is reviewed in the Appendix to this chapter.

B. PEPTIDE TEMPLATE MODELS FOR NONPEPTIDE LIGAND DESIGN: THE OVERLAY APPROACH

An increasingly common approach in the discovery of nonpeptide antagonists of peptide receptors utilizes radioligand binding assays to screen compound databases and natural products extracts in hopes of finding novel nonpeptide ligands.[2] Such studies have yielded receptor antagonists to CCK,[2] glucagon,[3] growth hormone releasing factor,[4] oxytocin,[5] vasopressin,[6] and Substance P.[7] The first nonpeptide angiotensin II receptor antagonists (e.g., compound *1*), were reported in a Takeda patent on benzylimidazoles designed as diuretic agents.[1]

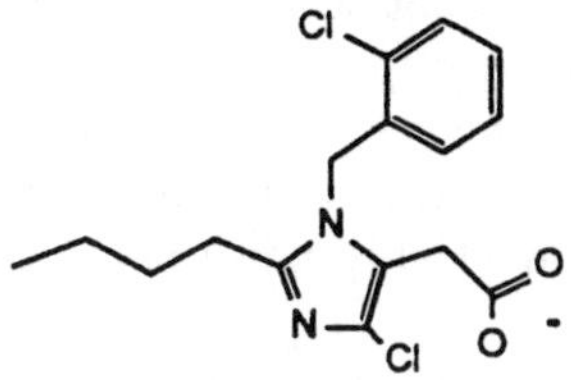

1

In most instances, the nonpeptide angiotensin II receptor antagonist leads have been modified empirically to yield high affinity analogs without the incorporation of structural information from the native peptide ligand. Potentially, three binding modes appear to be available for nonpeptide angiotensin II receptor ligands: (1) a mimetic mode, in which the ligand binds in the peptide "binding pocket," (2) a steric mode, in which the ligand binds partially or totally outside the binding pocket but sterically prevents native peptide from binding or, (3) an allosteric mode, in which the ligand binds at some site distant to the active site and thereby modifies receptor conformation in a way that reduces its avidity for the native peptide. It may be exceedingly difficult to discover which mode of binding is utilized by the receptor. Nonetheless, it is possible that analogs of nonnative ligands may bind in either steric or allosteric modes and may indeed be potent receptor antagonists. In fact, mutagenesis experiments with the NK-1 receptor have shown recently that the nonnative Substance P receptor antagonist, CP 96345, may be a steric or an allosteric inhibitor of Substance P, since residues in the NK-1 receptor that have been shown to be critical for nonpeptide receptor antagonist binding are not important for the native peptide.[8]

If, however, a nonpeptide ligand binds to the receptor in a mimetic mode, then the employment of structural and conformational information from the peptide in the design nonpeptide analogs may succeed in the discovery of analogs with enhanced affinity. This design approach has been utilized by several groups in the discovery of novel, potent nonpeptide angiotensin II receptor antagonists. In what has come to be known as the "overlay approach," models of angiotensin II have been employed to design of analogs of *1* with significant higher affinities for angiotensin II receptors.

II. NONPEPTIDE LIGAND DESIGN BASED UPON TEMPLATE MODELS OF ANGIOTENSIN II

A. THE OVERLAY APPROACH IN THE DESIGN OF BENZYLIMIDAZOLE ANALOGS

The overlay approach for the development of nonpeptide antagonists of peptide agonists involves three steps in the design of a nonpeptide ligand from peptide: (1) generation of hypothetical correlations between binding groups on the non-peptide ligand with the peptide ligand, (2) development of a three-dimensional model in which the nonpeptide ligand is superimposed upon the peptide ligand with the highest degree of overlap among the correlated binding groups, and (3) the generation of testable hypotheses (i.e., identification of potential new analogs of the nonpeptide ligand). The manner in which these aspects of the overlay approach were employed to identify novel nonpeptide angiotensin II receptor antagonist is detailed below.

If one assumes that certain groups in the prototypic angiotensin II receptor antagonist, *1*, assume the same three-dimensional relationship at the angiotensin II receptor as analogous groups on angiotensin II, then one can consider superimposing or overlaying benzylimidazole *1* onto angiotensin II in a way that positions these correlated groups in *1* into the same three-dimensional array as in angiotensin II. The low affinity of *1* relative to angiotensin II could be the result of (1) a less than optimal geometric relationship between receptor binding groups on *1* or (2) an insufficient number of important receptor binding groups on *1*. Angiotensin II contains an appreciable number of potential receptor binding groups (Figure 1, Chapter 2). If one correlates the binding groups in *1* with similar groups in angiotensin II, then examination of the three-dimensional relationship between these two ligands may either suggest a more suitable geometry for the binding groups in *1* (i.e., a new scaffold, altered chain lengths, etc.) or reveal positions on *1* where other binding groups could be appended based on consideration of the structure of angiotensin II.

In the process of creating an overlay of nonpeptide ligands with the native peptide, the chemist has the option of either superimposing the two molecules as rigid structures (a *rigid fit*) or allowing the peptide conformation to change in ways that maximize the degree of overlap of correlated putative binding groups (minimize the distance between correlated groups in a *flexible fit*).* In one instance described in this chapter, a rigid fit was employed when the nonpeptide ligand was fit to a particular angiotensin II conformational model (which was held "rigid" during the overlay so as to not disturb the conformation). The model of Fermandjian et al.[9] (described in the Appendix) was employed in this case in the design of the biphenylimidazoles.[10–13] In two other instances, a flexible fit was

*Note that the conformation of the nonpeptide ligand may change during the overlay process in both cases. Overlays that are attempted without modifying the nonpeptide ligand have been less fruitful.

employed in the overlay process, which resulted in novel conformations of angiotensin II. In this second type of overlay, a novel conformational model is derived from folding of angiotensin II around a nonpeptide ligand in the process of overlay. This type of model more closely reflects the true relationship between the nonpeptide ligand and the native peptide, since the common receptor binding elements can occupy the same positions in space. This second approach has been described by Samanen et al.[14] and Pierson and Freer[15] in the design of nonpeptide angiotensin II receptor antagonists. In the case of Samanen et al.,[14] only those conformations of angiotensin II were considered that were consistent with the "conformational structure–activity relationships" that were described in Chapter 2. It is important to emphasize that the only meaningful overlays are those that suggest new analogs of *1*.

B. ELEMENTS OF BENZYLIMIDAZOLE STRUCTURE THAT MIMIC ANGIOTENSIN II

Several investigator groups have assumed that the benzylimidazole *1* bound to the angiotensin II receptor in a mimetic mode (i.e., that certain functional groups in *1* mimicked functional groups in angiotensin II by binding to the same sites on the angiotensin II receptor). Table 1 contains a complete list of potential correlations between the functional groups on the benzylimidazole *1* and angiotensin II. These potential correlations are based upon the structure–activity relationships of both series. Different groups have explored various combinations of correlations between the functional groups in the benzylimidazole *1* and angiotensin II and have employed different conformational models of angiotensin II.

C. THE DESIGN OF BIPHENYLIMIDAZOLES FROM A POSITION SIX OVERLAY WITH THE ANGIOTENSIN II MODEL OF FERMANDJIAN

The first reported utilization of an overlay of the benzylimidazole *1* onto angiotensin II succeeded in the development of a series of potent biphenyl- and benzyl- substituted imidazoles (e.g., *2,* DuP 753, derived from *1*)[10–12,16]

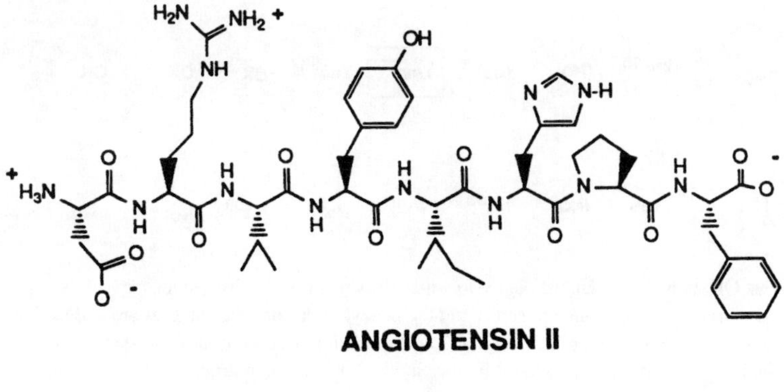

ANGIOTENSIN II

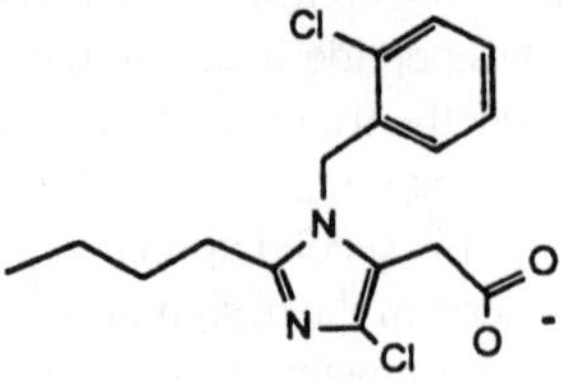

1

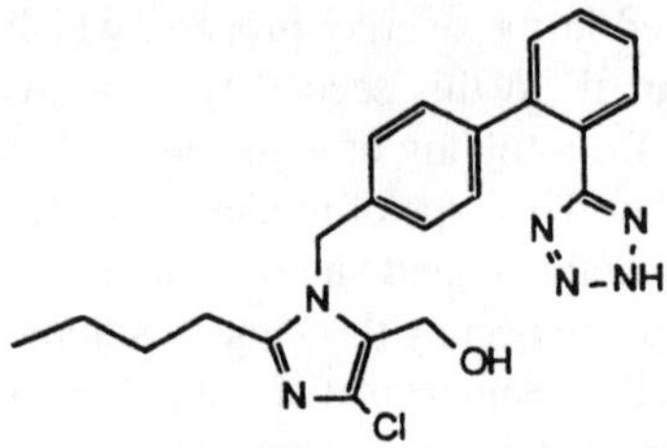

Dup 753 (Losartan)

2,

TABLE 1
Potential Correlations Between Angiotensin II and Benzylimidazole *1*

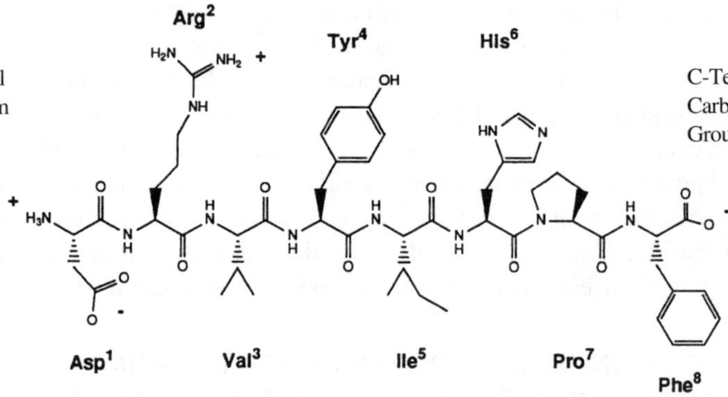

	Asp[1]	Arg[2]	Val[3]	Tyr[4]	Ile[5]	His[6]	Pro[7]	Phe[8]	COOH
	OK	(OK)	Yes	(No)	Yes (boxed)	(No)	Yes[SB]	Yes	(No)
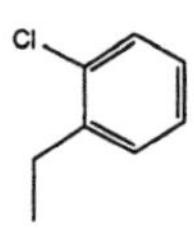	OK[SB]	(No)	Yes[SB]	Yes[SB] (boxed)	Yes[SB]	(OK)[SB]	OK[SB]	OK	(No)
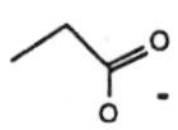	Yes	(No)	No[SB]	(No?)	No[SB]	No[SB]	No[SB]	No	Yes (boxed)

Note: Yes, OK, or No represent the degree to which the structure–activity relationships of these positions appear to correlate between angiotensin II and the benzylimidazoles. Parentheses are added if there are no particular data. SB = unpublished data from SmithKline Beecham, now appearing in Chapter 2. Boxes highlight the correlation favored by the SmithKline Beecham group.

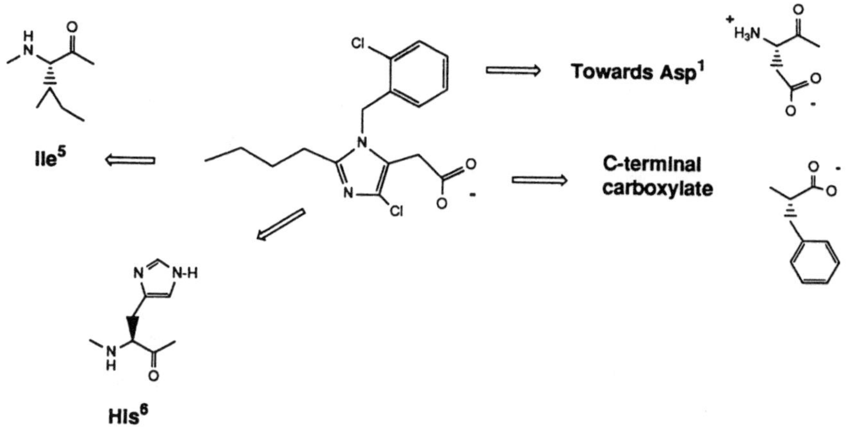

FIGURE 1. The correlation (Duncia et al.)[10] between angiotensin II and benzylimidazole *1*.

The rationale that has been presented[10,13] for the development of the biphenylimidazoles was to relate benzylimidazole *1* to angiotensin II through a rigid overlay with the model of angiotensin II described originally by Fermandjian et al.[9] (see Appendix). In this overlay, the central imidazole is related to [His[6]]-AII,* the [Bu[2]]-imidazole to [Ile[5]]-angiotensin II and the [$CH_2CO_2H^5$]-imidazole to the C terminal carboxylate of angiotensin II, as summarized in Figure 1. In this overlay, the [(o-Cl)Bzl[1]]-imidazole points toward the β-carboxyl group of [Asp[1]]-angiotensin II. The rationale for this initial orientation of the [(o-Cl)Bzl[1]]-imidazole group towards the carboxylate of [Asp[1]]-angiotensin II, as opposed to some other group on *1*, is not clear. The assumption that an appropriately extended carboxylate or tetrazole should mimic the Asp sidechain is tenuous, since the Asp sidechain carboxylate is not thought to be important for activity (see Chapter 1). Nonetheless, Duncia et al.[10] indicated that the disparity in the structure–activity relationships around the β-COOH group of the Asp of angiotensin II suggested that substitution at this position may have profound effects. These investigators extended the [(o-Cl)Bzl[1]]-imidazole group, ultimately resulting in compound *2*, Dup 753 (losartan). The biphenylimidazole *2* and related analogs represent an important series of potent, orally active angiotensin II receptor antagonists. Other investigators began with the biphenylimidazole *2* and discovered additional novel and potent analogs. The structure–activity relationships around this series of angiotensin II receptor antagonists is discussed in greater depth in Chapter 4.

Figure 2 displays an overlay of a hybrid molecule in which the [(o-Cl)Bzl[1]]-imidazole of *1* is replaced with an o-carboxy-biphenyl group to show that the

*In this chapter, peptide nomenclature will be borrowed for abbreviations of the sidechains in the imidazole *1*, namely that the butyl sidechain in the benzylimidazole *1* will be denoted by [Bu[2]]-imidazole, the (o-Cl)benzyl group will be denoted by [(o-Cl)Bzl[1]]-imidazole and the acetic acid group will be denoted by ($CH_2CO_2H^5$)-imidazole.

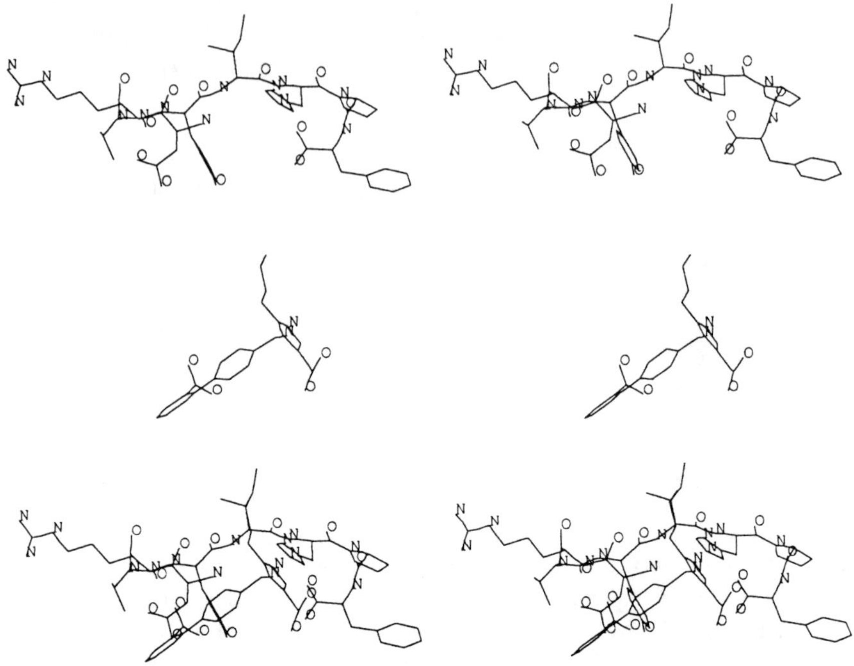

FIGURE 2. A position six overlay between the model of Fermandjian et al.[9] and benzylimidazole *1* *in stereo.*

original orientation of the $[CH_2CO_2H^5]$-imidazole can be maintained while the o-carboxybiphenyl group extends well into the region of the sidechain carboxylate of $[Asp^1]$-angiotensin II. It is important to note that some questions arise upon consideration of the overlay in Figure 2. First, Duncia et al.[10] acknowledge that their overlay would not predict that the $[CH_2CO_2H^5]$-imidazole group would be successfully replaced with a nonacidic group. Second, a visual comparison of the model of Fermandjian et al.[9] with the model of angiotensin II used by Duncia et al.[10] in their overlay with the benzylimidazole *1* suggests that the two angiotensin II models are rather different. For example, the Tyr^4 and His^6 sidechains point in opposite directions in the model displayed by Duncia et al.,[10] which differs from the model of Fermandjian et al.,[9] which emphasized a parallel positioning of the two aromatic groups. Other visual discrepancies are also notable. The reasons for these discrepancies are not apparent, since the desired overlay can be achieved with the original conformation proposed by Fermandjian et al.[9] (Figure 2). Finally, as noted above, the rationale for any position six overlay is weakened by the fact that $[(N\tau Me)His^6]$- and $[N\pi Me)His^6]$-angiotensin display diminished activity. The rationale that has been described for the design of the biphenylimidazoles, nonetheless, succeeded in the discovery of an important class of nonpeptide angiotensin

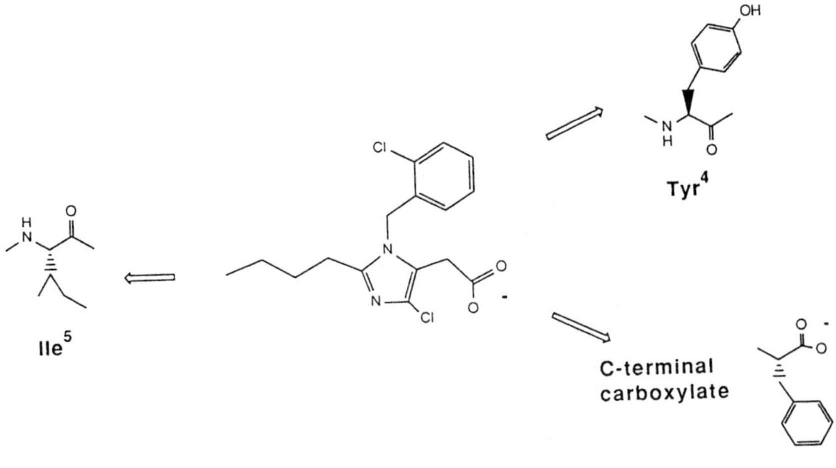

FIGURE 3. The correlation (Samanen et al.)[14] between angiotensin II and benzylimdazole *1*.

II receptor antagonists, and the biphenylimidazoles have served as the starting point for the synthesis of many other analogs (Chapter 4).

D. UTILIZATION OF MODEL II TO DEVELOP THE POTENT ANGIOTENSIN II RECEPTOR ANTAGONISTS, SK&F 107011 AND SK&F 108566

1. The Position Four-Eight Overlay

An overlay that attracted the attention of other workers[14] involved a correlation of [Bu^2]-imidazole with [Ile^5]-angiotensin II, [(o-Cl)Bzl^1]-imidazole with [Tyr^4]-angiotensin II and [$CH_2CO_2H^5$]-imidazole with the C terminal COOH of angiotensin II (Figure 3). The benzylimidazole *1* may be aligned in a position four-eight overlay with either the model of Fermandjian et al.[9] (Figure 4) or the model of Matsoukas et al.[17] (Figure 5). On inspection of these overlays, however, it is apparent that neither overlay readily suggested binding groups of angiotensin II that might be appended onto the benzylimidazole.

Although the angiotensin II Model I, described in the Appendix, may be desirable in its relative congruence with peptide structure–activity relationships, it cannot give a position four-eight overlay, due to the large distance between the Tyr^4 sidechain and the C terminal carboxylate (Figure 6).

2. Model II, Discovery and Use

The fact that all of the known models of angiotensin II are derived from physical studies of flexible linear peptides and not from highly constrained peptides with high affinity suggests that none of these models may serve as good templates for nonpeptide angiotensin II receptor antagonist design. The possibility

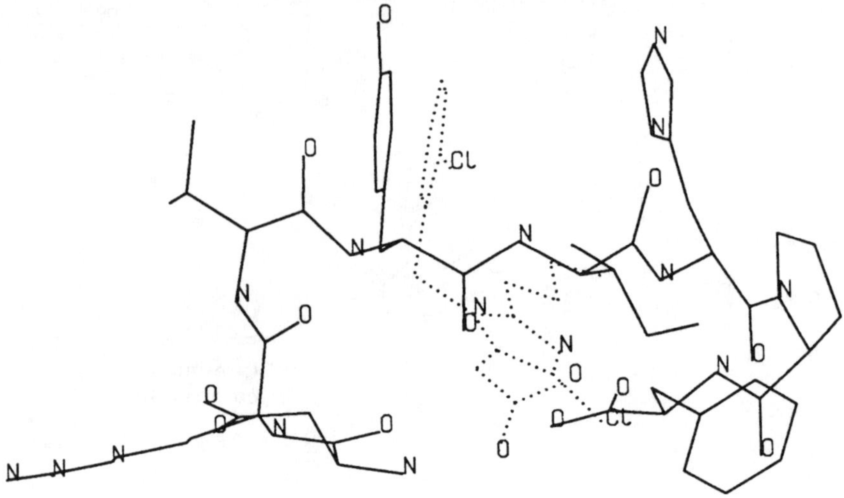

FIGURE 4A. A position four/eight overlay between the model of Fermandjian et al.[9] and benzylimidazole *1*.

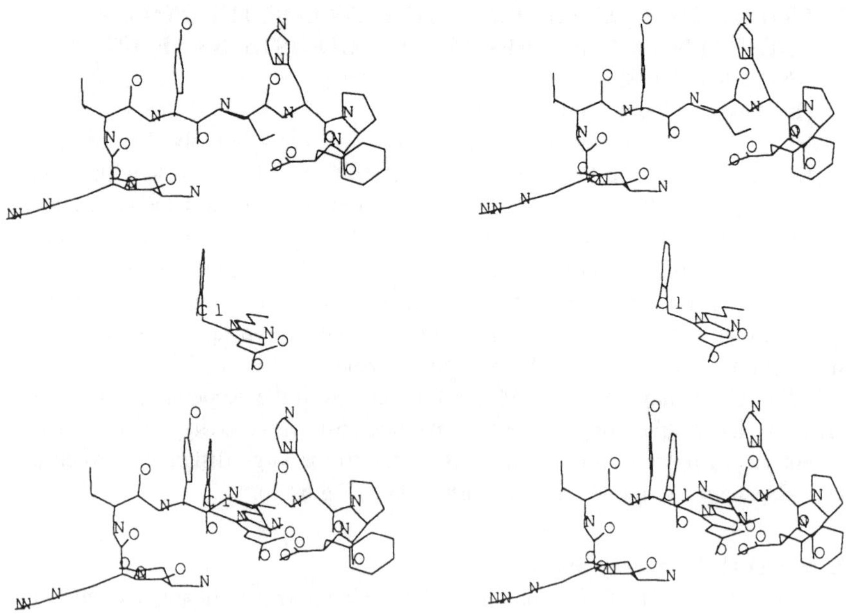

FIGURE 4B. A position four/eight overlay between the model of Fermandjian et al.[9] and benzylimidazole *1 in stereo.*

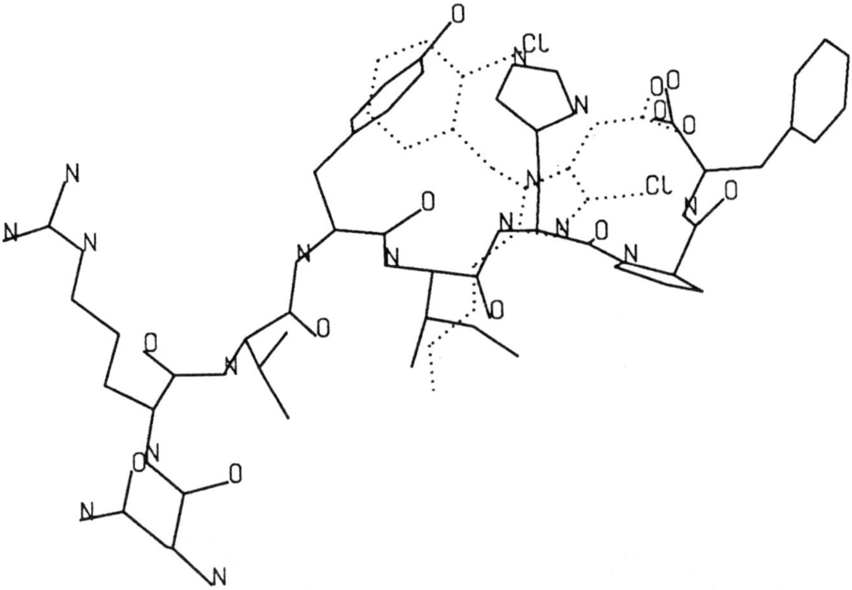

FIGURE 5A. A position four/eight overlay between a representation of the model of Matsoukas et al.[17] and benzylimidazole *1*.

that structural groups in benzylimidazole *1* may actually adopt the same three-dimensional relationship at the angiotensin II receptor as the complementary structural groups in angiotensin II led Samanen et al.[14] to propose that angiotensin II might be folded around *1* into a new conformation that might serve as a template model for nonpeptide angiotensin II receptor antagonist design. Since angiotensin II is rich in receptor binding groups, this model could help to identify positions in the benzylimidazole structure where putative binding groups in angiotensin II could be appended to yield compounds with enhanced affinity for angiotensin II receptors. This section describes the development of what is called angiotensin II Model II and the compounds that were designed from overlays between the benzylimidazole *1* and Model II. Through this process, compounds were discovered with affinities that are 40,000-fold greater than the benzylimidazole *1* and superior to the octapeptide, saralasin.[18,19]

Model I, which is described in the Appendix, was selected for folding around *1,* since it is the model that is most consistent with the activities of conformationally constrained linear peptides. Empirical manipulation of Model I (using Nicholson hand-held models) provided a rough version of Model II (Figure 7). A goal of the modeling exercise was to arrive at a conformation that allowed for a good overlay of angiotensin II with benzylimidazole *1* that was still consistent with the set of conformational constraints that can be imposed upon angiotensin II (Table 7 of the Appendix). Surprisingly, simple alteration of the Tyr^4 in Model I

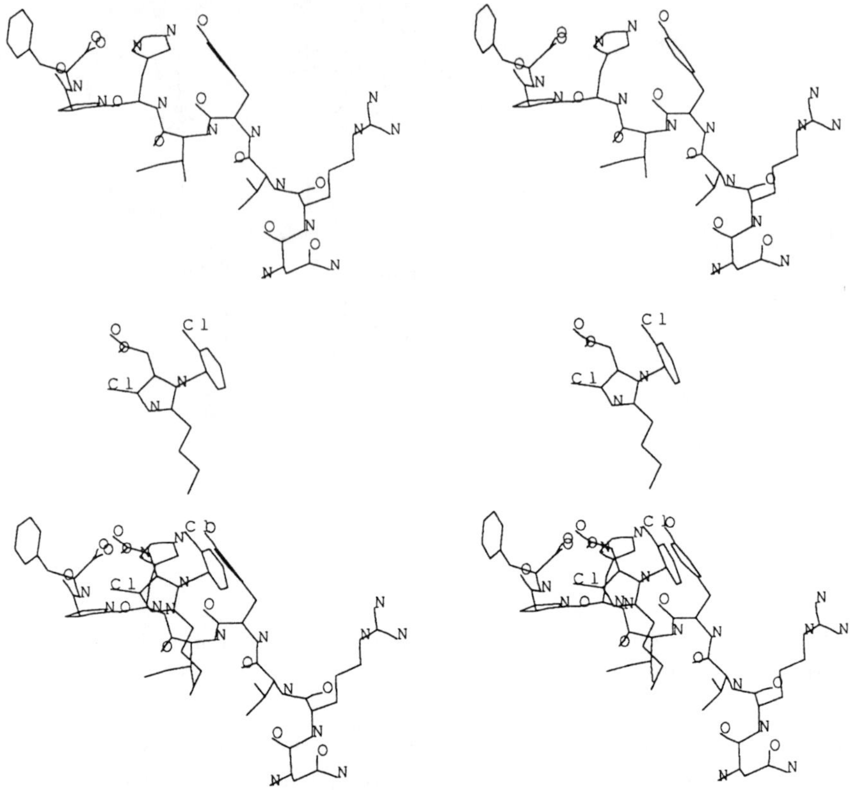

FIGURE 5B. A position four/eight overlay between a representation of the model of Matsoukas et al.[17] and benzylimidazole *1 in stereo.*

from an α_R to an α_L conformation was all that was necessary to obtain a conformation that would provide for a good overlay with the benzylimidazole and yet be consistent with an (αMe)Tyr substitution at position four. The rough revised model was reproduced in SYBYL 3.4, and a final fit between this model of angiotensin II and benzylimidazole *1* (Figure 8) was obtained through a mutual minimization of angiotensin II and *1* using the MAXIMIN program of SYBYL 3.4. It is of interest to note that Models I and II, as well as the models of Fermandjian et al.[9] and Matsoukas et al.,[17] in SYBYL give almost identical conformational energies.

Inspection of sidechain orientations of Model II and benzylimidazole *1* in SYBYL suggested that a lengthening of the acetic acid sidechain in *1* would allow for a better overlay between correlated groups. To determine if this better placement of the (o-Cl)Bzl group in relation to the carboxylate resulted in an increase in affinity for angiotensin II receptors, the propionic acid analogs *9* and

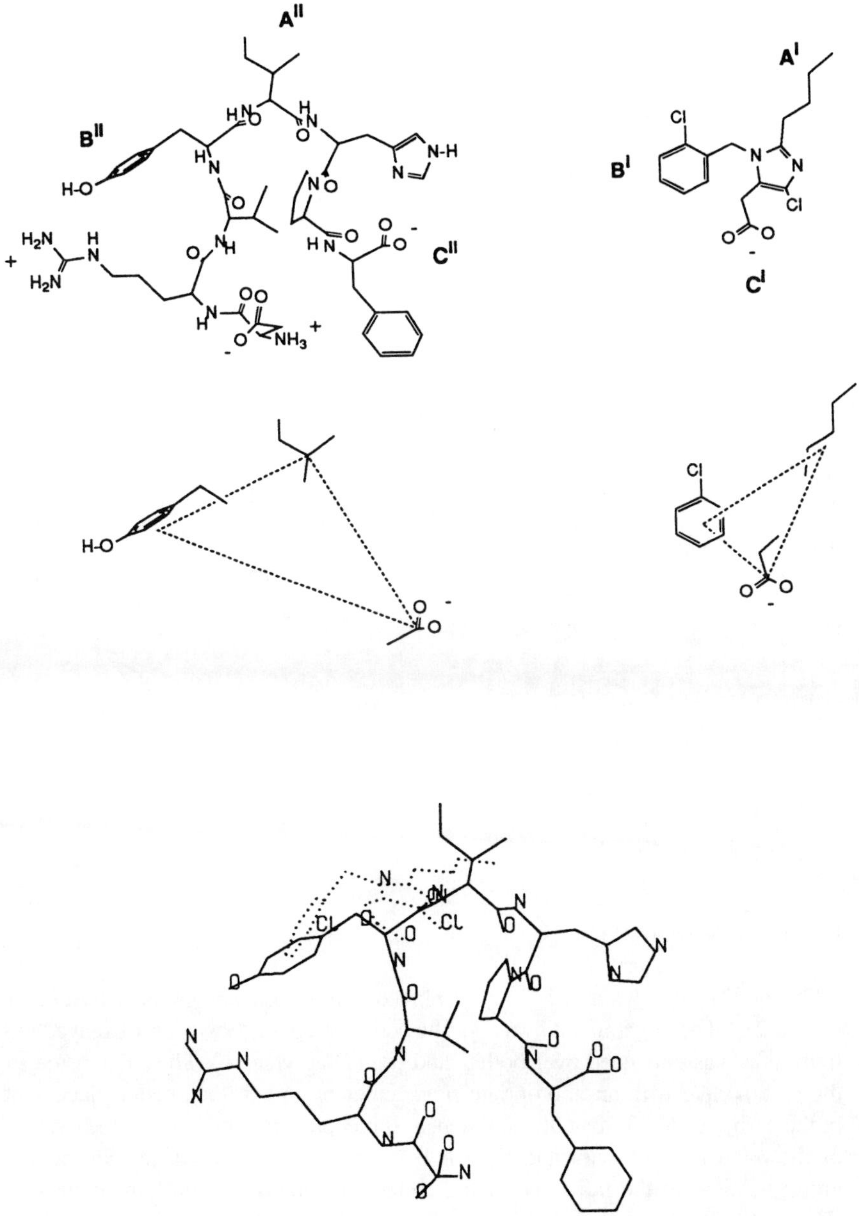

FIGURE 6. Model I cannot give a position four/eight overlay. The C-terminal carboxylate group in Model I is too far from the Tyr phenol and the Ile isobutyl to allow for a suitable overlay of the benzylimidazole.

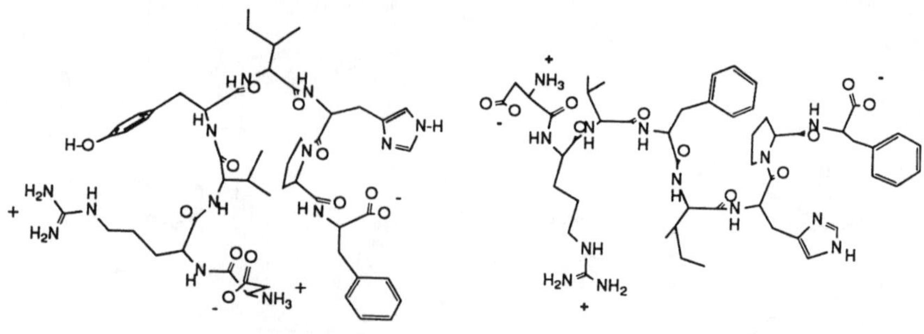

Amino Acid	MODEL I Torsion Angles φ	ψ	χ[1]		MODEL II Torsion Angles φ	ψ	χ[1]
Asp[1]	—	153	−174		—	150	−174
Arg[2]	−76	89	173		−74	89	−172
Val[3]	−130	146	180		−134	145	−172
Tyr[4]	−96	−45	−63		**75**	**88**	**−139**
Ile[5]	−93	−45	−63	⟶	−82	118	−177
His[6]	62	91	−160		67	72	−41
Pro[7]	−75	87			−59	119	
Phe[8]	−67	−28	−64		−86	—	−140

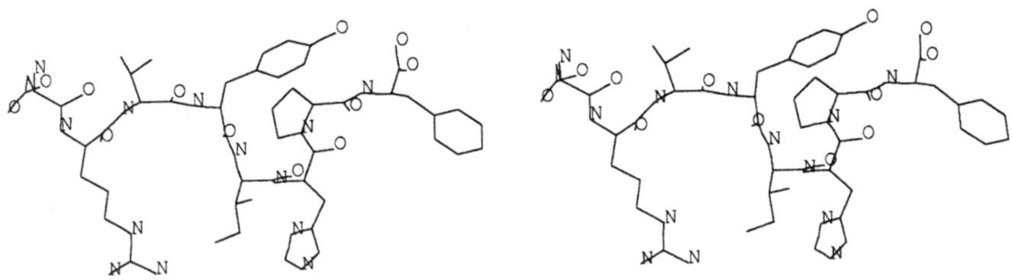

FIGURE 7. Conversion of model I to model II.

10 (Table 2) were evaluated. They displayed comparable antagonist activity to *1* (Table 2).* The constrained *trans*-acrylic acid analog *11,* however, displayed activity that was enhanced over both *1* and the *cis*-acrylate *12.* Thus, extension of the carboxylate with an appropriate rigid tether provided for a better placement of the carboxylate group at the angiotensin II receptor. Examination of the overlay of the *trans*-acrylate with angiotensin II suggested the appendage of a benzyl group to pick up the influence of the benzyl sidechain in [Phe[8]]-angiotensin II (Figure 9). It is well known that angiotensin II receptor antagonists that bear aromatic sidechains in position eight display enhanced activity over saralasin.[20]

*Although angiotensin II octapeptide analogs display comparable activities in the *in vitro* and radioligand-binding assays, many of the low potency benzylimidazoles display lower receptor affinity than *in vitro* potency. The reason for this discrepancy is not known.

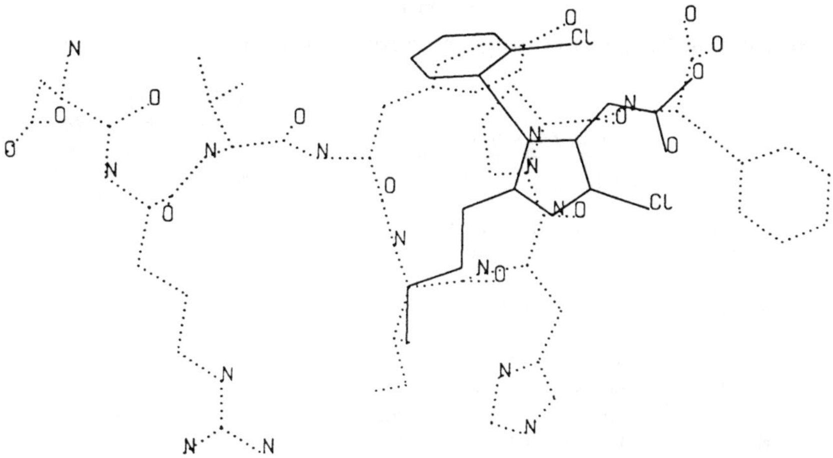

FIGURE 8A. A position four/eight overlay between model II and benzylimidazole *1*.

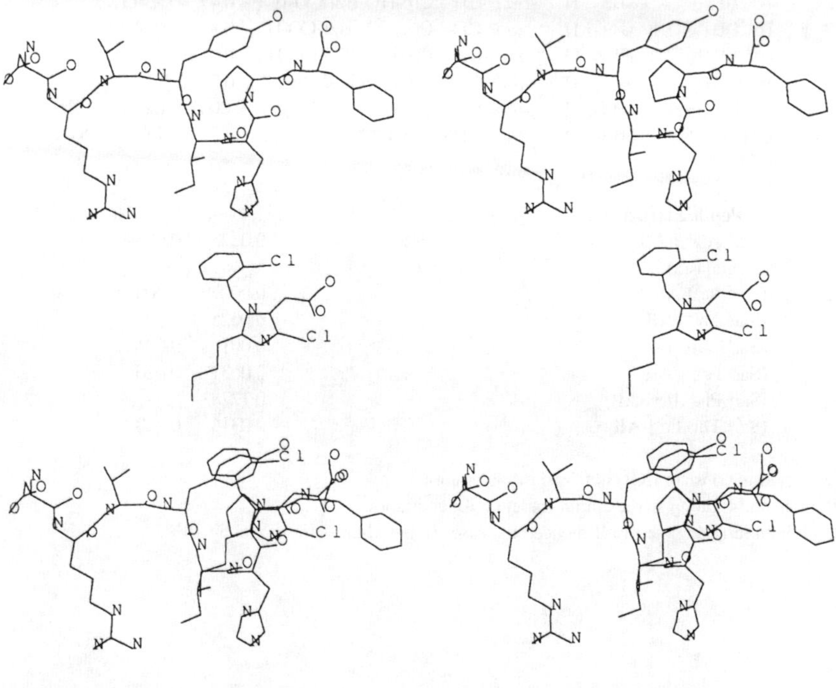

FIGURE 8B. A position four/eight overlay between model II and benzylimidazole *1 in stereo.*

TABLE 2
Activities of Peptide and Nonpeptide Angiotensin II Receptor Ligands

No.	Imidazole substituents 1	2	4	5	*In vitro* rabbit aorta[a] $K_B \mu M$	Binding rat Mes.[b] $IC_{50} \mu M$	*In vivo* rat BP[c] ID_{50}
1	-(σCl)Bzl	-Bu	Cl	CH_2CO_2H	2.70	43.0	30 ± 3.5
2	Dup 753-Ph-Ph(σCO_2H)	-Bu	Cl	-CHOH			
3	-(σCl)Bzl	-Bu	H	-CH_2CO_2H	1.90	12.0	22.5
4	-(σCl)Bzl	-SPr	H	-CH_2CO_2H	1.90	38.0	
5	-(σCl)Bzl	-SPr	H	-CH_2CH_2OH	71.0	98.0	
6	-(σCl)Bzl	-SPr	H	-CH_2CO_2Me	55.6	82.0	
7	-H	-SPr	H	-(R,S)-CH(CH_2Ph)CO_2H	>100	>1000	
8	-H	-Bu	H	-C(CO_2H)=CHPh	>100	>1000	
9	-(σCl)Bzl	-Bu	H	-$CH_2CH_2CO_2H$	2.76	16.5	
10	-(σCl)Bzl	-SPr	H	-$CH_2CH_2CO_2H$	8.50	55.0	
11	-(σCl)Bzl	-Bu	H	*trans*-CH=CHCO_2H	0.81	8.9	15 ± 0.72
12	-(σCl)Bzl	-Bu	H	*cis*-CH=CHCO_2H	14.0	124.0	
13	-(σCl)Bzl	-Bu	H	-*trans*-CH=C(CH_3)CO_2H	0.97	22.3	
14	-(σCl)Bzl	-Bu	H	*trans*-CH=C(Bzl)CO_2H	0.064	2.60	14 ± 3.3
15	-(σCl)Bzl	-Bu	H	-*cis*-CH=C(Bzl)CO_2H	8.70	37.60	
16	-(σCl)Bzl	-Bu	H	-*trans*-CH=C(p-MeO-Bzl)CO_2H	0.12	0.81	9.00
17	-(σCl)Bzl	-Bu	H	-*trans*-CH=C(p-HO-Bzl)CO_2H	0.47	**0.12**	
18	-(σCl)Bzl	-Bu	H	-*trans*-CH=C(p-NO_2-Bzl)CO_2H	0.31	**0.40**	
19	-(σCl)Bzl	-Bu	H	-*trans*-CH=C(CH_2-1-Thi)CO_2H	0.051	0.44	3.60
20	-(σCl)Bzl	-Bu	H	-*trans*-CH=C(CH_2imid)CO_2H	0.31	0.25	
21	-(pCO_2H)Bzl	-Bu	H	-*trans*-CH=C(CH_2-1-Thi)CO_2H	0.00021	0.001	0.08
22	-CH_2-Ph-Ph — (σCO_2H)	-Bu	H	-*trans*-CH=C(**Thi**)CO_2H		**1.6**	**6.6**
	Peptide structure						
23	[Sar[1],Ala[8];]-AII (saralasin)				0.0025	0.0034	
24	[Sar[1],Ile[8]]-AII				0.00079	0.00075	
25	[Sar[1],Thi[8]]-AII				0.0025	0.00011	
26	[Sar[1]]-AII				0.001	0.0013	
27	[Sar[1],His[8]]-AII				0.0631	0.026	
28	[Sar[1],Phe[6],Ile[8]]-AII				0.1778	—	
29	[Sar[1],Thi[6],Ile[8]]-AII				0.0316	0.140	

[a]Dissociation constant (KB) obtained in rabbit aorta.
[b]Radioligand-binding studies in rat mesentary membranes.
[c]Dose-inhibiting angiotensin II induced increases in blood pressure in rats by 50%.

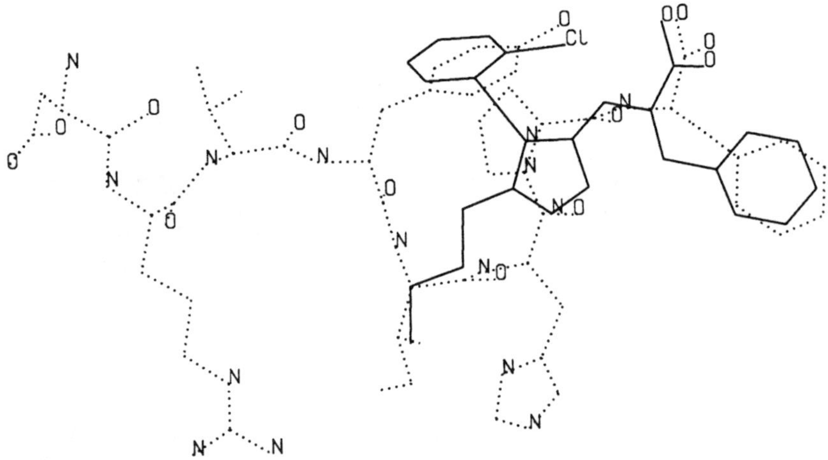

FIGURE 9A. A position four/eight overlay between model II and benzylimidazole *14*.

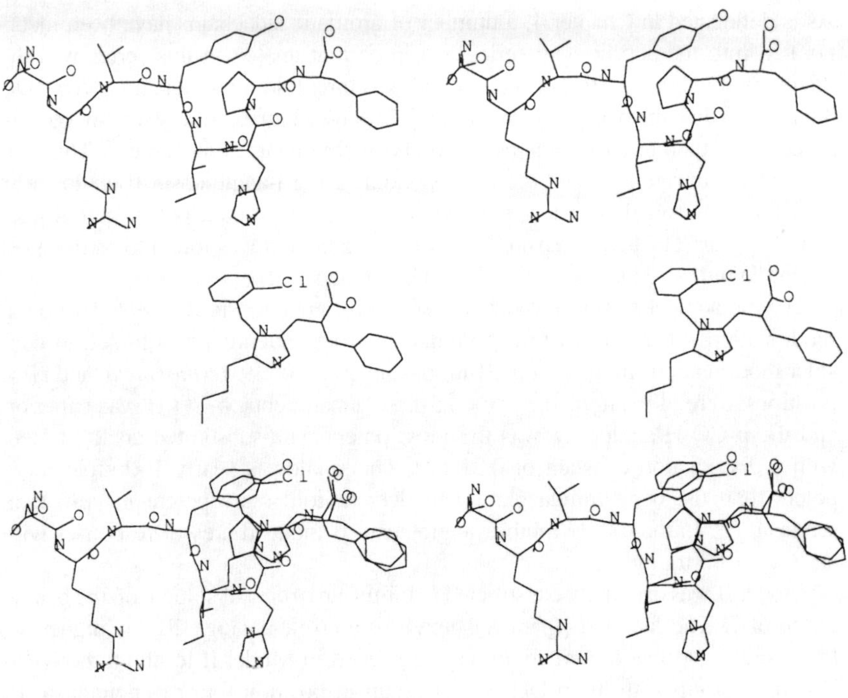

FIGURE 9B. A position four/eight overlay between model II and benzylimidazole *14 in stereo.*

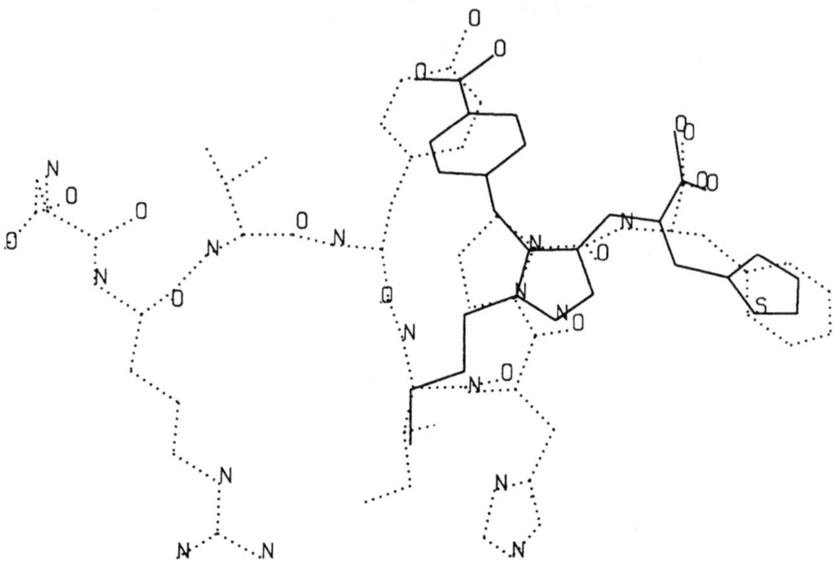

FIGURE 10A. A position four/eight overlay between model II and benzylimidazole *21*.

As is delineated in Chapter 4, a number of aromatic sidechains have been incorporated into the acrylic acid series. An important analog in this series was the thienylmethyl analog *19*. This compound was threefold more potent *in vivo* (ID_{50} 3.6 mg/kg) than the benzylacrylate *14* and displayed higher affinity for angiotensin II receptors than any of the substituted arylmethyl analogs in Table 2. This was particularly intriguing since the corresponding [Phe^8]-angiotensin II analog was [Thi^8]-angiotensin II, which was previously reported to be a potent partial agonist.[21–23] [Sar^1,Thi^8]-angiotensin II *25* was prepared and was found to be the most potent ligand tested to date (IC_{50} 0.11 n*M*), (Chapter 2).

Having achieved improvements in affinity for angiotensin II receptors through modifications of the [$CH_2CO_2H^2$]-imidazole group, attention was turned to consider modifications in the [$benzyl^1$]-imidazole group of *19. Ortho, meta,* and *para* positions were all examined.[24] Among these substituents, p-CO_2H was superior, and the p-CO_2H analog *21* was the most potent para-substituted analog tested, with a dissociation constant of 0.21 n*M.* This analog is nearly 1000-fold more potent than the benzylimidazole *1* and over tenfold more potent *in vitro* than saralasin *23.* In the rat, *21* inhibits angiotensin II-induced pressor responses with an IC_{50} of 80 μg/kg.

Model II was originally constructed to mimic an *ortho* substituent on the benzyl group of *1.* The fact that [(para-X)$benzyl^1$]-imidazole analogs display higher affinity suggested that a correction may be needed in Model II to allign the *para*-X-benzyl group with the p-OH-benzyl group in position four of angiotensin II. This could be accomplished by altering the χ^1 angle of Tyr from $-139°$ to $-88°$

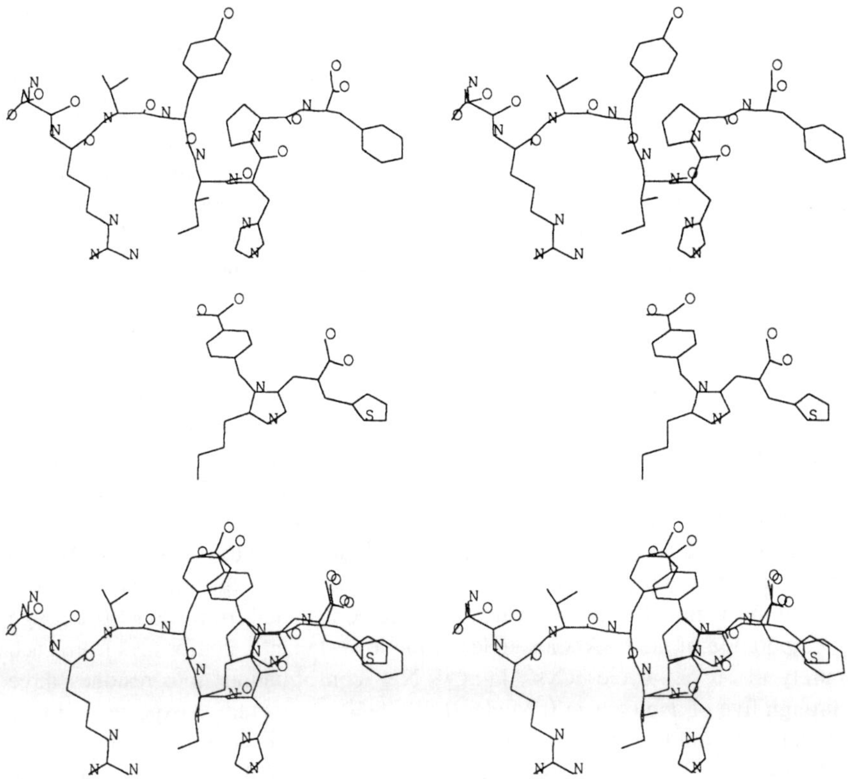

FIGURE 10B. A position four/eight overlay between model II and benzylimidazole *21 in stereo.*

(Figure 10). The degree to which the structure–activity relationships of the non-peptide benzyl substituents parallel the structure–activity relationships of the Tyr^4 in peptide analogs is currently being evaluated.

Model II was originally constructed with the Phe^8 benzyl group positioned over the Pro^7-Phe^8 amide, as suggested by Marshall[25] for agonists. Angiotensin II octapeptide antagonists, of course, may place the Phe^8 benzyl in other orientations. It is interesting to note that the benzylacrylates cannot be overlaid in the Phe^8 agonist conformation. If the Phe^8 x^1 angle is changed to $-140°$, however, then a good overlay of aryl groups can be achieved. Thus, the arylmethylacrylates mimic an antagonist conformation of angiotensin II.

3. Model III, A Refinement on Model II[26]

Subsequent to the development of the benzylacrylic acids through utilization of the overlay hypothesis, [hCys^{3,5}]-angiotensin II[27] and [Sar^1, hCys^{3,5}Ile^8]-angiotensin II[27,28] appeared in the literature. These analogs contain a cyclic structure and retain potent angiotensin II receptor agonist and antagonist activity, respectively. Upon examination of Model II, it was readily apparent that the model was not consistent with a disulfide bridge between a hCys in position three and a hCys in position five, since the sidechain in position three of Model II points above the plane of the molecule and the sidechain in position five of Model II points below the plane (Figure 11).[26] Because the spatial relationship of positions four and five to position eight is a critical component of Model II, an important test of the overlay hypothesis would be to attempt revision of Model II into a conformation that would be consistent with the hCys analogs and yet retain an acceptable overlay with the benzylimidazoles.

To evaluate the influence of the macrocycle on Model II, a full conformational search was conducted on the model tripeptide, Ac-S,S-cyclo-hCys-Ala-hCys-NH_2. A description of this search has been described recently.[26] Four types of conformation were found for this peptide as seen in Table 3. Conformer B was the most abundant type of conformer, and the alanine residue (which would correspond to Tyr in angiotensin II) was placed in an α_L conformation (Table 3). Likewise, Kataoka et al.[29] have postulated recently that Tyr^4 occupies an α_L conformation in the $hCys^{3,5}$ analogs. In all cases, the α to β carbon vectors of both $hCys^3$ and $hCys^5$ point roughly in the same direction, due to the disulfide bridge. Four analogous angiotensin II conformations were constructed from Model II through incorporation of the backbone angles of the lowest energy conformers from each family of Ac-S,S-cyclo-hCys-Ala-hCys-NH_2 conformations into residues three through five of angiotensin II Model II (Table 4). As would be expected, all four conformers could be closely aligned at their C terminal residues six through eight. Of importance, however, was the high degree of overlap that occurred among the Tyr^4 and $hCys^5$ sidechains with the corresponding sidechains in Model II. The topological arrangement of positions four, five, and eight in Model II that gave the important overlay with the benzylimidazole antagonists could be obtained in all four altered conformers. The major difference between Model II and this set of conformers, which is referred to as Model III collectively, is in the orientations of the Arg^2 alkylguanidine sidechain. As seen in Figure 12, each Model III conformer places the Arg^2 guanidine group much closer to the Tyr^4 phenol than in Model II. Because the overlays with benzylimidazole antagonists do not attempt to align with the Arg^2 guanidine group, neither Model II nor Model III should be expected to ''define'' the orientation of that group. The discovery of conformationally constrained analogs of angiotensin II, which define the relationship between positions one and two with the rest of the peptide, might help to ''increase the resolution'' of these models. As might be expected, [$hCys^{3,5}$]-angiotensin II can be built readily in Model III conformations (Figure 13). Most important is the fact that, since the conformation between Tyr^4 and Phe^8 has not changed, the overlay hypothesis is upheld with Model III, inasmuch as a good overlay may still be obtained with the benzylimidazole antagonists (Figure 14).

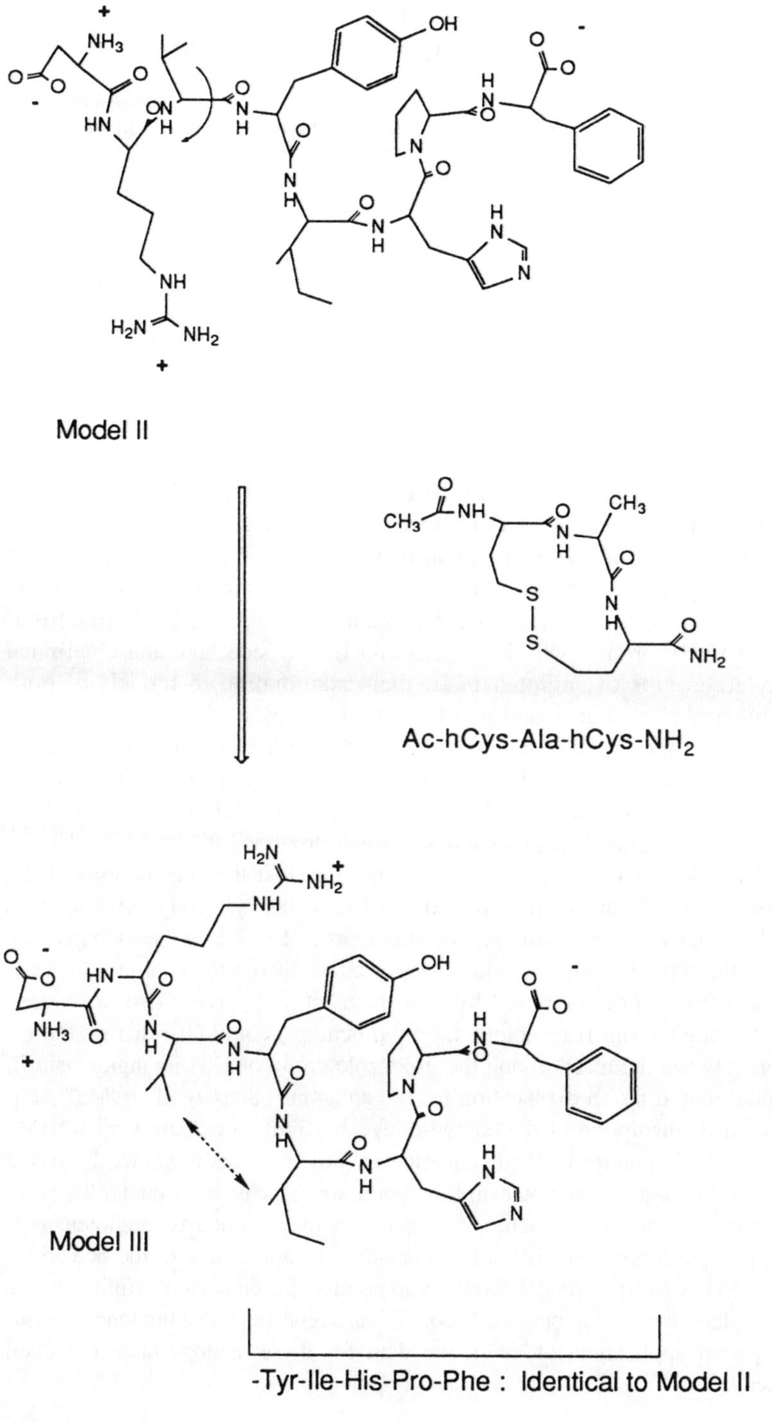

FIGURE 11. Conversion of model II to model III.

TABLE 3
Conformers of Ac-S,S-cyclo-hCys-Ala-hCys-NH_2

Family no.	Conformer[a] type	Sample conformer	Torsion Phi	Angle Psi	Lowest energy (kcal/mole)	#Conformers in lowest 10	20	100[a]
A	β_3	9	−130.8	99.8	4.86	1	1	25
B	α_L helix	317	58.5	78.2	5.69	6	10	30
C	α_R helix	454	−90.1	−30.1	5.77	2	6	23
D	$C7_{eq}$	262	−90.8	81.6	6.72	1	2	15

[a]One conformer was eliminated because it was not feasible for an L-amino acid.

E. AN ALTERNATE POSITION SIX OVERLAY OF THE BENZYLIMIDAZOLE WITH ANGIOTENSIN II

The presence of imidazole groups in both *1* and angiotensin II suggests that the imidazole group in *1* might occupy the same space as the imidazole group in the His[6] sidechain. If one includes the correlations of the imidazole (o-Cl)Bzl[1] and $(CH_2CO_2H)^5$ groups with the angiotensin II Tyr[4] sidechain and C terminal carboxylate groups of angiotensin II, then examination of models of both benzylimidazole *1* and angiotensin II leads to the suggestion that the Tyr[4] sidechain and C terminal carboxylate may cluster about the His[6] sidechain in angiotensin II. This suggests a model of angiotensin II that is similar to the Tyr[4]-His[6]-C terminal charge relay model of Matsoukas et al.,[17] in a rigid overlay with benzylimidazole *1* (Figure 15). (This model is also discussed in the Appendix.) A potential problem with this overlay lies in the fact that the substitutions of the imidazole of His[6] in angiotensin II with methyl or benzyl groups give analogs with substantially reduced affinities for angiotensin II receptors (see Chapter 2). Although these modifications could be expected to disrupt the charge-relay conformation of the model proposed by Matsoukas et al.,[17] giving rise to reduced affinity for angiotensin II receptors, the modifications could also disrupt the interaction between the receptor and the imidazole group of His[6] in angiotensin II. The model may offer an explanation for the antagonist activity of [(OMe)Tyr[4]]-angiotensin II (disruption of the charge-relay), but fails to explain why [(NτMe and NτMe)His[6]]-angiotensin II and angiotensin II methyl ester retain weak agonist activity and fail to act as angiotensin II receptor antagonists (see Chapter 2). More importantly, the overlay depicted in Figure 15 places putative angiotensin II binding groups in positions that are inaccessible for appendage to the benzylimidazole *1*. This overlay model would not predict the enhanced affinity of the benzylacrylates (described below). Moore[30] has suggested some analogs in a European patent application, although the data for these analogs have not been published.

TABLE 4
Conformers of Angiotensin II Model III

Position in angiotensin II	Torsion angle	Differences between conformers A–D and model II				
		Model II	A	B	C	D
Asp	ϕ	—				
1	ψ	150.0				
	θ	−178.6				
Arg	ϕ	−74.0				
2	ψ	89.0				
	θ	174.3				
Val	ϕ	−133.5		*−60*		
3	ψ	145.6	163.58	*−57.6*	*99.2*	133.9
Conf. type:		β_1	β_1	α_R	β_3	β_1
	θ	−175.4	164.9	179.8	175.9	180.0
Tyr						
4	ϕ	74.8	−130.8	58.5	−90.1	−90.8
	ψ	87.8	99.8	78.2	−30.1	81.6
Conf. type:		α_L	β_3	α_L	α_R	C_7^{eq}
	θ	175.8	195.1	175.9	184.9	181.3
Ile						
5	ϕ	−81.8	−73.9	−80.2	*69.2*	*−116.9*
	ψ	117.6			*60*	
Conf. type:		β_2	β_2	β_2	α_L	β_1
	θ	−177.4				
His						
6	ϕ	67.1				
	ψ	72.1				
	θ	174.7				
Pro	ϕ	−58.8				
7	ψ	118.6				
	θ	−171.9				
Phe	ϕ	−86.2				
8	ψ	—				
Distance						
Tyr OH to Arg guan:		14.4 A	10.9 A	2.44 A	8.79 A	8.6 A

Note: Torsion angles in conformers only shown if different from Model C.

F. THE TEMPLATE MODEL OF PIERSON AND FREER DERIVED A POSITION SIX OVERLAY

Pierson and Freer[15] recently reported the development of different angiotensin II template models derived from overlays with the DuP 753 and its structural analogs. This group began with a model described by De Coen and Ralston.[31] Pierson and Freer[15] altered the conformation to account for the macrocycle in the

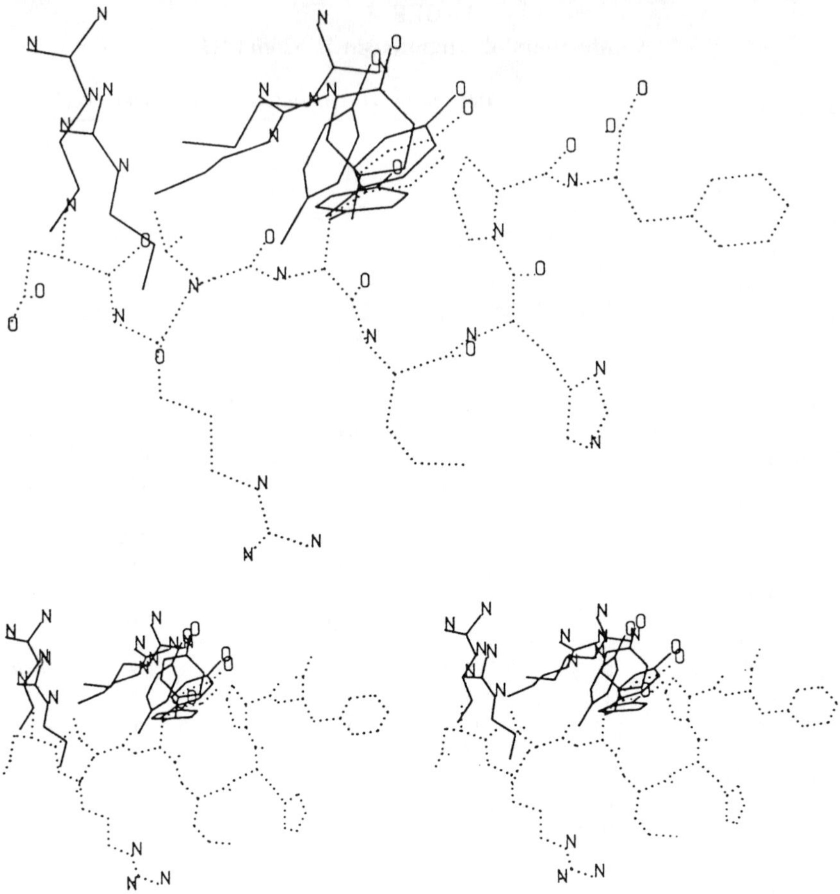

FIGURE 12. Model III conformers. *Relationship between Arg and Tyr sidechains in each model III conformer.* The Arg-alkylguanidine and Tyr phenol sidechains in model III conformers W–Z (solid lines) are superimposed upon model II (dots) to highlight the most important differences between model II and model III. Were the entire structures to be shown, complete overlap of the model III comformers with model II from the carbonyl of Tyr[4] to the c-terminus (Ile-His-Pro-Phe) would be evident, as well as the lack of overlay of the Asp[1] and Val[3] sidechains.

potent cyclic angiotensin II analog, [Sar[1], hCys[3,5], Ile[8]]-angiotensin II. Onto this model Dup 753 was superimposed, aligning the imidazole template with the His[6] imidazole, the distal phenyl of the biphenyl group with the phenyl group of Phe[8], the butyl group with the Ile[5] sidechain, and the acetic acid carboxylate with the C terminal carboxylate group. Both compounds were allowed to ''relax'' into conformations that optimized the overlay. This final conformation was used to study the overlay of several different nonpeptide angiotensin II receptor antagonists.

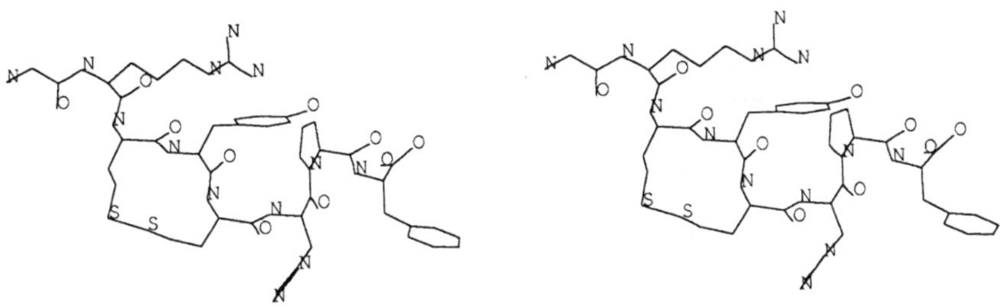

FIGURE 13. A model III conformer depicted in [hcys^{3,5}]-Angiotensin II *in stereo.*

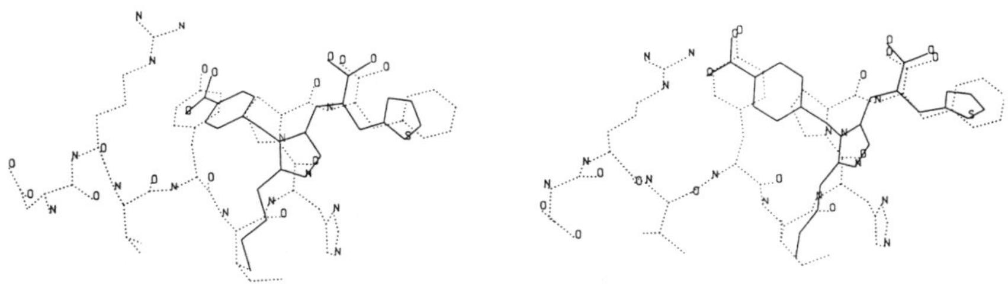

FIGURE 14. A position four/eight overlay between model III and benzylimidazole *14 in stereo.*

The original benzylimidazole *1* was reported to fit the model relatively well in the same overlay that was employed for Dup 753. Other small angiotensin II receptor molecule antagonists, Exp 6155 and S8308, which are relatively inactive, did not fit the model in the same overlay as Dup 753, perhaps due to a lack of appropriate binding groups. Pierson and Freer[15] reported that they are evaluating the model with conformationally constrained analogs of angiotensin II. Since no atomic coordinates or dihedral angles are reported for this model, it is not possible to compare the Pierson and Freer[15] model to the Model II or Model III of Samanen[14] or to the model of Fermandjian.[9] This overlay is also different from that employed by Duncia et al.,[13] who oriented the biphenylcarboxylate in the direction of the Asp[1] sidechain.

A potential problem with the model of Pierson and Freer[15] resides in the super-positioning of the template imidazole with the His[6] sidechain. As noted earlier and in Chapter 2, (NMe)His analogs of angiotensin II have been reported to display diminished activity, suggesting a negative steric interaction with the angiotensin II receptor. It would seem, therefore, that the benzylimidazole antagonists would consequently display negative steric interactions with the receptor, giving rise to lower affinity. As noted in the Appendix, the most probable model of Dc Coen and Ralston[31] is not consistent with the conformationally constrained angiotensin II analogs in the literature. Pierson and Freer[15] do not indicate whether these inconsistencies are eliminated in their modeling.

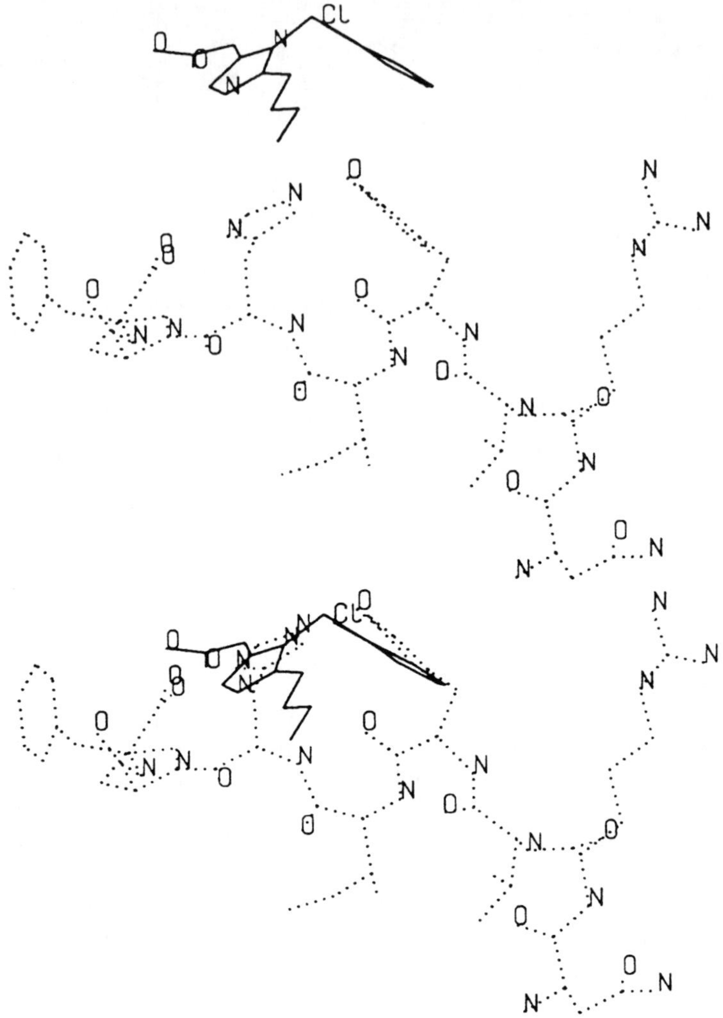

FIGURE 15A. An overlay between a representation of the model of Matsoukas et al.[17] and benzylimidazole *1*.

G. A POSITION FOUR OVERLAY WITH THE MODEL OF FERMANDJIAN

An overlay that attracted the attention of one group of investigators[17] is depicted in Figure 16. In this position four overlay, the [(o-Cl)Bzl[1]]-imidazole is related to [Tyr[4]]-angiotensin II, the [Bu[2]]-imidazole to [Ile[5]]-angiotensin II, and the [$CH_2CO_2H^5$]-imidazole lies within the amide backbone. Most models can be employed for this overlay. The rigid overlay shown in Figure 16 utilizes the model of Fermandjian et al.[9] This overlay does not place a particular emphasis on the

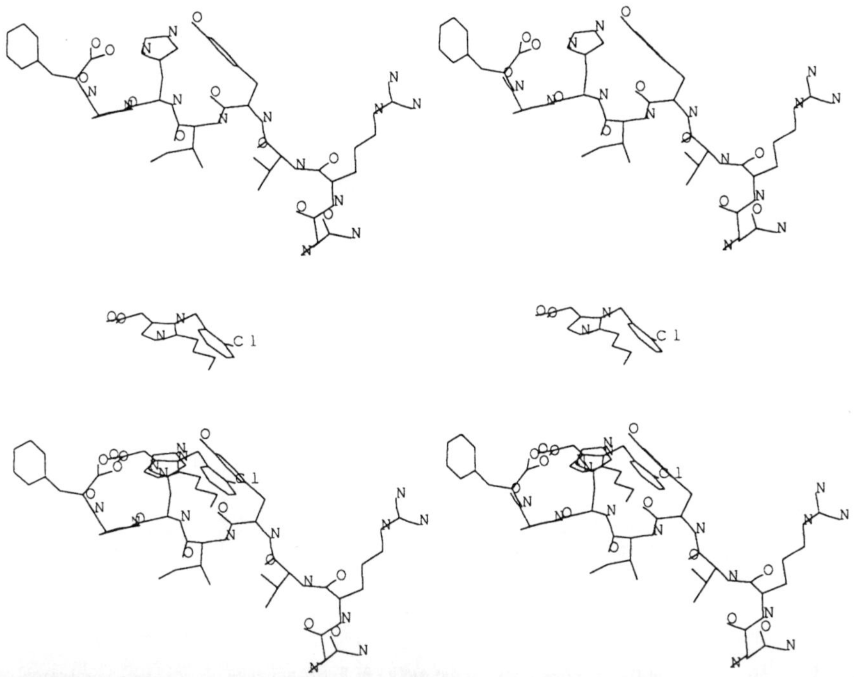

FIGURE 15B. An overlay between a representation of the model of Matsoukas et al.[17] and benzylimidazole *1 in stereo.*

benzylimidazole carboyxlate group, because it is related to portions of angiotensin II that do not require a carboxylate (The Asp^1 sidechain can be eliminated without loss of potency in agonists or antagonists; Chapter 2). As noted above, however, amide and ester analogs of *1* display reduced activities, suggesting to the contrary that the benzylimidazole carboxylate group is critical for activity. For this reason, the position four overlay was not pursued.

H. A POSITION EIGHT OVERLAY WITH ANGIOTENSIN II

Another early overlay that was considered related the benzylimidazole *1* to the C terminal phenylalanine in angiotensin II.[18] Marshall,[25] had noted that analogs of angiotensin II in which the Phe^8 sidechain is positioned over the Pro-Phe amide (e.g., [δ-Phe^8 and Thiq^8]-angiotensin II) retain agonist activity, whereas analogs of angiotensin II in which the Phe^8 sidechain is held out of this orientation (e.g., [Ind^8]-angiotensin II) lack agonist activity and are angiotensin II receptor antagonists, a so-called "agonist hypothesis."[25] In this "agonist conformation," one can place the imidazole group of *1* in the plane of the Pro-Phe amide group of angiotensin II (Figure 17). The [1-(o-Cl)Bzl]-imidazole overlays the phenyl ring of [Phe^8]-angiotensin II (or Thiq^8 in Figure 17), and the carboxylate group in imidazole *1* overlays the C-terminal carboxylate in angiotensin II. In this overlay,

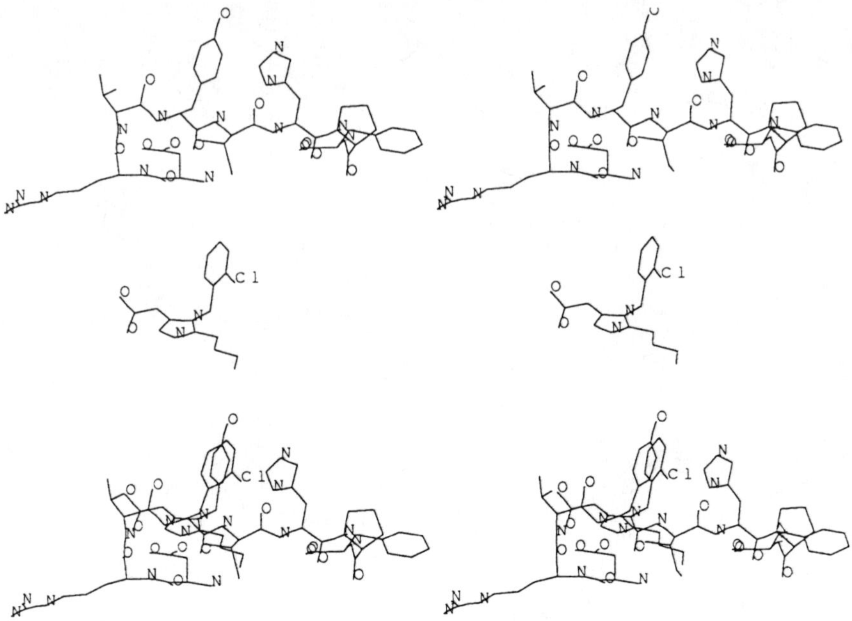

FIGURE 16. A position four overlay between the model of Fermandjian et al.[9] and benzylimidazole *1 in stereo.*

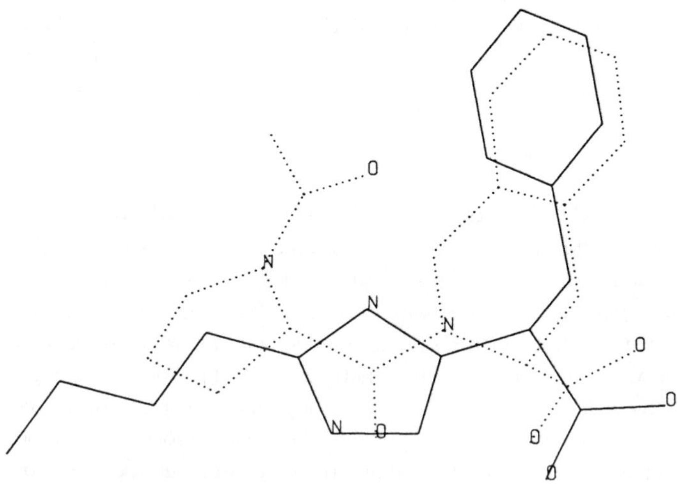

FIGURE 17. A position eight overlay. Overlay of benzylimidazole **I** onto -Pro-Thiq, the C terminal dipeptide of a constrained angiotensin II peptide [Thiq[8]]-AII.

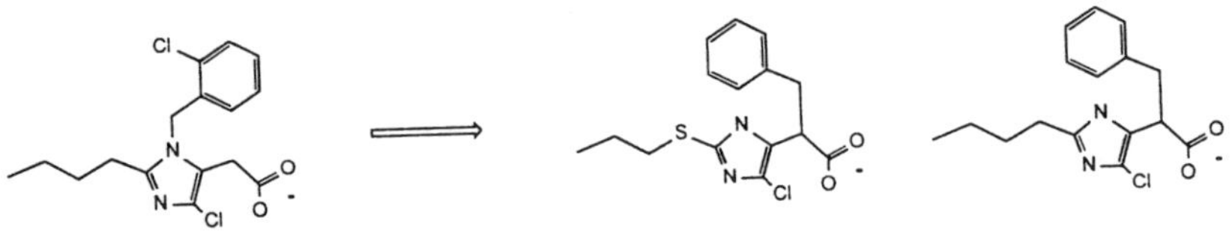

FIGURE 18. Derivation of benzylimidazole analogs from the position eight overlay.

the [2-Bu]-imidazole overlays the aliphatic sidechain of [Pro[5]]-angiotensin II. Since this overlay focuses upon the C terminus of angiotensin II, most models of angiotensin II suffice, as long as the [Phe[8]]-angiotensin II sidechain is in an "agonist" conformation. Apparent from this overlay was the possibility that the benzyl group on the imidazole of *1* could be translocated to the acetic acid sidechain (Figure 18), to give rise to analogs *7* and *8.* As seen in Table 2, however, analogs *7* and *8* are inactive. Also apparent from this overlay is the possibility that the weak angiotensin II receptor antagonist activity of benzylimidazole *1* may be due to its being a poor mimic of the C terminus or to the fact that it mimics only a small portion of angiotensin II and thereby lacks a sufficient number of receptor binding groups. Analog *372,* Boc-His-L-lactam-Phe (Chapter 2, Table 15) is derived from the C terminal tripeptide and contains a His residue, which is important in the octapeptide, as well as the L-lactam replacement for Pro that gave rise to potent angiotensin II receptor antagonist activity in the octapeptide.[32] Analogs *366* to *371,* Tyr-Ile-His-Pro-Phe, Ile-His-Pro-Phe, Boc-Pro-Phe, Boc-Pro-*D*-Phe, Boc-His-Pro-Phe, and Boc-His-Pro-*D*-Phe (Chapter 2), also add His and Tyr residues to the C terminal Phe, all important binding groups in the octapeptide. None of these compounds exhibited agonist or antagonist activity at concentrations up to 1 μ*M.* Due to these initial failures and the subsequent success of another overlay, the position eight overlay was not considered further.

I. COMPARISON OF THE BIPHENYLIMIDAZOLES WITH THE IMIDAZOLE ACRYLATES

1. Differences in Structure–Activity Relationships

The biphenylimidazoles (e.g., DuP 753; losartan) and the benzylacrylates (e.g., SK&F 108566) both contain alkyl groups in position two and an imidazole group as the basic template. However, these two classes of angiotensin II receptor antagonists differ in a number of ways. Whereas a carboxylate group in position five appears to be important in the acrylate series, the significance of this functional group is lost in the biphenylimidazole series.[10] A carboxylate group or tetrazole mimetic is an important substituent in the pendant biphenyl group, however. As is discussed in Chapter 4, replacement of the acrylate carboxylate in *21* with a tetrazole group gave an analog with reduced activity. Furthermore, incorporation

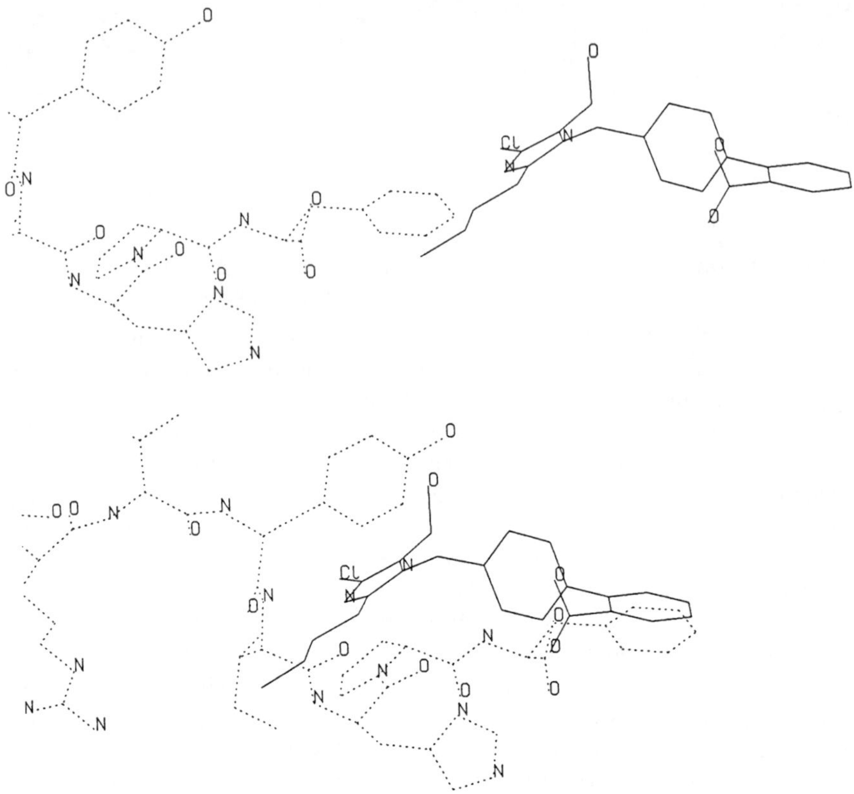

FIGURE 19A. A position four/eight overlay between model II and benzylimidazole *2*.

of the biphenyl carboxylate group into an acrylate analog *14* gave compound *22* with reduced activity (Table 2). These differences suggest that the two series of angiotensin II receptor antagonists may display different modes of binding to the angiotensin II receptor. Examination of the degree to which important binding groups in angiotensin II Model II and the biphenylimidazole *2* can be overlaid might shed light on these differences. Figure 19 displays one possible overlay, in which the alcohol group in *2* is aligned with the phenolic hydroxyl of Tyr^4, and the biphenyl of *2* is stretched through the lipophilic sidechains of Pro-Phe, placing both carboxylate groups in the same space. In this orientation, the 2-butyl group can reach Ile^5 as in other Model II overlays. This overlay places the *ortho* carboxylate (or tetrazole mimetic in DuP 753) in the same position as the C terminal carboxylate. This overlay further accounts for the increased affinity observed upon oxidation of the alcohol group on *2* to a carboxylate, since it is observed that a *para*-carboxybenzyl group provides enhanced affinity over the *para*-hydroxybenzyl group that mimics Tyr in the benzylacrylate series of angiotensin

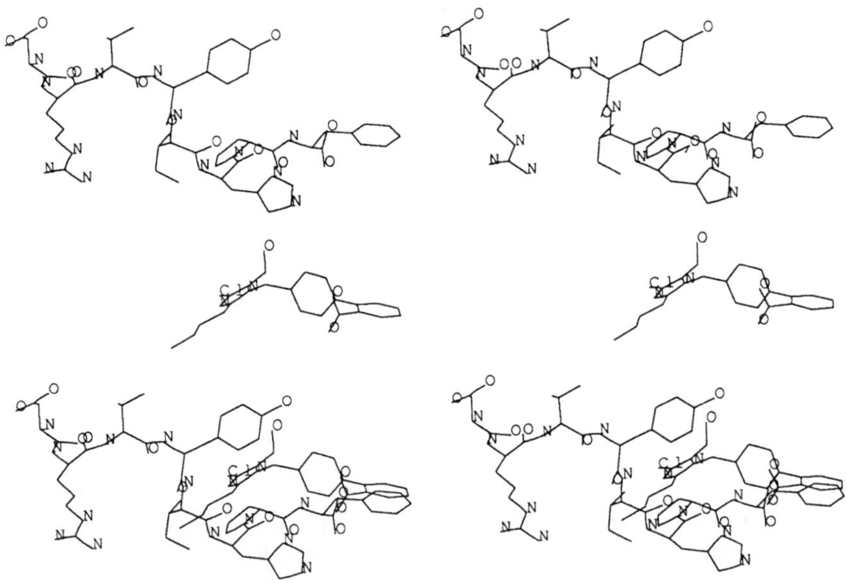

FIGURE 19B. A position four/eight overlay between model II and benzylimidazole *2 in stereo.*

II receptor antagonists. There is at present, however, no data to suggest that the biphenyl analog *2* mimics angiotensin II in this fashion.

J. COMPARISON OF THE AT_1 SELECTIVE BENZYLACRYLATES WITH THE AT_2 SELECTIVE SPINACINES: BOTH OVERLAY MODEL II

A series of nonpeptide bicyclic angiotensin II receptor antagonists that displays selective inhibition of the AT_2 receptor was recently described.[33] These spinacine derivatives share several structural features in common with the arylacrylates: (1) both classes of molecules contain a benzylimidazole structure, (2) the carboxylate group is connected to the imidazole through a 2-carbon chain, and (3) an aryl group is appended to the carboxylate tether. However, a comparison of PD 121981 (a spinacine derivative) with SK&F 108566 (a benzylacrylate) reveals some distinct differences: (1) PD 121981 lacks an alkyl substituent in position two, whereas the 2-butyl group in SK&F 108566 is critical, (2) the 2-carbon carboxylate tether in PD 121981 is part of a piperidine ring fused to the imidazole (as found in the natural product spinacine), whereas the 2-carbon tether in SK&F 108566 is an olefin, and (3) the substituent on the *para* position of the 1-benzyl group in PD 121981 is maximized in the electron donating groups $-OH$, $-OMe$, and $-NH_2$, whereas that substituent in SK&F 108566 is maximized in the electron withdrawing groups $-NO_2$ and $-CO_2H$ (Chapter 4).

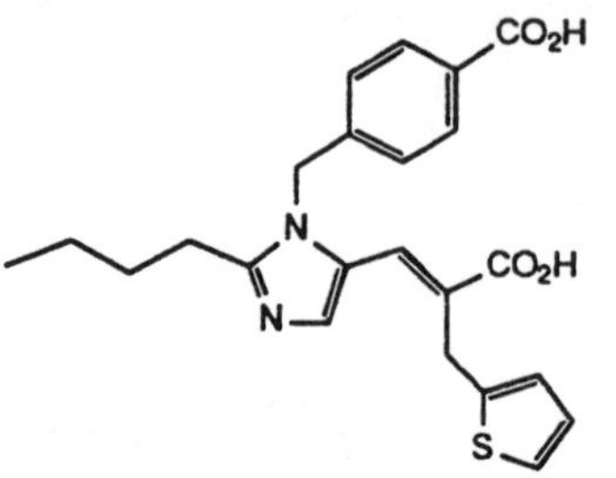

SK&F 108566

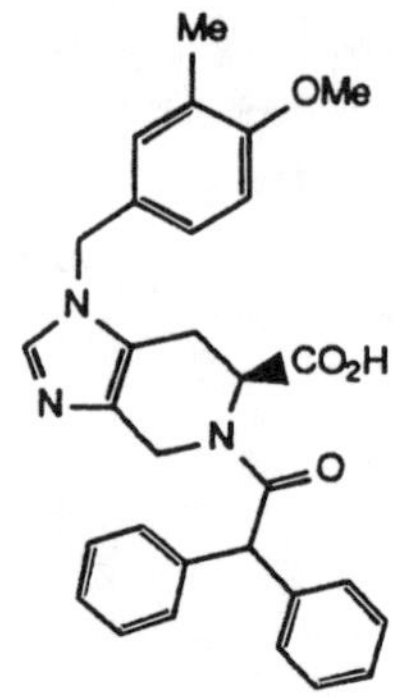

PD 121981

This unusual combination of differences appears to account for the fact that PD 121981 displays high affinity for AT_2 receptors and is inactive at AT_1 receptors, whereas SK&F 108566 displays alternate selectivity, has high affinity for AT_1 receptors, and is inactive at AT_2 receptors (Chapter 4). These differences suggest that the AT_2 pharmacophore is distinct from the AT_1 pharmacophore (as has recently been confirmed by peptide structure–activity relationships; Chapter 2).

To begin to address the question of whether the AT_2 receptor requires a different conformation for angiotensin II than the AT_1 receptor, an overlay of PD 121981 with angiotensin II Model II has been attempted. As seen in Figure 20, a high degree of overlap may be achieved between the [1-benzyl]-spinacine group and [Ile^5]-angiotensin II, between the [6-COOH]-spinacine group and the C terminal carboxylate group in angiotensin II, and one of the phenyl groups in [7-diphenylacetyl]-spinacine. Of interest is the fact that the other phenyl group in [7-diphenylacetyl]-spinacine could be overlaid with the [His^6]-angiotensin II sidechain, providing a rationale for the enhanced affinity observed with diphenylacetyl analogs over phenylacetyl analogs. Although this modeling exercise employed Model II, Model III should overlay with PD 121981 in a similar fashion, since both models are nearly identical from Tyr^4 through Phe^8. It has been observed that *para*-substituents on the diphenylacetyl group reduced affinity for angiotensin II receptors of the AT_2 subtype, in agreement with the structure–activity relationships for [Phe^8]-angiotensin II (refer to Chapter 2). One of the phenyl groups could be replaced with either cyclohexane or cyclopentane without loss of affinity for AT_2 receptors. Such a modification is also acceptable for [Phe^8]-angiotensin II. The overlay of the second phenyl group with [His^6]-angiotensin II does not need to be invoked to explain the higher affinity of diphenylacetyl over phenylacetyl, since [Sar^1,Diphenylalanine^8]-angiotensin II is more potent than [Sar^1]-angiotensin II.[34] This overlay suggests that the AT_2 receptor does not necessarily require a different conformation of angiotensin II for high affinity interaction.

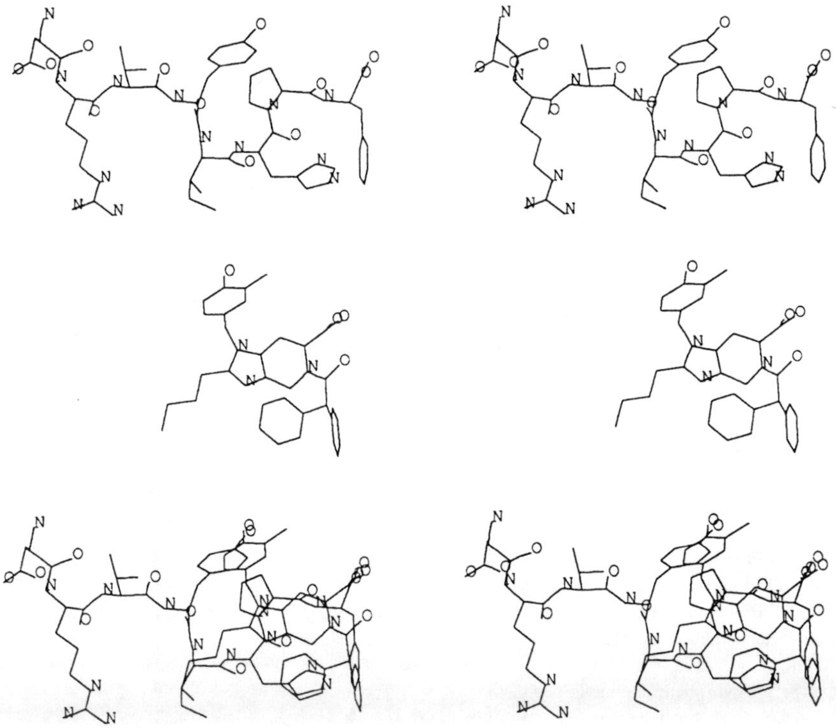

FIGURE 20. An overlay between model II and PD 121981 *in stereo.*

K. FACTORS THAT COMPLICATE THE OVERLAY APPROACH

The fact that analog design from Model II has successfully led to the identification of potent angiotensin II receptor antagonists is not proof that compounds such as SK&F 108566 mimic angiotensin II at the receptor in the manner depicted by the overlay. As noted earlier in this chapter, nonpeptides can interact potentially with a peptide receptor in a mimetic, steric, or allosteric mode. It is tempting to speculate that SK&F 108566 might bind in a mimetic mode, whereas DuP 753 might bind in a steric or allosteric mode. The extent to which the structure–activity relationships of correlated binding groups in these series parallel the structure–activity relationships of the correlated binding groups in angiotensin II would lend support to the hypothesis.

A comparison of the structures of the acrylates with angiotensin II raises the possibility of the arylmethyl group on the acrylate sidechain mimicking [His^6]-angiotensin. Accordingly, three acrylate analogs bearing thienylmethyl, benzyl, and imidazolemethyl groups, compounds *18, 14,* and *20,* were compared with the corresponding [Sar^1,L-NH($CHCH_2X$)CO^6,Ile^8]-angiotensin II and [Sar^1,L-NH($CHCH_2X$)CO_2H^8]-angiotensin II analogs (Table 5). The comparison of Table 5 is hampered by the fact that some of the compounds in the comparison are

TABLE 5
Comparison of Position Six and Eight Angiotensin II Octapeptide Analogs with Benzylimidazole Analogs

	Position six angiotensin II analogs [Sar^1,-NH($CHCH_2R$)CO-)^6, Ile^8]-angiotensin II			Position eight angiotensin II analogs [Sar^1, -$NHCHCH_2RCO_2H$)^8]-angiotensin II		
Rabbit aorta[a]	Imid[d] 0.79	> Thi 31.6	> Phen 177.8	Phen 1.00[e] ~	Thi 2.55	> Imid 63.1
Rat mesentery[b]	Imid 0.75	> Thi 140		Thi 0.11	> Phen 1.3	> Imid 26.0
Bovine adrenal cortex[c]	Imid 3.0	> Phen 300	> Thi 950	Phen 3.0 ≥	Thi 10	> Imid 120

	Position five substituted 1-oClBzl-2-butylimidazole t-CH=C(CH_2R)-CO_2H		
Rabbit aorta[a]	Thi 50	= Phen 64	> Imid 310
Rat mesentery[b]	Imid 250 ~	Thi 440	> Phen 2600
Bovine adrenal cortex[c]	Thi 2000	> Phen 3700	> Imid 6400

[a](10^{-9}*M*) *In vitro,* rabbit aorta.
[b](10^{-9}*M*) Binding, rat mesenteric artery.
[c](10^{-9}*M*) Binding bovine adrenal cortex (note the twofold difference in listed potencies and affinities of the octapeptides compared to the benzyl acrylic acids).
[c]Abbreviations: Imid = imidazole (histidine in peptides); Thi = thiophene (thienylalanine in peptides); Phen = phenyl (phenylalanine in peptides).
[d]EC_{50} n*M,* agonist activity, since [Sar^1]-AII is a pure agonist.

agonists, while others are antagonists, making direct comparisons difficult to interpret. Furthermore, as discussed in Chapter 2, cross-species comparisons cannot be made. Suffice it to say that the data does not provide a firm correlation and is at best equivocal. It may be more appropriate to compare three dissimilar sidechains, such as ionic, aliphatic, and aromatic, since one can imagine that relatively minor alterations in receptor topography could alter receptor preferences for similar ligand sidechains. This analysis highlights the difficulties in attempting to validate an overlay hypothesis.

The abundance of lipophilic binding groups on angiotensin II suggests that the receptor binding pocket may be rich in lipophilic binding sites. It is feasible that a small nonnative ligand could bind to the receptor via several energetically similar but physically distinct binding modes. A nonnative ligand could also interact with functional groups on the receptor that are inaccessible to the peptidic native ligand. Such a novel receptor site (exo-site) has been employed in a comparison of the interaction of thrombin and hirudin with the thrombin receptor.[35,36] Receptor affinity derives from the sum of several different modes of interaction. Thus, a particular binding group in a small, nonnative ligand may not always mimic a corresponding binding group in a large native peptide.

A substituent may also play a slightly different role, both sterically and electronically, on different ligands. The direction from which a functional group in the nonnative ligand approaches a receptor-binding group may differ from the native ligand, leading to a reduced interaction or a negative steric interaction. As a result, the energy of interaction can be different, especially if the substituent is involved in hydrogen bond formation with the receptor. A substituent may play a conformational role in one ligand and not another.

Of the three types of tissues employed in the comparison of Table 5, the rabbit aortic smooth muscle is known to contain AT_1 receptors but to lack AT_2 receptors, the rat adrenal cortex is known to contain both AT_1 and AT_2 receptors, while the rat mesentery contains only AT_1 receptors. The variations in activities of the benzylacrylates in the three assays systems may reflect the relative abundance of the AT_1 vs. other subtypes in the three tissues. The discovery of AT_1 isoforms, AT_{1A} and AT_{1B}, in the rat[37,38] suggests the potential for an even greater degree of complexity.

III. IDENTIFYING THE BIOACTIVE CONFORMATION: WHAT'S MISSING?

A. SUFFICIENTLY CONSTRAINED LIGANDS

As discussed in Chapter 2, despite the efforts of many groups, there has yet to be identified a high affinity peptide or semipeptide analog that is sufficiently constrained that its NMR spectra might be expected to reveal a discrete conformation. It may be that a certain degree of flexibiity is required in octapeptide ligands. Despite its novelty, the peptide CGP 4211A[39] (see Chapter 2) is also too flexible to reveal discrete conformational information. Until sufficiently constrained peptides or semipeptides are discovered, it is unlikely that the conformations of a ligand in solution will provide useful conformational information regarding the receptor-bound conformation.

B. STRUCTURE OF RECEPTOR-BOUND LIGAND

In cases where the receptor can be obtained in soluble form, it may be possible to obtain useful information about the receptor-bound ligand by NMR spectroscopy. A recent example is that of the analogue of FK 506 and its binding protein.[40] Because the angiotensin II receptor is a membrane-bound glycoprotein composed of seven helical transmembrane-spanning domains, it is highly unlikely that a soluble form of the receptor will be obtained. Also unlikely is the possibility of an X-ray crystal structure of the receptor-ligand complex.

C. THE NEED TO DEFINE CONFORMATIONAL FEATURES THAT DISTINGUISH AT_1 RECEPTORS FROM AT_2 RECEPTORS

As noted in Chapter 2, most of the structure–activity studies relating to the peptide analogs of angiotensin II predate the discovery of angiotensin II receptor

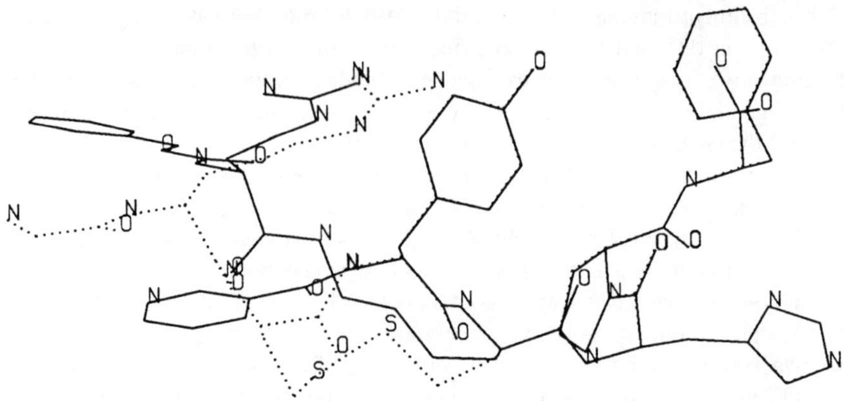

FIGURE 21A. An overlay between model II and CGP 42112A.

subtypes. Although the peptides are being reinvestigated in radioligand binding assays that discriminate between AT_1 and AT_2 receptors, at present there are no known conformational differences between peptide ligands that are selective for AT_1 or AT_2 receptors. Because angiotensin II does not distinguish between AT_1 and AT_2 receptors, it may not be possible to identify differences between the receptor- bound conformations of angiotensin II at AT_1 and AT_2 receptors. Because the AT_2 selective semipeptide, CGP 42112A (Chapter 2), can be overlaid onto Model III quite readily (Figure 21), it would be reasonable to propose that the binding pocket of AT_2 receptors displays the same topological arrangement as AT_1 receptors but that AT_2 receptors contain different residues in that pocket, which may be identified by analogs of angiotensin II. CGP 42112A, however, is expected to be too flexible to provide useful conformational information in solution. As noted earlier, the AT_2-selective ligand, PD 121981, can also be overlaid onto Model II. Thus, there does not as yet appear to be any difference in the conformational requirements for AT_1 selectivity over AT_2 selectivity.

IV. CONCLUSION

In the Appendix, a variety of conformational models of angiotensin II have been discussed. Ultimately, none of the models may bear much relevance to the receptor-bound conformation of angiotensin II. Through the development of template models, however, the definition of a receptor-bound conformation of angiotensin II has not been an obligatory prerequisite for the design of nonpeptide angiotensin II receptor antagonists. A detailed examination of the discovery of nonpeptide angiotensin II receptor antagonists and their respective structure–activity relationships follows in Chapter 4.

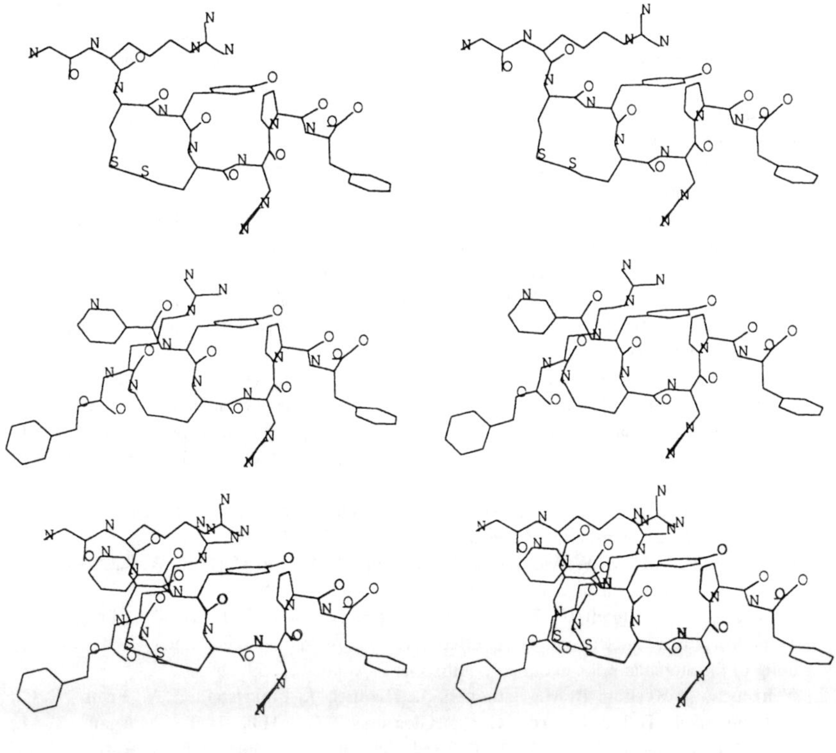

FIGURE 21B. An overlay between model II and CGP 42112A *in stereo.*

REFERENCES

1. **Furukawa, Y., Kishimoto, S., and Nishikawa, K.,** Hypotensive Imidazole-5-Acetic Acid Derivatives, U.S. Patent 4, 355, 040, 1982.
2. **Freidinger, R. M.,** Non-peptide ligands for peptide receptors, *Trends Pharmacol. Sci.,* 10, 270, 1989.
3. **Collins, J. L., Dambek, P. J., Goldstein, S. W., and Faraci, W. S.,** CP-99,711: a nonpeptide glucagon receptor antagonist, *Bioorg. Med. Chem. Lett.,* 2, 915, 1992.
4. **Valentine, J. J., Nakanishi, S., Hageman, D. L., Snider, R. M., Spencer, R. W., and Vinick, F. J.,** GRP nonpeptide antagonist, *Bioorg. Med. Chem. Lett.,* 2, 333, 1992.
5. Oxytocin Antagonists, U.S. Patent 4, 1, 412, 1989.
6. **Yamamura, Y., H., O., Chihara, T., Kondo, K., Onogawa, T., Nakamura, S., Mori, M., Tominaga, M., and Yuuchi, Y.,** OPC-21268, an orally effective nonpeptide vasopressin V1 receptor antagonist, *Science,* 252, 572, 1991.
7. **Snider, M., Constantine, J., Lowe, J., Longo, K., Lebel, W., Woody, H., Drozda, S., Desai, M., Vinick, F., Spencer, R., and Hess, H.,** Nonpeptide subst. P antagonist, *Science,* 251, 435, 1991.
8. **Gether, U., Johansen, T. E., Snider, R. M., Lowe J. A., III, Nakanishi, S., and Schwartz, T.,** Different binding epitopes on the NK_1 receptor for substance P and a non-peptide antagonist, *Nature,* 362, 345, 1993.
9. **Fermandjian, S., Lintner, K., Haar, W., Fromageot, P., Khosla, M. C., Smeby, R. R., and Bumpus, F. M.,** Conformation–function relationship of angiotensin II and analogues, in *Peptides, 1976, Proc. 14th Eur. Peptide Symp.,* Loffet, A., Ed., University of Brussels, Brussels, 1976, 339.

10. **Duncia, J. V., Chiu, A. T., J., C. D., Gregory, G. B., Johnson, A. L., Price, W. A., Wells, G. J., Wong, P. C., Calabrese, J. C., Timmermans, P. B. M. W. M.,** The discovery of potent nonpeptide angiotensin II receptor antagonists: a new class of potent antihypertensives, *J. Med. Chem.,* 33, 1312, 1990.
11. **Carini, D. J., Duncia, J. V., Johnson, A. L., Chiu, A. T., Price, W. A., Wong, P. C., and Timmermans, P. B. M. W. M.,** Nonpeptide angiotensin II receptor antagonists: N-[(benzyloxy) benzyl] imidazoles and related compounds as potent antihypertensives, *J. Med. Chem.,* 33, 1330, 1990.
12. **Wong, P. C., Price, W. A., Jr., Chiu, A. T., Wong, N. T., Duncia, J. V., Carini, D. J., Johnson, A. L., and Timmermans, P. B. M. W. M.,** Losartan (DuP 753), an orally-active nonpeptide angiotensin II receptor antagonist, *Cardiovasc. Drug Rev.,* 9, 317, 1991.
13. **Duncia, J. V., Carini, D. J., Chiu, A. T., Johnson, A. L., Price, W. A., Wong, P. C., and Timmermans, P. B. M. W. M.,** The discovery of DuP 753, a potent, orally active nonpeptide angiotensin II receptor antagonist, *Med. Res. Rev.,* 12, 149, 1992.
14. **Samanen, J., Weinstock, J., Hempel, J., Keenan, R. M., Hill, D. T., Ohlstein, E. H., Weidley, E. F., Aiyar, N., and Edwards, R.,** A molecular model of angiotensin II (Ang II) for rational design of small Ang II antagonists with increased potency, in *Peptides: Structure and Biology, Proc. 12th Am. Peptide Symp.,* Rivier, J. E. and Smith, J. A., Eds., ESCOM, Leiden, Holland, 1992, 386.
15. **Pierson, M. E. and Freer, R. J.,** Analysis of the active conformation of angiotensin II and nonpeptide AII antagonists, *Int. J. Peptide Protein Res.,* 5, 102, 1992.
16. **Timmermans, P. B. M. W. M., Wong, P. C., Chiu, A. T., and Herblin, W. F.,** Nonpeptide angiotensin II receptor antagonists, *Trends Pharm. Sci.,* 12, 55, 1991.
17. **Matsoukas, J. M., Bigam, G., Zhou, N., and Moore, G.,** ^{1}H-HMR studies of [Sar]-angiotensin II conformation by nuclear overhauser effect spectroscopy in the rotating frame (ROESY): clustering of the aromatic rings in dimethylsulfoxide, *Peptides,* 11, 359, 1990.
18. **Weinstock, J., Keenan, R. M., Samanen, J., Hempel, J., Finkelstein, J. A., Franz, R. G., Gaitanopoulos, D. E., Girard, G. R., Gleason, J. G., Hill, D. T., Morgan, T. M., Peishoff, C. E., Aiyar, N., Brooks, D. P., Frederickson, T. A., Ohlstein, E. H., Ruffolo, R. R., Stack, E. J., Sulpizio, A. C., Weidley, E. F., and Edwards, R. M.,** 1-(Carboxybenzyl)imidazole-5-acrylic acids: potent and selective angiotensin II receptor antagonists, *J. Med. Chem.,* 34, 1514, 1991.
19. **Keenan, R. M., Weinstock, J., Finkelstein, J. A., Franz, R. G., Gaitanopoulos, D. E., Girard, G. R., Hill, D. T., Morgan, T. M., Samanen, J. M., Hempel, J. C., Eggleston, D., Aiyar, N., Griffin, E., Ohlstein, E. H., Stack, E. J., Weidley, E. F., and Edwards, R. M.,** Imidazole-5-acrylic acids: potent nonpeptide angiotensin II receptor antagonists designed using a novel peptide pharmacophore model, *J. Med. Chem.,* 35, 3858, 1992.
20. **Samanen, J., Narindray, D., Adams, W., Cash, T., Yellin, T., and Regoli, D.,** Effects of D-amino acid substitution on antagonist activities of angiotensin II analogues, *J. Med. Chem.,* 31, 510, 1988.
21. **Khosla, M. C., Hall, M. M., Smeby, R. R., and Bumpus, F. M.,** Agonist and antagonist ationships in 1- and 8- substituted analogs of angiotensin II, *J. Med. Chem.,* 17, 1156, 1.
22. **Saltman, S., Baukal, A., Waters, S., Bumpus, F. M., and Catt, K.,** Competitive binding activity of angiotensin II analogues in an adrenal cortex radioligand-receptor assay, *J. Endocrinol.,* 97, 275, 1975.
23. **Saltman, S., Fredlund, P., and Catt, K.,** Actions of angiotensin II antagonists upon aldosterone production by isolated adrenal glomerulosa cells, *J. Endocrinol.,* 98, 894, 1976.
24. **Keenan, R. M., Weinstock, J., Finkelstein, J. A., Franz, R. G., Gaitanopoulos, D. E., Girard, G. R., Hill, D. T., Morgan, T. M., Samanen, J., Peishoff, C. E., Tucker, L. M., Aiyar, N., Griffin, E., Ohlstein, E. H., Stack, E. J., Weidley, E. F., and Edwards, R. M.,** Potent nonpeptide angiotensin II receptor antagonists. II. (1-Carboxybenzyl)imidazole-5-acrylic acids, *J. Med. Chem.,* 36, 1880, 1993.
25. **Marshall, G. R.,** Determination of the receptor bound conformation of angiotensin, *Dtsch. Apoth.-Ztg.,* 85, 2783, 1986.

26. **Samanen, J. M., Peishoff, C. E., Keenan, R. M., and Weinstock, J. M.,** Refinement of a molecular model of angiotensin II (AII) employed in the discovery of potent nonpeptide antagonists, *Bioorg. Med. Chem. Lett.,* 3, 909, 1992.
27. **Sugg, E. F., Dolan, C. A., Patchett, A. A., Chang, R. S. L., Faust, K. A., and Lotti, V. J.,** Cyclic disulfide analogues of [Sar^2, Ile^8]-angiotensin II, in *Peptides: Chemistry, Structure, and Biology,* Rivier, J. E. and G. R., M., Eds., ESCOM, Leiden, Holland, 1990, 305.
28. **Spear, K. L., Brown, M. S., Reinhard, E. J., McMahon, E. G., Olins, G. M., Paloma, M. A., and Patton, D. R.,** Conformational restriction of angiotensin II: cyclic analogues having high potency, *J. Med. Chem.,* 33, 1935, 1990.
29. **Kataoka, T., Beusen, D. D., Clark, J. D., Yodo, M., and Marshall, G. R.,** The utility of side-chain cyclization in determining the receptor-bound conformation of peptides: cyclic tripeptides and angiotensin II, *Biopolymers,* 32, 1519, 1992.
30. **Moore, G. J.,** Methods for modelling tertiary structures of biologically active ligands including agonists and antagonists thereto and novel synthetic antagonists based on angiotensin, *WO 91/10140,* July 11, 1991.
31. **De Coen, J. L. and Ralston, E.,** Theoretical conformational analysis of Asn^1,Val^5 Angiotensin II, *Biopolymers,* 16, 1929, 1977.
32. **Samanen, J., Cash, T., Narindray, D., Brandeis, E., Adams, W., Weideman, H., Yellin, T., and Regoli, D.,** An investigation of angiotensin II agonist and antagonist analogues with 5,5-dimethylthiazolidine-4-carboxylic acid and other constrained amino acids, *J. Med. Chem.,* 34, 3036, 1991.
33. **Blankley, C. J., Hodges, J. C., Klutchko, S. R., Himmelsbach, R. J., Chucholowski, A., Connolly, C. J., Neergaard, S. J., Van Nieuwenhze, M. S., Sebastian, A., Quin, J., Essenburg, A. D., and Cohen, D. M.,** Synthesis and structure–activity relationships of a novel series of non-peptide angiotensin II receptor binding inhibitors specific for the AT_2 subtype, *J. Med. Chem.,* 34, 3248, 1991.
34. **Hsieh, K., LaHann, T. R., and Speth, R. C.,** Topographic probes of angiotensin and receptor: potent angiotensin II agonist containing diphenylalanine and long-acting antagonists containing biphenylalanine and 2-indan amino acid in position 8, *J. Med. Chem.,* 32, 898, 1989.
35. **Hartman, G. D., Egbertson, M. S., Halszenko, W., Laswell, W. L., Duggan, M. E., Smith, R. L., Naylor, A. M., Manno, P. D., Lynch, R. J., Zhang, G., Chang, C. T. C., and Gould, R. J.,** Non-peptide fibrinogen receptor antagonists. I. Discovery and design of exosite inhibitors, *J. Med. Chem.,* 35, 4640, 1992.
36. **Vu, T., Wheaton, V. I., Hung, D. T., Charo, I., and Coughlin, S. R.,** Domains specifying thrombin-receptor interaction, *Nature,* 353, 674, 1991.
37. **Kakar, S. S., Riel, K. K., and Neill, J. D.,** Differential expression of angiotensin II receptor subtype mRNAs (AT-1A and AT-1B) in the brain, *Biochem. Biophys. Res. Commun.,* 185, 688, 1992.
38. **Kitami, Y., Okura, T., Marumoto, K., Wakamiya, R., and Hiwada, K.,** Differential gene expression and regulation of Type-1 angiotensin II receptor subtypes in the rat, *Biochem. Biophys. Res. Commun.,* 188, 446, 1992.
39. **Whitebread, S., Mele, M., Kamber, B., and de Gasparo, M.,** Preliminary biochemical characterization of two angiotensin receptor subtypes, *Biochem. Biophys. Res. Commun.,* 163, 284, 1989.
40. **Petros, A. M., Gampe, R. T., Jr., Gemmecker, G., Neri, P., Holzman, T. F., Edalji, R., Hochlowski, J., Jackson, M., McAlpine, J., Luly, J. R., Pilot-Matias, T., Pratt, S., and Fesik, S. W.,** NMR studies of an FK-506 analogue, [U-^{13}C]ascomycin, bound to FKBP: conformation and regions of ascomycin involved in binding, *J. Med. Chem.,* 34, 2925, 1991.

Appendix

THEORETICAL AND PHYSICAL STUDIES OF ANGIOTENSIN II CONFORMATION

TABLE OF CONTENTS

0-8493-8545-8/94/$0.00+$.50

© 1994 by CRC Press, Inc.

I. INTRODUCTION

In 1972, Marshall and Bosshard[1] delineated the differences among three physical states of a peptide hormone in which the conformation of the peptide may be examined: (1) the *crystalline state,* in which the peptide interacts with neighboring peptide molecules (intermolecular interactions) subject to crystal-packing forces and with solvent molecules that happen to cocrystallize with the peptide, (2) the *solution state,* in which the conformation of the peptide results from intramolecular interactions and from interactions with solvent, and (3) the biologically active conformation or *receptor-bound conformation,* determined by interactions with functional groups (binding groups) within the receptor, as well as intramolecular and solvent interactions. These authors made a compelling case, using angiotensin II as their subject, for the spectroscopic evaluation of peptide analogs bearing conformational constraints that retained high receptor affinity as a useful approach toward understanding receptor-bound conformation. Evidence supporting this proposal subsequently became available. Premilat and Maigret[2] in 1980, Marcionini et al.[3] in 1983, and more recently, Nikiforovich et al.[4] in 1987 employed statistical methods for generating families of conformations for angiotensin II that could be related to spectroscopic data. In particular, Premilat and Maigret[2,3] showed that H_N-$H_{C\alpha}$ NMR coupling constants calculated from a weighted mixture of the conformational families matched the experimental coupling constants of angiotensin II in acid solution. In essence, these investigators showed that angiotensin II in solution can occur in a variety of conformations, varying from fully extended to relatively folded, compact conformations. As such, work that has evaluated the solution conformation of native angiotensin II or its highly flexible linear analogs, such as saralasin, [Sar^1,Ala^8]-angiotensin II, [Sar^1,(OMe)Tyr^4]-angiotensin II or [Sar^1,Ile^8]-angiotensin II, could no longer be taken as relevant to any discrete conformation, especially the receptor-bound conformation. Unfortunately, the bulk of conformational studies on angiotensin II have led to proposals of single conformational models on the basis of spectroscopic studies of flexible linear angiotensin II analogs in solution. None of these models can adequately reflect the dynamics of angiotensin II in solution. However, such models may at least partially describe the receptor-bound conformation. For this reason, these models, as well as the totally theoretical models, are examined in this Appendix. The possibility of induced fit at the receptor,[5] however, implies that the receptor-bound conformation may be different from these solution conformations.

II. THEORETICAL CONFORMATIONS

A. THE CONFORMATIONS OF DE COEN AND RALSTON[6–8]

Conformations were generated for [Asn^1,Val^5]-angiotensin II employing Ala^6-Pro-Ala to obtain a set of 116,640 backbone structures. Their energies were evaluated using semiempirical potential energy calculations. Similar conformations

TABLE 1
Ten Most Probable Conformations, De Coen and Ralston[8]

No.	Conformer[a]	$d(C_\alpha^1–C_\alpha^8)$ (Å)	Energy (kcal/mol)	Relative probability (%)	No. of sidechain arrangements
1	bdebacac	7.7	−52.9	44.2	240
2	bbebacab	7.1	−52.9	36.5	151
3	bdebacab	7.7	−52.4	12.1	149
4	beaeecad	17.2	−51.5	1.8	87
5	bdebacad	7.7	−51.3	1.8	148
6	bbebabdb	11.0	−50.3	0.9	520
7	baebebed	9.0	−51.4	0.8	36
8	bbebecab	10.2	−50.5	0.7	163
9	bdebecab	10.1	−50.8	0.7	49
10	bbebeceb	9.4	−51.0	0.5	53

[a] See De Coen and Ralston for the stereochemical code defining each conformer.[8]

were pooled together into 18 sets of conformations. Sidechains then were added to the ten sets of conformations bearing the 1421 lowest energy conformers, and all of these conformers were minimized. The ten resulting families of conformations that were "most probable" are displayed in Table 1. The three most probable families of conformation represent folded and compact structures, rather similar in shape, varying little in conformational energy. As might be expected, a wide variety of sidechain orientations were found. The backbone torsion angles for the most probable conformer is listed in Table 2. The authors stressed that angiotensin II appears to be a polymorphic molecule (i.e., it is likely that its conformation in solution is best described by a combination of families of conformation rather than a single conformation). They calculated peptide backbone NMR coupling constants from the mixture of conformations. Unfortunately, comparison of the calculated coupling constants with experimentally determined coupling constants revealed few similarities.

B. THE CONFORMATIONS OF PREMILAT AND MAIGRET[2,3]

These investigators studied five families of conformations that were determined with a Monte-Carlo procedure,[2,3] (Table 3). Unlike the set of conformations determined by De Coen and Ralston,[6,8] which were all folded conformations, these types of conformation varied between fully extended and folded structures. Peptide backbone coupling constants were calculated for each family, for the whole Monte-Carlo sample, and from the mixture of five families (Table 4). As might be expected, none of the individual families gave calculated coupling constants that matched experimentally determined coupling constants, but those determined from the whole Monte-Carlo sample or from the mixture of families came close to the experimental data, much closer than the earlier set determined by De Coen

TABLE 2
Most Probable Conformation, De Coen and Ralston[8]

Amino acid	Letter	Backbone angles		Sidechain angles			
		φ	ψ	χ1	χ2	χ3	χ4
Asn[1]	b	52	163	−163	−114		
Arg[2]	b	−133	150	−162	−178	77	−173
Val[3]	e	−80	81	−182			
Tyr[4]	b	−161	162	−155	−118	−84	
Val[5]	a	−63	−36	−184			
His[6]	c	−145	71	−178	−127		
Pro[7]	a	−60	−38				
Phe[8]	b	−155	161	66	−98		

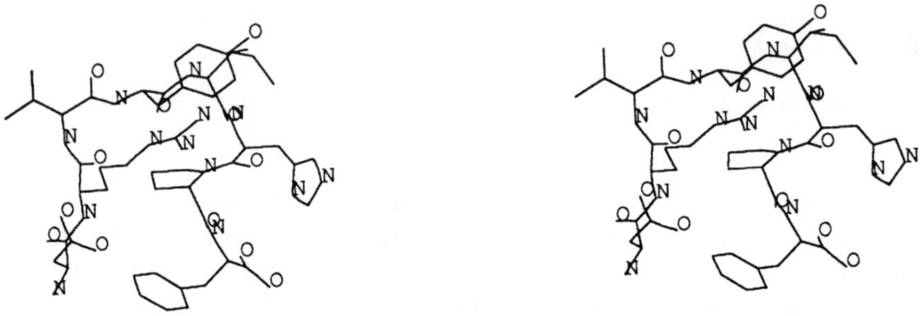

TABLE 3
Five Most Probable Conformations, Marchionini et al.[33]

Cluster	Asn[1] φ,ψ	Arg[2] φ,ψ	Val[3] φ,ψ	Tyr[4] φ,ψ
1	—,135	−145,120	−70,150	−65,−45
2	—,180	−165,130	−70,155	−85,145
3	—,180	−160,125	−90,180	−85,145
4	—,180	−165,130	−75,150	−85,145
5	—,175	−165,115	−135,150	−85,145

Cluster	Ile[5] φ,ψ	His[6] φ,ψ	Pro[7] φ,ψ	Phe[8] φ,ψ
1	−90,150	50,90	−60,110	−140,—
2	−65,135	−160,90	−60,−30	−100,—
3	−100,55	−170,170	−60,135	−135,—
4	−80,130	70,90	−60,130	−100,—
5	−140,150	−155,90	−60,150	−40,—

TABLE 4
Angiotensin II NMR J (H_N-$H_{C\alpha}$) Coupling Constants Calculated from Clusters of Conformations Determined by Marchionini et al.[3]

		${}^{J}H_N{-}H_{C\alpha}$						
Cluster	**Asn[1]**	**Arg[2]**	**Val[3]**	**Tyr[4]**	**Ile[5]**	**His[6]**	**Pro[7]**	**Phe[8]**
1	—	9.7	5.4	4.5	9.0	5.5	—	10.3
2	—	6.4	5.4	8.2	4.5	7.3	—	10.3
3	—	7.3	9.0	8.2	10.3	5.4	—	10.8
4	—	6.4	6.4	8.2	7.3	5.5	—	10.3
5	—	6.4	10.8	8.2	190.3	8.2	—	1.0
Calculated from whole Monte-Carlo sample	—	6.6	7.9	7.6	7.2	6.1	—	8.0
Calculated from weighted typical conformation	—	6.6	7.8	8.0	8.0	6.7	—	7.5
Experimental data								
Glickson[a]	—	6.5	7.9	7.2	8.0	6.0	—	7.3
Bleich[b]	—	6.7	7.6	8.0	8.5	7.0	—	7.5
Lenkinski[c]	—	—	—	8.1	8.5	6.6	—	—

[a] Glickson et al.[54]
[b] Bleich et al.[55]
[c] Lenkinski et al.[26]

and Ralston.[6,8] The end-to-end distance and the distance between the Tyr and Phe sidechains determined for the whole sample were in good agreement with those distances calculated by Schiller[9] in fluorescence-energy transfer studies. Thus, Premilat and Maigret[2,3] have substantiated what De Coen and Ralston[6,8] proposed, that the conformation of angiotensin II is best described by a combination of families of conformation.

C. THE CONFORMATIONS OF NIKIFOROVICH ET AL.[4]

Nikiforovich et al.[4] took an approach that was similar to Premilat and Maigret.[2,3] This group employed the Monte-Carlo procedure, but rather than sample the entire conformational space, they generated Monte-Carlo statistical samples in the vicinity of the previously determined 12 low-energy conformers of an N-terminal spin-labeled analog of [Asn[1]]-AII. Backbone NMR coupling constants, the values of singlet-singlet energy transfer between the Phe and Tyr aromatic sidechains, as well as the spin-lattice relaxation times for several protons, were calculated for each of the 12 Monte-Carlo samples and matched with the corresponding experimental data in solution by independent selection of their statistical weights. It was shown that for the flexible molecule angiotensin II good agreement between calculated and experimental data cannot be achieved without taking into account at least five "indispensible" low-energy backbone conformers.

III. PHYSICAL STUDIES OF ANGIOTENSIN II CONFORMATION

A number of physicochemical techniques have been utilized to study the conformation of angiotensin II analogs in solution.[10] As will become apparent, these studies provide an abundance of enticing and often conflicting data. Given the probable lack of relevance of these studies to receptor-bound conformation, they will only be mentioned briefly.

A. FLUOROMETRY

With the presence of chromophores in positions four (tyrosine) and eight (phenylalanine), several investigators have attempted to determine the distance between these groups and N terminal chromophores in angiotensin II analogs via fluorometry.[9,11–13] From these studies, an average end-to-end distance of approximately 15 Å was determined in the pH range of 1.5 to 8.5, indicating that the conformations of angiotensin II in solution are as compact as some of the models that have been proposed for angiotensin II.[9]

B. TITRATION EXPERIMENTS

Given the abundance of titratable groups in angiotensin II analogs, a number of titration experiments have been performed, both titrimetric[11,14,15] and spectroscopic,[16,17] on angiotensin II and a number of analogs in an attempt to identify unusual pKas that may indicate that certain groups possibly are involved in intramolecular interactions, indicative of solution conformation. Table 5 lists the pKas of [Asn^1]-angiotensin II.

C. NMR SPECTROSCOPY

Two detailed studies exist providing an abundance of information relevant to the conformation of angiotensin II in solution.[16,18] In both studies, nearly complete assignments of hydrogens in AII analogs allowed the evaluation of chemical shift variation with pH (Table 6). A number of the variations could be ascribed to the inductive effects upon deprotonation of the ionizable groups in AII[19] shown in Table 5,[14,16,17] such as the Phe^8 $C^{\alpha}H$ and $N^{\alpha}H$ shift upfield upon elevation of the pH from 2 to 5, which results from deprotonation of the C terminal carboxylate group. Several variations in chemical shift (starred in Table 6), however, cannot be ascribed to inductive effects but could be due to shielding or deshielding by aromatic groups that adopt proximal locations to the altered hydrogen atom. These latter variations may be due to variations in conformation with pH.

The spin-lattice relaxation rates induced by a N terminal spin label in an angiotensin II analog were employed in a conformational study that suggested a variety of folded and unfolded conformations.[4] The presence of proline in position seven of angiotensin II has suggested the presence of a γ turn about position seven. A γ turn in position seven would entail a hydrogen bond between the NH of Phe^8 and the carbonyl of His.[6]

TABLE 5
pKas of Angiotensin II

Titrated group				
Phe^8	CO_2^-	pKa 3.07 (1)		
		pKa 3.50 (3)		
		pKa 3.34 (4)		
His^6	imidazole	pKa 6.26 (1)		
			pKa^D 6.82 (2)	
			pKa^D 6.94 (3)	
			pKa 6.30 (4)	
Tyr^4	phenol			pKa 10.2 (1)
				pKa^D 10.5 (2)
				pKa^D 10.85 (3)
				pKa 10.0 (4)
Asn^1	α-NH_2		pKa^D 6.98 (1)	
			pKa 6.92 (4)	

Notes: (1) pKa determined by PMR in H_2O at 23° C, Glickson et al.[16]
(2) pKa^D determined by PMR in D_2O at 4° C, Glickson et al.[16]
(3) pKa^D determined by ^{13}C NMR in D_2O at 27° C, Zimmer et al.[17]
(4) pKa determined by direct titration in 0.1 *M* KCl at 25° C, Paiva et al.[14]
All inductive effects due to deprotonation of ionizable groups result in upfield chemical shifts.[19]

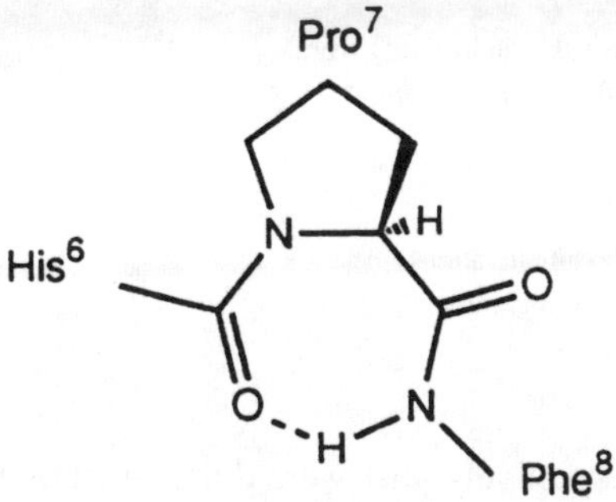

Proposed C7 γ turn about Pro[7] in angiotensin II

Weinkam and Jorgenson[20,21] sought evidence for a proline γ turn in the NMR spectrum of angiotensin II analogs. These authors chose to study the NMR spectrum of [des-Asp[1],Gly[2]]-angiotensin II, since it lacked overlapping signals that existed in the spectrum of angiotensin II, preventing complete assignment of signals to protons in angiotensin II. In fact, a thorough study by Glickson et al.[16] of the NMR of angiotensin II was still incapable of complete NH signal assignment. Samanen et al.[22] previously demonstrated that substitution of Dtc for proline in position seven of [Sar[1]]-angiotensin II and [Sar[1],Ile[8]]-angiotensin II gave the analogs [Sar[1],Dtc[7]]-angiotensin II and [Sar[1],Dtc[7],Ile[8]]-angiotensin II with agonist and

TABLE 6
Observed Alterations of Angiotensin II Structure in the NMR Spectrum

pH Change:		**2 → 5**	**5 → 8**	**8 → 12**
			Variation in chemical shift	
	Group	**pKa ~ 3**	**pKa ~ 6.5**	**pKa ~ 10.2**
Phe^8	C$^\alpha$H	upfield (1,2)[a]	upfield (1,2)*	
	N$^\alpha$H	upfield (1)[a]	upfield (1)*	
Pro^7	1 — C$^\delta$H		upfield (1,2)*	
His^6	C^{2}H		upfield (1,2)[b]	
	C^{4}H		upfield (1,2)[b]	
	C$^\beta$H		upfield (1)[b]	
	C$^\alpha$H		upfield (1,2)[b]	
Ile^5	N$^\alpha$H		"upfield" (1)*	
Tyr^4	C^{m}H		upfield (1,2)*	upfield (1,2)[c]
	C^{o}H			upfield (1,2)[c]
	C$^\beta$H			upfield (1)[c]
	C$^\alpha$H		upfield (1,2)*	
	N$^\alpha$H		"upfield" (1)*	
Asn^1	C$^\beta$H		upfield (1)[d]	
	C$^\alpha$H		upfield (1,2)[d]	
	1° $CONH_2$	upfield (1)*		
Asp^1 or Sar^1	C$^\alpha$H		upfield (2)[d]	

Notes: Upfield shifts are due either inductively from
- [a] Deprotonation of the C-terminal carboxylate,
- [b] The His^6 imidazole,
- [c] The Tyr^4 phenol,
- [d] The N-terminal amine,
- * Due to conformational alterations.

"Upfield" = based on our own assignments.
(1) Glickson et al.[16]
(2) Piriou et al.[18]

antagonist activities, respectively, that were diminished by two orders of magnitude (Chapter 2). They also showed that a C_7 or γ turn conformation is destabilized in model peptides bearing Dtc through the use of solution spectra and molecular modeling.[23,24] These results support the hypothesis that proline in position seven in angiotensin II receptor agonists and antagonists adopts a γ turn in the receptor-bound conformation. Although attempts have been made to obtain evidence in solution for the presence of a γ turn about proline in these Pro^7 analogs and for the absence of a γ turn in the Dtc^7 analogs, unfortunately, simple supportive evidence, such as a downfield shift in the amide hydrogen on Phe^8 in the Pro^7 analogs relative to the same signal in the Dtc^7 analogs, could not be found, although the changes in chemical shift patterns were indicative of changes in conformation.[25] In fact, a study[26,27] of peptide amide hydrogen exchange rates of angiotensin II in water showed a slower than predicted exchange for the His^6 NH (and not the Phe[8] NH). Lenkinski et al.[26,27] suggest that the conformation of angiotensin II in solution involves a conformation that may differ from the receptor-bound conformation (i.e., one with an Ile[5]-His[6] C_7^{eq}-C_5 bend).

The amide bond in linear peptides tends to be the *trans* isomer. But the amide bond preceding proline in peptides can also be found as the *cis* isomer. In solution, the Pro-Phe amide bond in angiotensin II analogs occurs mostly as the *trans* isomer, although some *cis* peptide (16%) has been identified in several studies.[28] If the receptor-bound conformation involves a *cis* His-Pro amide group, the rate of *cis-trans* isomerization in angiotensin II is sufficiently rapid as not to present a rate limiting step in receptor binding.[29]

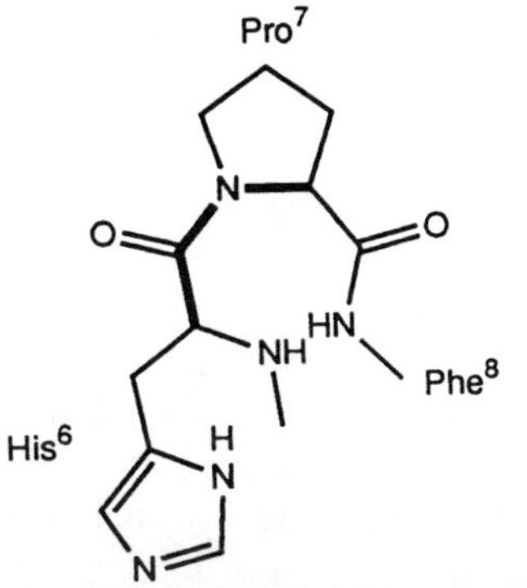

cis-amide

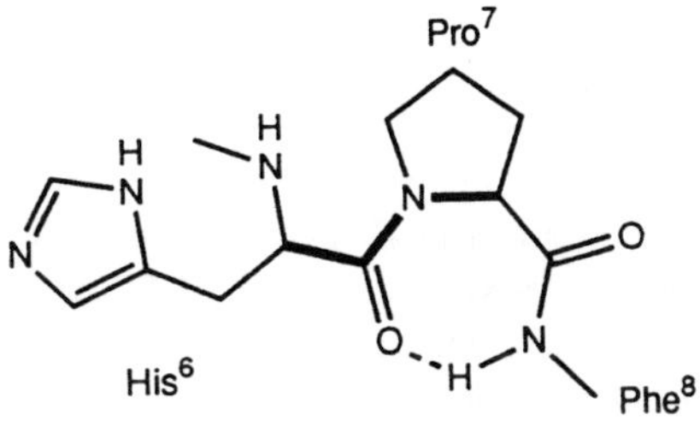

trans-amide

In Chapter 2, evidence was cited that supports the role of the His[6] imidazole either in binding to the receptor and/or in maintaining an intramolecular hydrogen bond or other intramolecular interaction critical to the receptor-bound conformation. Basosi et al.[30] have found that the interaction between Mn(II) ions and angiotensin II gives rise to paramagnetic enhancement of nuclear relaxation rates for the protons neighboring the Mn(II) interactions.[30] They also found that the binding of Mn(II) ions to angiotensin II is reversible and pH dependent (~6.8) and enhances the imidazole 2-H and 4-H relaxation rates relative to other aromatic signals. Thus, at neutral pH, the His sidechain is available for Mn(II) interaction. If the sidechain is also involved in an intramolecular interaction, the equilibrium between the exposed and unexposed state does not appear to favor the unexposed state.

Studies of bulk in the position five sidechain of angiotensin II receptor agonists, led to the postulate that β-branched sidechains served to enhance agonist activity (Chapter 2). Fermandjian et al.[31–33] proposed that variations in certain ^1H NMR parameters of various angiotensin II analogs and fragments suggested that bulk in the position five sidechain had an influence upon the solution conformation of angiotensin II analogs.

Aumelas et al.[34,35] have also proposed that the steric volume of the position eight sidechain is important for both agonist and antagonist activity, citing the variation in ^3J HC_α-NH as well as the chemical shift of the C_aH with sidechain structure in position eight of several angiotensin II analogs. It should be noted that they were led to propose that the benzyl sidechain resides over the C terminal carboxylate in the receptor-bound conformation of angiotensin II receptor agonists, which is not supported by the high agonist activities of [ΔPhe^8]-angiotensin II and [L-Thiq^8]-angiotensin II (see Chapter 2).

Matsoukas et al.[36–38] have studied the NMR spectra of [Sar,(pOMe)Phe^4]-angiotensin II that suggest a conformation in which the Tyr hydroxyl, the His imidazole, and the C terminal carboxylate groups are involved in a so-called ''charge relay'' system.

D. CIRCULAR DICHROISM (CD) STUDIES

Fermandjian et al.[32,33,39–41] have conducted CD studies on angiotensin II analogs and fragments, determining pH-dependent variations in the CD spectra of the angiotensin II analogs in both water and triflurorethanol. Among the pH-dependent variations in the CD spectra, the change in amplitude at 275 nm between acid and alkaline pH was proportional to changes in analog structure that influenced biological activity. These authors showed, for example, that the steric bulk of the sidechain in position five of angiotensin II analogs influenced both the change in amplitude at 275 nm with pH and biological activity. It is important to note, however, that the amplitude changes at 275 nm are quite minor compared to other transitions (i.e., much larger transitions were observed at 230 and 240 nm[41]). The focus on the amplitude changes at 275 nm was based on the fact that this alteration results from the electronic transitions of the phenolic hydroxyl group of tyrosine. Because the changes in amplitude between 230 and 240 nm may be influenced by n-π^* and π-π^* electronic transitions of the peptide bonds as well as from electronic transitions of the tyrosine phenolic ring,[42] changes in amplitude in the 230 to 240 nm region can also be interpreted with care to yield valuable information regarding the solution conformation of angiotensin II analogs.

E. LASER RAMAN SPECTROSCOPY

Fox and Tu[43] examined in laser raman spectroscopy the amide I and amide III bands of a series of angiotensin II analogs in both the solid state and in solution. The spectral data suggested that the angiotensin II conformations were more flexible in solution. The positions and intensities of these bands were indicative of β turn structure. The 850/830 cm^{-1} tyrosine doublet suggested that the sidechain was exposed to solvent. Thus, as with the His^6 sidechain, if the Tyr^4 sidechain is also involved in an intramolecular interaction, the equilibrium between the exposed and unexposed state does not appear to favor the unexposed state. As such, a conformation involving an intramolecular interaction with the Tyr^4 hydroxyl group cannot be a predominant conformation in solution.

F. X-RAY CRYSTALLOGRAPHY

The first X-ray crystal structure of angiotensin II was reported only recently.[44] X-ray crystal structures of linear peptides can be quite different from conformations determined in solution, and X-ray structures of crystals grown in different solvents may differ. Ideally, one would prefer the crystal structure of angiotensin II bound to its receptor. However, the fact that the angiotensin II receptors are membrane-bound glycoproteins diminishes the prospect for the determination of such a structure. The X-ray structure was not solely of angiotensin II but was of an angiotensin II-antibody complex; this work will be discussed below.

IV. MODELS THAT RATIONALIZE PHYSICAL DATA AND ACTIVITIES OF SIDECHAIN-MODIFIED ANALOGS

A. MODELS THAT RATIONALIZE PHYSICAL DATA ALONE

As discussed above, a number of groups have investigated the conformations of angiotensin II in solution through a variety of techniques. Although these studies have provided some useful information regarding the behavior of angiotensin II and its analogs in solution, interpretation of these data have led to speculation relating to specific models of the receptor-bound conformation. It should be emphasized, however, that the likelihood that any of these conformations actually represents the receptor-bound conformation is extremely low.

1. ORD: The first model proposed for angiotensin II was an α *helix* by Smeby et al.[45] on the basis of ORD studies.
2. Titration/Single Model: Shortly after the publication of Smeby et al.,[45] Paiva et al.[14] proposed a *random coil* conformation.
3. NMR/Two Models: Printz et al.[46] concluded that NMR, titration experiments and conformational energy calculations potentially would support two models of angiotensin II, bearing either a γ turn involving Val-Tyr-Ile or a β turn involving Val-Tyr-Ile-His.
4. Laser Raman Spectroscopy/Two Models: Fox and Tu[43] determined the laser raman spectrum for angiotensin I, angiotensin II, angiotensin III, and [Sar[1],Ala[8]]-angiotensin II. The positions and intensities of the amide I and III bands suggested the presence of a β turn conformation in all four compounds.[43] Lagant et al.[47] found that the laser raman data matched the frequencies that would be predicted for both a β turn conformation involving Val-Tyr-Ile-His and a C_7C_5 conformation involving Tyr-Ile-His.

B. A MODEL OF ANGIOTENSIN II BOUND TO AN ANTI-IDIOTYPIC ANTIBODY

Protein chemists have found that an antibody raised against an antibody that in turn was raised against a peptide hormone can display binding behavior similar to the parent hormone, as was shown recently with antibodies against angiotensin II.[48] Such antibodies are called anti-idiotypic antibodies. Garcia et al.[44] have taken an anti-idiotypic antibody to angiotensin II and raised anti-idiotypic antibodies that still bind angiotensin II with high affinity, in spite of the fact that the antibodies were not raised initially against angiotensin II itself. A crystal structure was obtained ultimately for the complex formed between this antibody, called MAb 131, and angiotensin II. The fact that the most buried residues on angiotensin II are also the most immunogenic, namely -Tyr-Ile-His-Pro-, and that modifications to these residues in angiotensin II analogs abolish binding to MAb 131, led the authors to suggest that the structure of the Fab-bound hormone may be relevant to the bioactive conformation. However, the stereoviews provided for the Fab-bound angiotensin II appear to point Val^3 and Ile^5 in opposite directions from each other, which is an orientation that cannot be adopted by one of the high affinity $hCys^{3,5}$ analogs in Chapter 2. Thus, it does not appear that the conformation adopted by angiotensin II upon binding to MAb 131 represents a bioactive conformation that binds to angiotensin II receptors.

C. MODELS THAT RATIONALIZE BOTH PHYSICAL DATA AND ACTIVITIES OF SIDECHAIN MODIFIED ANALOGS

Two groups have attempted to bring together diverse sets of information, a variety of spectroscopic data, and the structures and activities of a number of analogs to create models of the receptor-bound conformation of angiotensin II. These two groups have devised two rather different model conformations.

1. The Fermandjian Model

Extensive studies of the conformation of angiotensin II and analogs by NMR and CD have been conducted. From the extensive body of information that exists, one model conformation has been proposed (Figure 1).[40] Despite the voluminous spectroscopic data[10,18,31–33,35,39,40,49] that are reported, the details relating to how this model was actually determined have yet to be reported. It would appear, however, that the model agrees with the spectroscopic data in a topological sense: the two-turns in the model agree with evidence in the CD and IR for a β-turn. CD/titration and NMR/titration data have also been interpreted to suggest relationships between sidechains and between sidechains and the backbone. However, this model does not account for the retained activities of certain analogs bearing backbone modifications, most notably [(α Me) Tyr^4]-angiotensin II, which was evaluated by Khosla et al.[49] The Fermandjian model has been used in the design of small molecule angiotensin II receptor antagonists.[50]

2. The Jorgensen/Moore Model

A number of studies have been conducted on the solution conformation of angiotensin II and analogs.[36–38] Moore and co-workers[36] have proposed a model of the receptor-bound conformation of angiotensin II, reportedly on the basis of

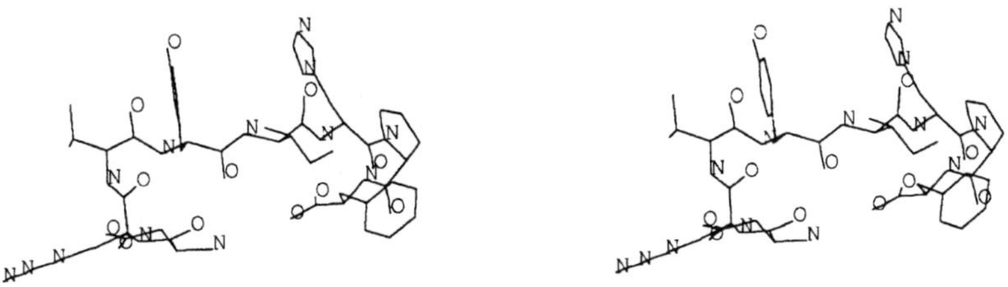

(Torsion angle not shown implies that a specific value has not been specified)

Amino Acid	Backbone Angles		Sidechain Angles			
	φ	ψ	χ¹	χ²	χ³	χ⁴
Asp[1]	—	180	−155			
Arg[2]	−80	−50	−157	−172	180	180
Val[3]	−90	30	66			
Tyr[4]	−155	150	−121	14		
Ile[5]	−160	120	−141	−160		
His[6]	−160	−150	−75	28		
Pro[7]	−60	−60				
Phe[8]	−90	−160	−80	89		

FIGURE 1. The model of Fermandjian et al.[40]

their spectral data and structure–activity relationships. Unfortunately, atomic coordinates or dihedral angles for the model have not been published, although a picture of the model has appeared. The chief features of the Moore model reside in the so-called "charge-relay system", which the authors believe is formed by interaction of the Tyr[4]and His[6] sidechains with the C terminal carboxylate group. The His sidechain and C terminal carboxylate interactions were first proposed by Jorgensen and Weinkham.[21] Although a third interaction between the tyrosine and histidine sidechains was not specifically mentioned, the Tyr sidechain was placed near the His sidechain in the model displayed by Jorgensen and Weihkham. An attempt to create a similar conformation is described below (see Figure 2).

V. MODELS THAT RATIONALIZE ACTIVITIES OF CONFORMATIONALLY CONSTRAINED PEPTIDES

A. CONFORMATIONAL HYPOTHESES FROM CONSTRAINED ANGIOTENSIN II ANALOGS

A number of cyclic and linear conformationally constrained analogs of angiotensin II were described in Chapter 2. Unfortunately, these studies have failed to discover a high affinity analog of angiotensin II that is sufficiently constrained to

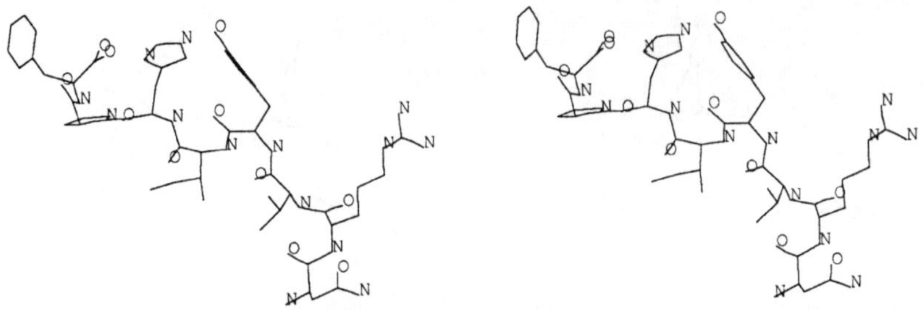

Amino Acid	Backbone Angles		Sidechain Angles			
	ϕ	ψ	χ^1	χ^2	χ^3	χ^4
Asp[1]	—	154	−176			
Arg[2]	−76	91	−171	−156	180	180
Val[3]	−134	148	174			
Tyr[4]	−89	−44	151	−88		
Ile[5]	−55	95	−178			
His[6]	−65	151	27	62		
Pro[7]	−62	113				
Phe[8]	72	71	−170	−93		

FIGURE 2. A Representation of the model of Matsoukas et al.[36]

provide useful conformational information from spectroscopic studies in solution. Nonetheless, this information has been applied to evaluate proposed models of the receptor-bound conformation.

B. CONFORMATIONAL HYPOTHESES DERIVED FROM CONSTRAINED ANGIOTENSIN II ANALOGS

The activities of linear angiotensin II analogs bearing conformational constraints were presented in Chapter 2. Figure 8 in Chapter 2 compared the conformational space (ϕ,ψ) that is accessible to Ac-XX-NHMe, in which XX is one of the conformationally constrained amino acids. Marshall and Bosshard[1] have proposed that the local conformation about such constrained amino acids, XX, in angiotensin II analogs would be found in some portion of the accessible space determined for the corresponding Ac-XX-NHMe. If a series of constrained amino acids were evaluated in angiotensin II analogs, those analogs with the highest biological activity would indicate which type of constraint was tolerated by the angiotensin II receptor. Assuming that the constraint had no positive or negative receptor interaction, one could postulate that the accessible space about XX in the AII analog would fall within the accessible space determined for Ac-XX-NHMe. This hypothesis leads to the results in Table 7, which may help to define receptor-bound conformations of angiotensin analogs. Although certain constrained amino

TABLE 7
Angiotensin II Backbone Conformations Consistent with Conformationally Constrained AII Analogs

Amino acid	Acceptable constraint[a]	Backbone angles[b]	
		φ	ψ
Asn[1]			
Arg[2]			
Val[3]	(NMe) Ala	−120 ± 30	30 to 210
Tyr[4]	(αMe)	−60 ± 20	−60 ± 40
	Tyr	60 ± 20	60 ± 40
Val[5]	Dtc—agonists and antagonists	−60 ± 20	−20 to −90, 110 to 180
	Cle—antagonists	−60 ± 20	−60 ± 40
		60 ± 20	60 ± 40
His[6]			
Pro[7]	C_7-Pro	−60 ± 20	60 ± 20
Phe[8]	(αMe)	−60 ± 20	−60 ± 40
	Phe	60 ± 20	60 ± 40

[a] Amino acid substituted in the position shown for the agonist [Sar[1]]-angiotensin II or the antagonist [Sar[1],Ile[8]]-angiotensin II with retention of activity.

[b] Taken from the references in Chapter 6 due to the reduced agonist activity in [Sar[1],Dtc[7]]-angiotensin II or the reduced antagonist activity in [Sar[1],Dtc[7],Ile[8]]-angiotensin II.

acids are indeed favored in different positions in angiotensin II analogs, when two or more of these receptor-favored constrained amino acids are incorporated into angiotensin II analogs, the resulting compounds display reduced activity (Table 11, Chapter 2). Possible interpretations that may explain this anomaly are multiple binding modes for angiotensin II analogs or a need for inherent flexibility, which is thwarted by the incorporation of two or more constrained amino acids. The former hypothesis would suggest that Table 7 cannot define the ultimate receptor-bound conformation, whereas the latter hypothesis would suggest that Table 7 may define the ultimate receptor-bound conformation. For lack of more precise information, it has been assumed that Table 7 most likely relates to the ultimate receptor-bound conformation, and as such, template model conformations that were consistent with Table 7 have been proposed. Another factor that must also be applied to the consideration of receptor-bound conformations of angiotensin II analogs is the fact that hCys[3,5] analogs are potent angiotensin II receptor agonists and antagonists. Although the S,S-cyclo-hCys-Tyr-hCys ring is quite flexible, the ring does impose one important constraint upon angiotensin II analogs: the vector drawn from the α carbon to the β carbon in hCys[3] and hCys[5] are roughly parallel to each other and point roughly in the same direction.

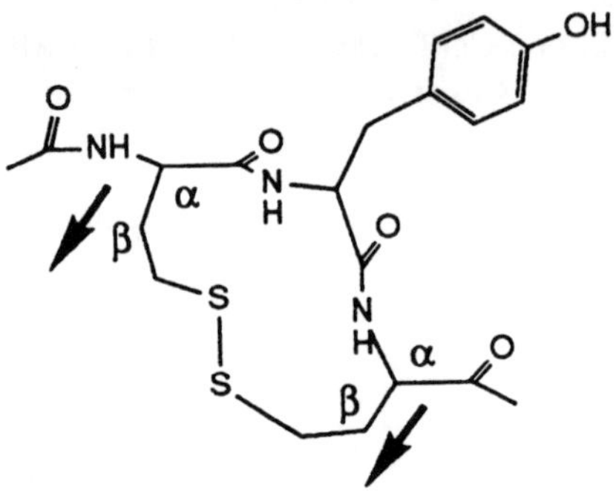

-hCys-Tyr-hCys-ring in [hCys]^{3,5}-AII

This relatively simple criterion may be important in distinguishing potential models of receptor-bound conformation.

Kataoka et al.[51] have found that *cis*- and *trans*-3-mercaptoproline can be substituted for Cys or hCys in positions three and five in angiotensin II analogs to give cyclic analogs in which the disulfide-bridged ring is more constrained than in the parent hCys^{3,5} analogs. In such analogs, it was concluded that a receptor-bound conformation in which Tyr assumes the backbone conformation associated with a left-handed α-helix was consistent with the data in the literature.[51]

The major factor that weighs against conformational interpretation of biologically active analogs bearing conformational constraints is that the functional groups that comprise the constraint can undergo receptor interactions along with other portions of the molecule. These interactions, positive or negative, cannot be distinguished readily from other interactions. Thus, one can only consider conformational interpretations about these analogs as hypothetical. To date, no analogs have been found that retain a high degree of activity and that are sufficiently constrained to give rise to relatively discrete solution conformations. (Notably, the more constrained Cys^{3,5} analogs were less active.)[51]

C. EVALUATION OF PREVIOUSLY PROPOSED MODELS

The set of linear conformationally constrained angiotensin II analogs and the cyclic hCys^{3,5} analogs suggest that the angiotensin II receptor is capable of recognizing angiotensin II analogs in which the backbone is constrained in certain types of conformation (Table 7). As seen in Table 8, some of the models concur at certain positions in the sequence (shown in bold face), but none of the previously proposed models of conformation are fully consistent with this set of conformationally constrained analogs.

D. MODEL I

Of the literature models summarized in Table 8, the lowest energy model of Premilat and Maigret[2] came the closest to being compatible with the set of

TABLE 8
Comparison of Literature Models of Conformation with Angiotensin II Backbone Conformations[a]

Amino acid	Constrained analogs[b] ϕ/ψ	De Coen and Ralston Model[c] ϕ/ψ	Fermandjian Model[d] ϕ/ψ	Premilat and Maigret Model[e] ϕ/ψ	Model I[f] ϕ/ψ
Asn/Asp^1		52	—	−40	—
		163	180	120	153
Arg^2		−133	−80	−60	−76
		150	−50	60	89
Val^3	−120 ± 30/	−80	**−90**	**−120**	**−130**
	30 to 210	**81**	**30**	**160**	**146**
Tyr^4	−60 ± 20/	−161	−155	−100	−96
	−60 ± 40	162	150	**−60**	−45
	or				
	60 ± 20/				
	60 ± 40				
Ile^5	−60 ± 20/	**−63**	−160	**−80**	**−93**
	−20 to −90,110	**−36**	120	**100**	**108**
	to 180				
	or				
	60 ± 20/				
	60 ± 40				
His^6		−145	−160	60	62
		71	−150	80	91
Pro^7	−60 ± 20/	**−60**	**−60**	**−60**	**−75**
	60 ± 20	−38	−60	140	**87**
Phe^8	−60 ± 20	−155	−90	−180	**−67**
	−60 ± 40	161	−160	80	**−28**
	or				
	60 ± 20/				
	60 ± 40				
Consistent with hCys^{3,5} Analogs[g]		No	No	No	No

[a] The backbone torsion angles may give topologically similar conformations with ϕ,ψ angles ± 10°. Torsion angles for residues that concur with the constrained analogs are shown in boldface.

[b] Based upon Samanen et al.,[22] see Table 7.

[c] De Coen and Ralston.[6]

[d] Fermandjian et al.[40]

[e] Premilat and Maigret.[2]

[f] See text for derivation.

[g] Capable of being applied to the [hCys^{3,5}]-angiotensin II analogs without a significant distortion of the conformation.

conformationally constrained analogs, since it did contain an extended Val[3] consistent with (NMe)Ala, a nearly helical Tyr[4] consistent with (αMe)Tyr, and a conformation about Ile consistent with Dtc[5]. It did not contain a γ turn about Pro[7], however, nor did it contain an α helical conformation about Phe[8], which would be consistent with (αMe)Phe. These deviations could be corrected simply by rotating the plane of the Pro-Phe amide bond. This minor modification, along with further minimizations via Simplex, gave rise to Model I (Table 8 and Figure 3).[52] The model was used to determine whether a model conformation could be employed in the design of analogs to the benzylimidazole angiotensin II receptor antagonist described by Furukawa et al.[53]

E. A VERSION OF THE MODEL OF MATSOUKAS ET AL.[36]

Despite the absence of a set of atomic coordinates or a set of dihedral angles for the model of Matsoukas et al.,[36] it is possible to derive a conformation of angiotensin II that could give a similar charge-relay system between the Tyr[4] and His[6] sidechains and the C terminal carboxylate groups (Figure 2), which is consistent with the set of conformationally constrained analogs in Table 7. Visual comparison of this version of the Matsoukas et al.[36] model with the picture of the model in the literature suggests that these are different conformations. The model in Figure 2 was also evaluated for use in the design of analogs to the benzylimidazole angiotensin II receptor antagonist described by Furukawa et al.[53]

VI. CONCLUSION

The enormous effort that has been expended over decades of research has failed to determine the conformation of angiotensin II in solution and at the angiotensin II receptor. This failure stems from the fact that angiotensin II is a flexible linear peptide and that analogs have yet to be identified that are sufficiently constrained to provide NMR spectra with discrete conformations and which still retain high affinity for angiotensin II receptors.

REFERENCES

1. **Marshall, G. R. and Bosshard, H. E.,** Studies on the biologically active conformation, *Circ. Res.,* 30 and 31, (Suppl. II), 143, 1972.
2. **Premilat, S. and Maigret, B.,** Statistical molecular models for angiotensin II and enkephalin related to NMR coupling constants, *J. Phys. Chem.,* 84, 293, 1980.
3. **Marchionini, C., Maigret, B., and Premilat, S.,** Models for the conformational behavior of angiotensin II in acidic aqueous solutions, *Biochem. Biophys. Res. Commun.,* 112, 339, 1983.
4. **Nikiforovich, G. V., Vesterman, B., Betins, J., and Podins, L. J.,** The space structure of a conformationally labile oligopeptide in solution: angiotensin, *Biomol. Struct. Dyn.,* 4, 1119, 1987.

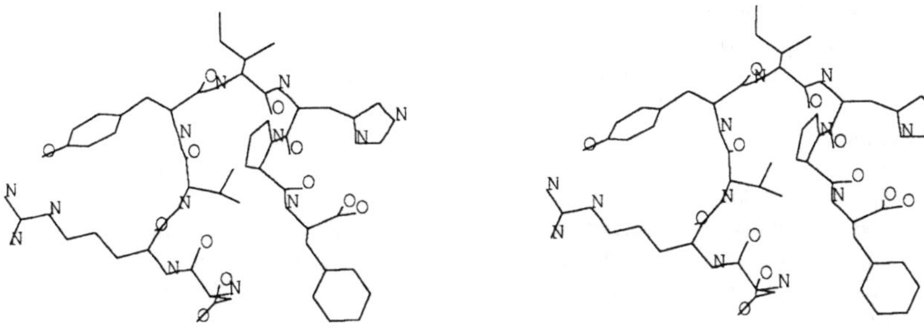

(Torsion angle not shown implies that a specific value has not been specified)

Amino Acid	Backbone Angles φ	ψ	Sidechain Angles χ^1	χ^2	χ^3	χ^4
Asp[1]	—	153	−174			
Arg[2]	−76	89	173	76	180	
Val[3]	−129	146	180	−66	66	
Tyr[4]	−96	−45	−63	106	−178	
Ile[5]	−93	108	178	−65	65	
His[6]	62	91	−160	−59		
Pro[7]	−75	87				
Phe[8]	−67	−28	−64	−75		

FIGURE 3. Angiotensin II model I.

5. **Burgen, A. S. V., Roberts, G. C. K., and Feeney, J.,** Binding of flexible ligands to macromolecules, *Nature,* 253, 753, 1975.
6. **De Coen, J. L. and Ralston, E.,** Theoretical conformational analysis of Asn[1],Val[5] angiotensin II, *Biopolymers,* 16, 1929, 1977.
7. **De Coen, J. L. and B., P.,** Correlation between conformational properties and biological activities of angiotensin II analogues, in *Peptides 1978, Proc. 15th Eur. Peptide Symp.,* Siemon, I. Z. and Kupryszewski, G., Ed., Wroclaw University Press, Wroclaw, Poland 1979, 425.
8. **De Coen, J. L. and Ralston, E.,** Theoretical conformational analysis of Asn[1], Val[5] angiotensin II, *Biopolymers,* 16, 1929, 1977.
9. **Schiller, P. W.,** Determination of the average end-to-end distance of two angiotensin II analogs by resonance energy transfer, *Int. J. Peptide Protein Res.,* 15, 259, 1980.
10. **Smeby, R. R. and Fermandjian, S.,** Conformation of angiotensin II, in *Chemistry and Biochemistry of Amino Acids, Peptides and Proteins,* Vol. 5, Weinstein, B., Ed., Marcel Dekker, New York, 1978, 117.
11. **Juliano, L., Laluce, C., Oliveira, M. C. F., and Paiva, A. C. M.,** Titration and fluorometric studies of the tyrosine side chain of angiotensin II and related peptides, *Biopolymers,* 18, 1793, 1979.
12. **Schiller, P.,** Conformational comparison of [Val[5],Trp[8]]-angiotensin II and [Val[4],Trp[7]]-angiotensin III by fluorescence measurements, *Can. J. Biochem.,* 57, 402, 1979.
13. **Schiller, P. W.,** Synthesis and solution conformation of [Trp[1],Val[5]]-angiotensin II, *Can. J. Biochem.,* 55, 75, 1977.
14. **Paiva, T. B., Paiva, A. C. M., and Scheraga, H. A.,** The conformation of angiotensin in aqueous solution, *Biochemistry,* 2, 1327, 1963.

15. **Juliano, L. and Paiva, A. C. M.,** Conformation of angiotensin II in aqueous solution. Titration of several peptide analogs and homologs, *Biochemistry,* 13, 2445, 1974.
16. **Glickson, J. D., Cunningham, W. D., and Marshall, G. R.,** Proton magnetic resonance study of angiotensin II (Asn^1Val^5) in aqueous solution, *Biochemistry,* 12, 3684, 1973.
17. **Zimmer, S., Haar, W., Maurer, W., Ruterjans, H., Fermandjian, S., and Fromageot, P.,** Investigation of the structure of angiotensin II using ^{13}C nuclear magnetic resonance spectra, *Eur. J. Biochem.,* 29, 80, 1972.
18. **Piriou, F., Lintner, K., Fermandjian, S., Fromageot, P., Khosla, M. C., Smeby, R. R., and Bumpus, F. M.,** Amino acid side chain conformation in angiotensin II and analogs: correlated results of circular dichroism and ^1H nuclear magnetic resonance, *Proc. Natl. Acad. Sci. U.S.A.,* 77, 82, 1980.
19. **Wutrich, K.,** *NMR in Biological Research: Peptides and Proteins,* North Holland/American Elsevier, Amsterdam, 1976, 39.
20. **Weinkam, R. J. and Jorgensen, E. C.,** Angiotensin analogs. IX. Conformational studies of angiotensin II by proton magnetic resonance, *J. Am. Chem. Soc.,* 93, 7038, 1971.
21. **Jorgensen, E. C. and Weinkam, R. C.,** The conformation of angiotensin II in Solution—Physicochemical and Synthetic Studies, in *Peptides, 1971, Proc. 11th Eur. Peptide Symp.,* Nesvadba, H., Ed., North-Holland, Amsterdam, 1973, 311.
22. **Samanen, J., Cash, T., Narindray, D., Brandeis, E., Adams, W., Weideman, H., Yellin, T., and Regoli, D.,** An investigation of angiotensin II agonist and antagonist analogues with 5,5-dimethylthiazolidine-4-carboxylic acid and other constrained amino acids, *J. Med. Chem.,* 34, 3036, 1991.
23. **Samanen, J., Cash, T., Eggleston, D., and Saunders, M.,** Conformational analysis of 5,5-dimethylthiazolidine-4-carboxylic acid (Dtc), a readily available proline analog, in *Peptides, Chemistry and Biology, Proc. 10th Am. Peptide Symp.,* Marshall, G., Ed., ESCOM, Leiden, Holland, 1987, 81.
24. **Samanen, J., Zuber, G., Bean, J., Romoff, T., Kopple, K., and Saunders, M.,** 5,5-Dimethylthiazolidine-4-carboxylic acid (Dtc) as a proline analog with restricted conformation, *Int. J. Peptide Protein Res.,* 35, 501, 1990.
25. **Samanen, J.,** unpublished results.
26. **Lenkinski, R. E., Stephens, R. L., and Krishna, N. R.,** Conformation of angiotensin II. Evidence for a specific hydrogen bonded conformation, *Biochemistry,* 20, 3122, 1981.
27. **Lenkinski, R. E., Stephens, R. L., and Krishna, N. R.,** The conformation of angiotensin II. II. The rates of peptide NH exchange with solvent for [Asn^1, Val^5]angiotensin II, angiotensin III and saralasin, *Biochim. Biophys. Acta,* 667, 157, 1981.
28. **Bleich, H. E., Freer, R. J., Stafford, S. S., and Galardy, R. E.,** Correlation of the biological activity and solution conformation of [Asp^1,Ile^5]- and [Phe^4,Tyr^8]angiotensin II, *Proc. Natl. Acad. Sci. U.S.A.,* 75, 3630, 1978.
29. **Galardy, R. E. and Liakopoulou-Kyriakides, M.,** The rate of s-cis/s-trans isomerization in angiotensin II is at least 70-fold greater than in His-Pro and is not rate limiting in receptor binding, *Int. J. Peptide Protein Res.,* 20, 144, 1982.
30. **Basosi, R., Gaggelli, E., and Valensin, G.,** The effect of binding divalent metal ions on the conformation of human angiotensin II in solution as probed by nuclear and electron paramagnetic resonance investigations, *J. Inorg. Biochem.,* 20, 263, 1984.
31. **Fermandjian, S., Sakarellos, C., Piriou, F., Juy, M., Toma, F., Lam Thanh, H., Lintner, K., Khosla, M. C., Smeby, R. R., and Bumpus, F. M.,** The key role of residue 5 in angiotensin II, *Biopolymers,* 22, 227, 1983.
32. **Fermandjian, S., Piriou, F., Sakarellos, C., Lintner, K., Khosla, M. C., and Bumpus, M. F.,** Conformation and activity in angiotensin II peptides, *Biopolymers,* 20, 1971, 1981.
33. **Sakarellos, C., Lintner, K., Piriou, F., and Fermandjian, S.,** Conformation of the central sequence of angiotensin II and analogs, *Biopolymers,* 22, 663, 1983.

34. **Aumelas, A., Sakarellos, C., Lintner, K., Fermadjian, S., Khosla, M. C., Smeby, R. R., and Bumpus, F. M.,** Studies on angiotensin II and analogs: impact of substitution in position 8 on conformation and activity, *Proc. Natl. Acad. Sci. U.S.A.,* 82, 1881, 1985.
35. **Aumelas, A., Sakarellos, C., Sakarellos-Daitsiotis, M., Lintner, K., Fermandjian, S., Khosla, M. C., Smeby, R. R., and Bumpus, F. M.,** Angiotensin II X^8-analogs with inhibitory properties: conformation of C-terminal moiety and affinity, in *Peptides 1984, Proc. 18th Eur. Peptide Symp.,* Ragnarsson, U., Ed., Almqvist and Wiksell International, Stockholm, 1984, 521.
36. **Matsoukas, J. M., Bigam, G., Zhou, N., and Moore, G.,** I. ^{1}H-NMR studies of [Sar]-angiotensin II conformation by nuclear overhauser effect spectroscopy in the rotating frame (ROESY): clustering of the aromatic rings in dimethylsulfoxide, *Peptides,* 11, 359, 1990.
37. **Matsoukas, J. M., Yamdagni, R., and Moore, G. J.,** II. ^{1}H-NMR studies of sarmesin and [Des^1]sarmesin conformation in dimethylsulfoxide by nuclear overhauser effect (NOE) enhancement spectroscopy: folding of the N- and C-terminal domains, *Peptides,* 11, 367, 1990.
38. **Matsoukas, J. M. and Moore, G. J.,** NMR Studies on angiotensin II: histidine and phenylalanine ring stacking and biological activity, *Biochem. Biophys. Res. Commun.,* 122, 434, 1984.
39. **Fermandjian, S., Lintner, K., Piriou, F., Toma, F., Thanh, L., and Fromageot, P.,** Biological peptides: conformation and activity, in *Peptides, 1978, Proc. 15th Eur. Peptide Symp.,* Wroclaw University Press, Poland, 1979, 449.
40. **Fermandjian, S., Lintner, K., Haar, W., Fromageot, P., Khosla, M. C., Smeby, R. R., and Bumpus, F. M.,** Conformation–function relationship of angiotensin II and analogues, in *Peptides, 1976, Proc. 14th Eur. Peptide Symp.,* Loffet, A., Ed., University of Brussels, Brussels, 1976, 339.
41. **Lintner, K., Fermandjian, S., Fromageot, P., Khosla, M. C., Smeby, R. R., and Bumpus, F. M.,** Circular dichroism studies of angiotensin II analogs: effects of primary sequence, solvent, and pH on the side-chain conformation, *Biochemistry,* 16, 806, 1977.
42. **Adler, A. J., Greenfield, N. J., and Fasman, G. D.,** Circular dichroism and optical rotatory dispersion of proteins and polypeptides, *Meth. Enzymol.,* 27D, 675, 1973.
43. **Fox, J. W. and Tu, A. T.,** Laser raman spectroscopic analysis of angiotensin peptides' conformation, *Arch. Biochem. Biophys.,* 201, 375, 1980.
44. **Garcia, K. C., Ronco, P. M., Verroust, P. J., Brunger, A. T., and Amzel, L. M.,** Three-dimensional structure of an angiotensin II-Fab complex at 3 Å. Hormone recognition by an anti-idiotypic antibody, *Science,* 257, 502, 1992.
45. **Smeby, R. R., Arakawa, K., M., B. F., and Marsh, M. M.,** A proposed conformation of isoleucyl^5-angiotensin II, *Biochim. Biophys. Acta,* 58, 550, 1962.
46. **Printz, M. P., Nemethy, G., and Bleich, N.,** Proposed models for angiotensin II in aqueous solution and conclusions about receptor topography, *Nat. New Biol.,* 237, 135, 1972.
47. **Lagant, P., Vergoten, G., Fleury, G., and Loucheux-Lefebvre, M. H.,** Normal modes of vibration of a model peptide adopting the C_7C_5 structure, *Int. J. Peptide Protein Chem.,* 24, 543, 1984.
48. **Couraud, P. O. J.,** Anti-angiotensin II anti-idotypic antibodies bind to angiotensin II receptor, *J. Immunol.,* 138, 1164, 1987.
49. **Khosla, M. C., Stachowiak, K., Smeby, R. R., Bumpus, F. M., Piriou, F., Lintner, K., and Fermandjian, S.,** Synthesis of [α-methyltyrosine-4]-angiotensin II: studies of its conformation, pressor activity and mode of enzymatic degradation, *Proc. Natl. Acad. Sci. U.S.A.,* 78, 757, 1981.
50. **Duncia, J. V., Chiu, A. T., J., C. D., Gregory, G. B., Johnson, A. L., Price, W. A., Wells, G. J., Wong, P. C., Calabrese, J. C., and Timmermans, P. B. M. W. M.,** The discovery of potent nonpeptide angiotensin II receptor antagonists: a new class of potent antihypertensives, *J. Med. Chem.,* 33, 1312, 1990.
51. **Kataoka, T., Beusen, D. D., Clark, J. D., Yodo, M., and Marshall, G. R.,** The utility of side-chain cyclization in determining the receptor-bound conformation of peptides: cyclic tripeptides and angiotensin II, *Biopolymers,* 32, 1519, 1992.

52. **Samanen, J., Weinstock, J., Hempel, J., Keenan, R. M., Hill, D. T., Ohlstein, E. H., Weidley, E. F., Aiyar, N., and Edwards, R.,** A molecular model of angiotensin II (Ang II) for rational design of small Ang II antagonists with increased potency, in *Peptides: Structure and Biology, Proc. 12th Am. Peptide Symp.,* Rivier, J. E. and Smith, J. A., Eds., ESCOM Science, Leiden, Holland, 1992, 386.
53. **Furukawa, Y., Kishimoto, S., and Nishikawa, K.,** Hypotensive Imidazole-5-acetic acid Derivatives, U.S. Patent 4, 355, 040, 1982.
54. **Glickson, J. D., Dadok, J., and Marshall, G. R.,** Proton magnetic double-resonance study of angiotensin II (Asn^1Val^5) in aqueous solution employing correlation spectroscopy. Assignment of peptide NH resonances and transfer of saturation from water, *Biochemistry,* 13, 11, 1974.
55. **Bleich, H. E., E., G. R., Printz, M. P., and Craig, L. C.,** Conformational studies of angiotensin peptides in aqueous solution by proton magnetic resonance, *Biochemistry,* 20, 4950, 1973.

Chapter 4

STRUCTURE–ACTIVITY RELATIONSHIPS OF NONPEPTIDE ANGIOTENSIN II RECEPTOR ANTAGONISTS

Joseph Weinstock and Richard M. Keenan

TABLE OF CONTENTS

I. INTRODUCTION

The development of potent and selective nonpeptide angiotensin II receptor antagonists can be traced to the discovery of angiotensin II antagonist activity at Takeda in a series of imidazoles originally investigated as diuretics.[1–3] Several of the most potent examples were subsequently reprepared at DuPont[4] and

0-8493-8545-8/94/$0.00+$.50
© 1994 by CRC Press, Inc.

SmithKline Beecham,[5] and the original observations were confirmed and extended. The key compound, CV 2947 (**1**), was then used as a chemical lead, as shown in Figure 1, by the DuPont Group to arrive at losartan (**5**)[4] and its active metabolite **6** and by the SmithKline Beecham group to discover SK&F 108566 (**10**).[5]

The DuPont workers[4] based their chemical design on an overlap of CV 2947 with the Fermandjian solution conformation of angiotensin II, as shown in Figure 2. The imidazole of CV 2947 was overlapped with the His^6 imidazole of angiotensin II, and the acetic acid carboxyl group was pointed toward the C-terminal carboxylic acid of angiotensin II. The butyl group at the 2-position of CV 2947 was pointed at the Ile^5 sidechain of angiotensin II, and the N-benzyl group was pointed at the N-terminus of angiotensin II. The *para*-position of the benzyl was chosen as the site to attach functionality to enlarge CV 2947 in a manner to better mimic angiotensin II. The β-carboxyl group of Asp^1, and the phenol of Tyr^4 in angiotensin II suggested introducing a carboxyl function as a first step, which gave rise to **2,** which has 25-fold higher affinity for the angiotensin II receptor than **1.** Modifying the imidazole-carboxyl relationship gave **3,** which, although a monocarboxylic acid, has an order of magnitude higher affinity for the angiotensin II receptor than **2.** The possibility that the terminal aromatic ring may contribute to binding must be considered. Compounds of this class are not orally active, possibly due to rapid metabolic cleavage to the diacid, followed by biliary excretion, but replacing the amide linkage with a single bond gave **4,** which was orally active. A further order of magnitude increase in binding affinity, with retention of oral activity, resulted when the biphenyl carboxyl moiety was replaced by the bioisosteric tetrazole to give **5.** One of the major metabolites of **5** in rats and humans is the carboxylic acid **6,** which has an order of magnitude higher affinity for the angiotensin II receptor than **5.** The changes that resulted in progressing from compound **1** to **6** were an approximate 30,000-fold increase in binding affinity to the angiotensin II receptor. The key pharmacophore for these nonpeptide angiotensin II receptor antagonists appears to be the 2′-tetrazolylbiphenylylmethyl group, which when incorporated into a wide variety of structural classes, yields compounds that are selective angiotensin II receptor antagonists.

The investigators at SmithKline Beecham[5] also based their initial drug design strategy on overlays of **1** on the Fermandjian solution model of angiotensin II. In one version, the benzyl of **1** overlaps the phenol of Tyr^4, while in another, both the N-benzyl sidechain and the carboxyl group mimicked Phe.[8] When compounds based on these overlays showed no significant improvement in binding affinity to the angiotensin II receptor compared to **1,** a new overlay, shown in Figure 3, was postulated, in which the N-benzyl and carboxyl groups of **1** correspond respectively to the aromatic sidechain of Tyr^4 and the carboxyl group of Phe^8 of angiotensin II. This required a new conformation of angiotensin II to be proposed, which brought these two groups into relatively close proximity. Consideration of this overlay led first to extending the sidechain in a conformationally restricted manner to better mimic the Tyr^4/Phe^8 separation to give **7,** and then adding the α-benzyl group to better mimic the Phe^8 side chain to give **8,** which has 15-fold

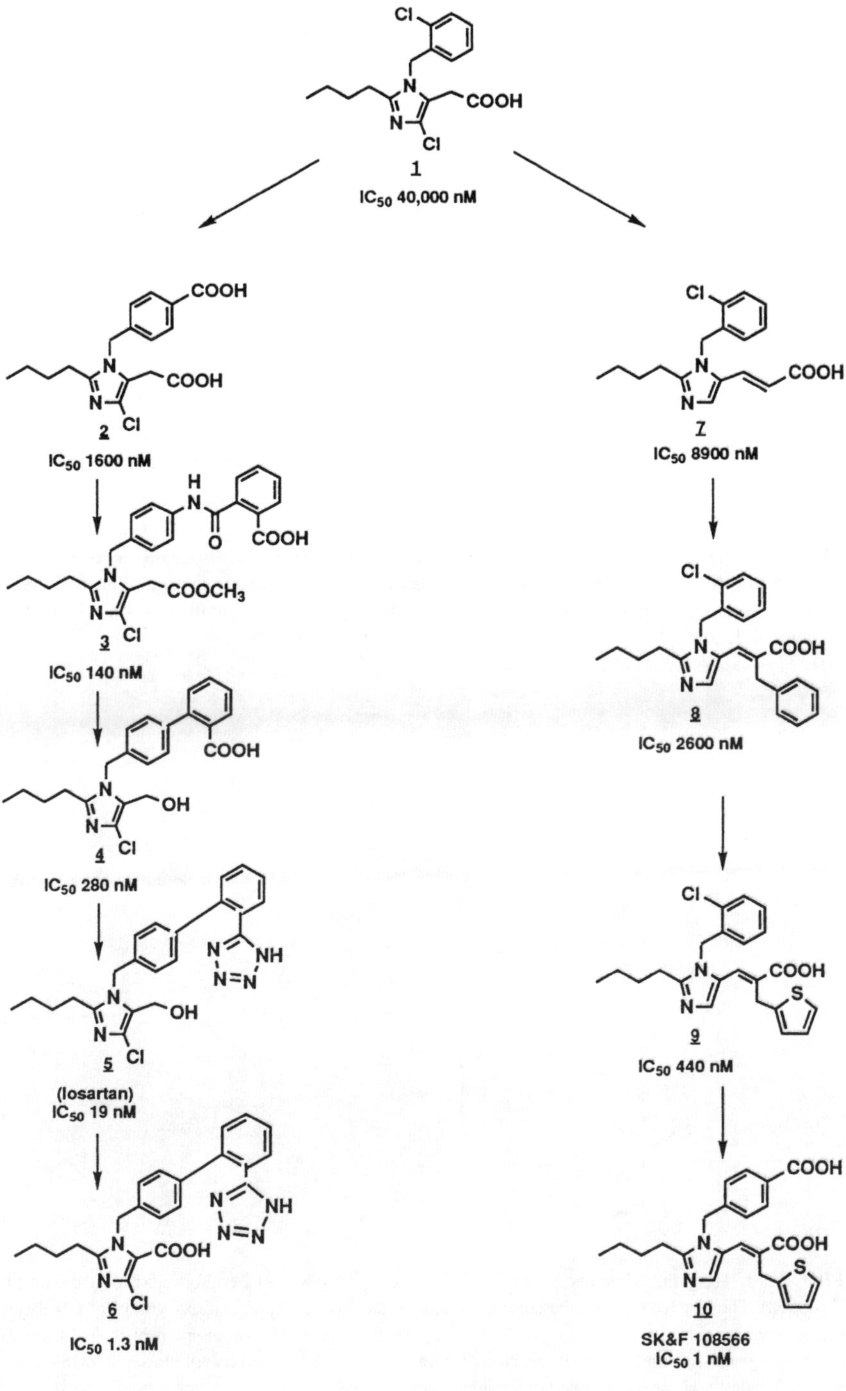

FIGURE 1. Genesis of losartan and SK&F 108566 from CV 2947.

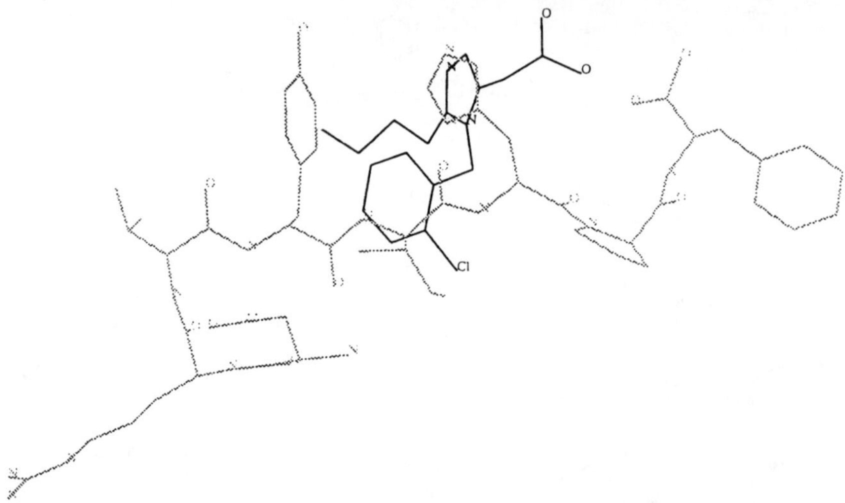

FIGURE 2. The DuPont overlay of CV 2947 (solid) on the Fermandjian solution conformation of angiotensin II (dotted). The key features are the superposition of the imidazole rings of the two structures, the pointing of the CV 2947 carboxyl toward the angiotensin II C-terminus carboxyl, and the pointing of the CV 2947 N-benzyl toward the angiotensin II N-terminus.

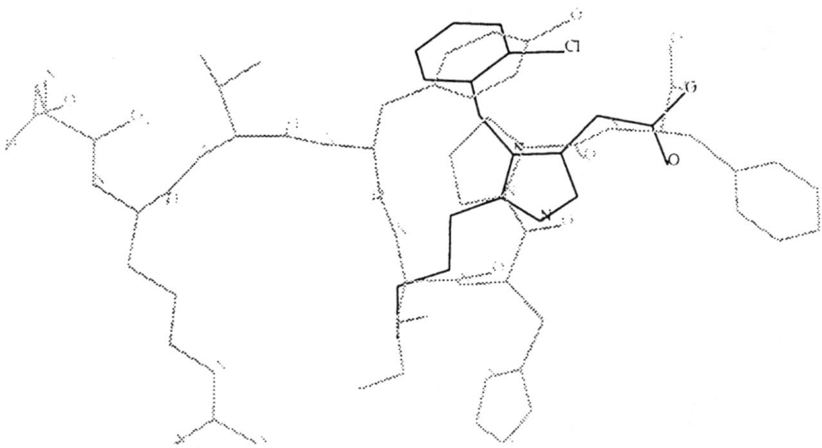

FIGURE 3. The SmithKline overlay of CV 2947 (solid) on the Model 2 conformation of angiotensin II (dotted). The key features are the correspondance of the N-benzyl and carboxy groups of CV 2947 to the aromatic side chain of Tyr^4 and the Phe^8 carboxyl, respectively, of angiotensin II. In Model 2 these two groups of angiotensin II are brought into close enough proximity to allow this, while preserving generally accepted peptide bond angles.

higher affinity for the angiotensin II receptor than **1.** Replacement of the α-benzyl group with a thienylmethyl function suggested by the structure–activity relationships of peptide angiotensin II receptor antagonists (Chapter 2 of Volume II) gave **9,** which has sixfold greater affinity than **8** and which is also orally active as an antagonist of angiotensin II-induced hypertension in the rat. Replacement of the 2-chlorobenzyl group of **9** with the 4-carboxybenzyl group to better mimic the acidic phenol of Tyr^4 gave **10,** which has 400-fold higher affinity for the angiotensin II receptor than **9** and retains oral activity. An overlay of **10** with the new overlay (Model II of Chapter 3 of Volume II) as shown in Figure 4. The change obtained in progressing from compound **1** to **10** resulted in a 40,000-fold increase in binding affinity to the angiotensin II receptor, and gave a compound whose affinity approximates the affinity observed with the potent peptide angiotensin II receptor antagonists.

In the remainder of this chapter, the structure–activity relationships of a number of diverse classes of nonpeptide angiotensin II receptor antagonists will be reviewed. Most of the major advances have been based on the developments made by the DuPont and SmithKline Beecham groups.

In this chapter, the IC_{50} values reported in the tables represent concentrations that inhibit radioligand binding to angiotensin II receptors and are not usually discussed in the context of the source of the angiotensin II receptor. Although different sources of the angiotensin II receptor may yield different IC_{50} values, in general, relative potencies are similar and typically do not vary much more than the IC_{50} values from different laboratories using the same receptor source. Furthermore, the IC_{50} values may be distorted by nonspecific binding to bovine serum albumin which is usually present in the assay mixture and so may not represent actual affinity.[6] In some cases where extensive structure–activity relationships are available, only limited data are included in the tables in order to show trends. The original literature may be consulted for more detailed, comprehensive data.

II. ANGIOTENSIN AT_1 RECEPTOR ANTAGONISTS

A. N-BENZYLIMIDAZOLES

The structure–activity relationships for several typical 1-benzylimidazole-5-acetic acids are summarized in Table 1. The original lead angiotensin II receptor antagonists[1–3] represented various 2-aryl analogs of **1–1,** but even the best of these antagonists are less potent than **1–1** in inhibiting the response to angiotensin II in rabbit aorta. Compounds with longer 2-alkyl groups have increased potency ($C_6>C_5>C_4>C_3$). The best 1-benzyl substituents reported in these patents are 2-chloro, 2-nitro, 4-butoxy, and 3-methyl-4-methoxy. Replacement of the acetic acid moiety by nonacidic functionalities, such as esters, substantially reduces potency. The quantitative data in Table 1 reported by others confirm that the 2-chloro and 2-nitro compounds are more active than the unsubstituted benzyl analog. However, the most potent compounds are the 4-carboxy (**1–5**) and the 4-(5-tetrazolyl)

TABLE 1
Imidazole Acetic Acids and Analogs

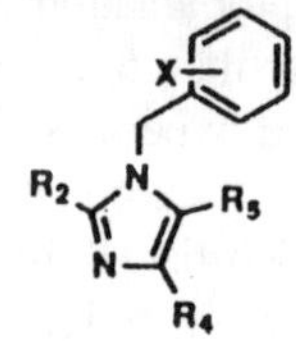

No.	X	R_2	R_4	R_5	IC_{50} (n*M*)	Ref.
1–1	H	n-Bu	Cl	CH_2COOH	150,000	30
1–2	2-Cl	n-Bu	Cl	CH_2COOH	40,000	30
1–3	2-Cl	n-Bu	CH_2COOH	Cl	>100,000	30
1–4	2-NO_2	n-Bu	Cl	CH_2COOH	15,000	30
1–5	4-COOH	n-Bu	Cl	CH_2COOH	1,200	30
1–6	2-Cl	n-Bu	H	CH_2COOH	12,000	5
1–7	2-Cl	PrS	H	CH_2COOH	38,000	5
1–8	2-Cl	PrS	H	$CH(CH_3)COOH$	54,000	5
1–9	2-Cl (CH$_3$ on benzyl CH$_2$)	PrS	H	CH_2COOH	880,000	5
1–10	3-COOH	Bu	Cl	CH_2COOH	100,000	30
1–11	2-COOH	Bu	Cl	CH_2COOH	38,000	30
1–12	4-CN_4H[a]	Bu	Cl	CH_2CN_4H[a]	1,200	30
1–13	2-Cl	Bu	H	CH_2CH_2COOH	16,500	31
1–14	2-Cl	Bu	H	$CH_2CH_2CH_2COOH$	>10,000	31
1–15	2-Cl	Bu	H	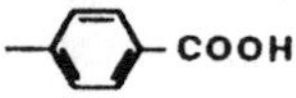	23,500	31

[a] 5-Tetrazolyl.

(**1–12**) analogs. The 2- and 3-carboxy analogs have lower affinity for the angiotensin II receptor than **1–5,** indicating the importance of regiochemistry in this series. Placement of the acetic acid sidechain in the 4-position (**1–3**) reduces affinity for the angiotensin II receptor.

Molecular modeling and NMR studies revealed that a 2-substituent on the phenyl ring of the benzyl moiety causes the imidazole and phenyl rings to favor an orthogonal relationship. Interfering with this relationship by introduction of a methyl group on the benzyl methylene, as in **1–9,** causes a loss of affinity, whereas introduction of a methyl group on the acetic acid methylene, as in **1–8,** does not affect the orthogonal relationship and does not greatly affect affinity for the angiotensin II receptor.

Consideration of the overlay model in which the benzyl group was pointed toward but not directly over the acidic β-carboxyl group of Asp[1] or the phenolic hydroxyl of Tyr[4] suggested that increasing the distance between the phenyl ring and its carboxyl group may allow for a better mimicry of angiotensin II (Table 2). This was accomplished by acylation of the 4-aminobenzyl intermediate with a cyclic anhydride. The resulting compounds, **2–3** to **2–5,** have lower affinity for the angiotensin II receptor than **1–5,** with **2–5** being somewhat better than the

TABLE 2
2-Butyl-1-(carboxybenzyl)-4-chloroimidazoles

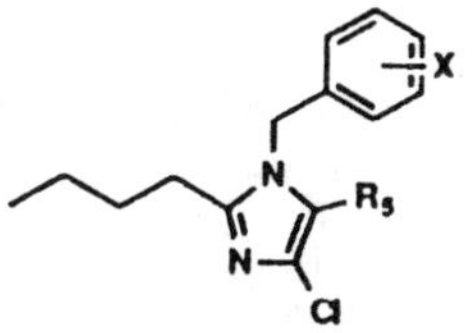

	X	R_5	IC_{50} (n*M*)	Ref.
1–5	4-COOH	CH_2COOH	1,200	30
2–1	4-COOH	CH_2OH	1,700	30
2–2	4-CH_2COOH	CH_2COOH	13,000	30
2–3	4-NHCO$(CH_2)_2$COOH	CH_2COOCH_3	46,000	30
2–4	4-NHCO$(CH_2)_3$COOH	CH_2COOCH_3	32,000	30
2–5	4-NHCOCH═CHCOOH (*cis*)	CH_2COOCH_3	11,000	30
2–6	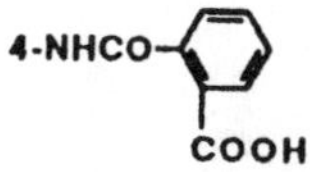	CH_2OH	140	30
2–7	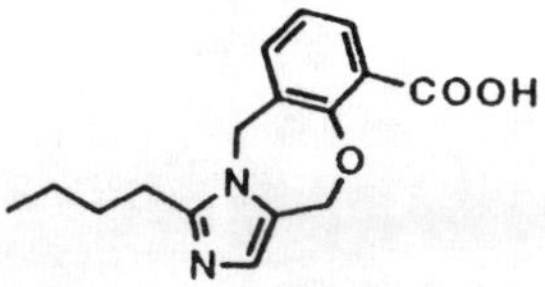		15,000	32
2–8	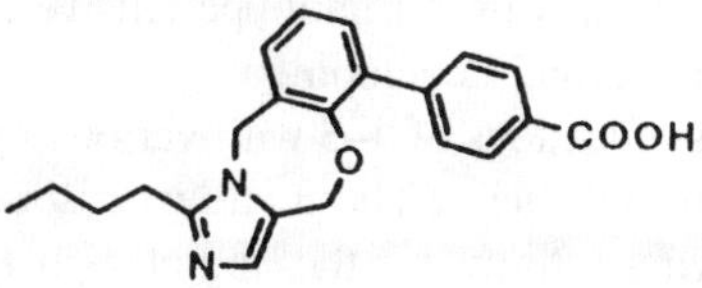		8,500	32

saturated analogs. However, the 2-carboxybenzamide **2–6,** an aromatic version of **2–5,** has approximately tenfold higher affinity for the angiotensin II receptor than **1–5** and became an important chemical lead. Compound **2–6** is a competitive antagonist, inhibiting angiotensin II-induced contraction in rabbit aorta with a pA_2 (i.e., $-\log K_D$) of 7.2. The compound lowers blood pressure in renal hypertensive rats following intravenous administration at a dose of 10 mg/kg, i.v., but not after oral administration at a dose of 100 mg/kg.

Because metabolic cleavage of the amide bond could inactivate **2–6,** other linkers of the phenyl groups were investigated. As observed in Table 3, compounds with short links (**3–1** to **3–5**) have affinity for the angiotensin II receptor comparable to that of **2–6,** and compounds with longer links (**3–6** to **3–8**) are substantially less active. All of the compounds in Table 3, with the exception of **3–5,** have poor oral activity. Compound **3–5** has ED_{30} of 3 mg/kg, i.v., and an oral ED_{30} of 11 mg/kg in the renal hypertensive rat. In the case of **3–9,** the absence of the election withdrawing 4-chlorine makes the imidazole more basic, and as such, the compound exists as a zwitterion and is therefore poorly absorbed. Since

TABLE 3
2-Butyl-1-(4-(2-carboxphenyl-x)benzyl-5-hydroxymethylimidazoles[33,30]

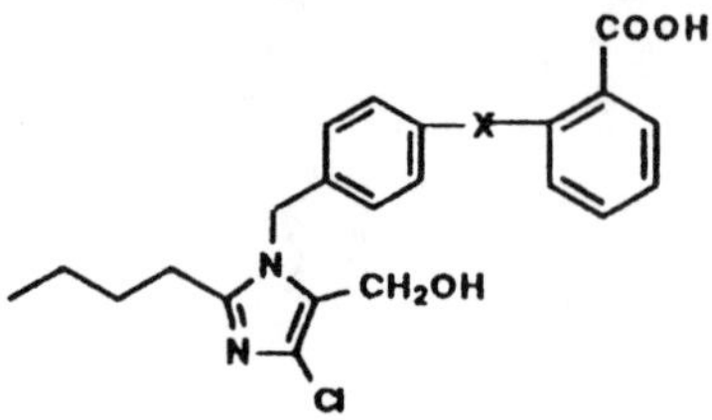

No.	X	IC_{50} (n*M*)
2–6	NHCO	140
3–1	O	400
3–2	S	400
3–3	C=O	160
3–4	OCH_2	920
3–5	bond	230
3–6	CH=CH (E)	5,400
3–7	CH=CH (Z)	2,300
3–8	NHCONH	2,400
3–9	bond (4-H)	270

oral activity is an absolute requirement for a useful antihypertensive agent, **3–5** became an important chemical lead for further development.

The structure–activity relationships of analogs of **3–5** with various substituents in the terminal phenyl ring are shown in Table 4. The 3-carboxy isomer of **3–5** (i.e., compound **4–1**), although showing affinity for the angiotensin II receptor comparable to **3–5,** is substantially less active *in vivo.* This observation suggests that the more sterically hindered 2-carboxylic acid may be less susceptible to metabolism and/or clearance. The 4-carboxy analog, **4–2,** has 50-fold lower affinity than **3–5,** illustrating the regiosensitivity of affinity for angiotensin II receptors. Various carboxylic acid replacement analogs of **3–5** are also shown in Table 4. In general, affinity for the angiotensin II receptor parallels acidity of the carboxylic acid replacement. The most interesting compound is the tetrazole analog, **4–12,** which has approximately one order of magnitude greater affinity for angiotensin II receptors than **3–5** and exhibits competitive antagonism of angiotensin II in the rabbit aorta.[4] Compound **4–12** is also approximately one order of magnitude more potent with regard to oral antihypertensive activity in the renal hypertensive rat, with a duration of action greater than 24 h. In the rat, compound **4–12** has an oral/intravenous dose ratio of <1, a bioavailability of 33%, and a biphasic antihypertensive response following intravenous dosing. This suggests the involvement of at least one active metabolite, and indeed in the rat, the 5-carboxylic acid, **7–15** (Table 7), which is approximately tenfold higher in

TABLE 4
1-(Biphenyl-4-yl)methyl-2-butyl-4-chloro-5-hydroxymethylimidazoles

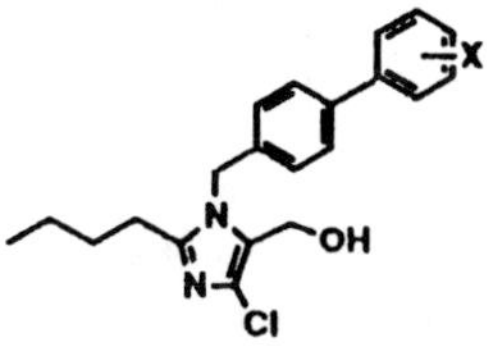

No.	X	IC_{50} (n*M*)	*In vivo* activity[a] i.v.	*In vivo* activity[a] p.o.	Ref.
3–5	2-COOH	230	3	ED_{30} = 11	33
4–1	3-COOH	490	45	>100	33
4–2	4-COOH	11,000	>10	>100	33
4–3	3-CH_3–2-COOH	1,900	3	10	33
4–4	6-CH_3O-2-COOH	3,000	10	100	33
4–5	2-CH_2COOH	1,500	—	—	34
4–6	2-SO_3H	79	—	—	34
4–7	2-$CONH_2$	35,000	—	—	33
4–8	2-$CONHOCH_3$	2,900	10	>100	33
4–9	2-$CONHSO_2C_6H_5$	140	>3	>30	33
4–10	2-$NHCOCF_3$	6,300	10	>100	33
4–11	2-$NHSO_2CF_3$	83	10	100	33
4–12	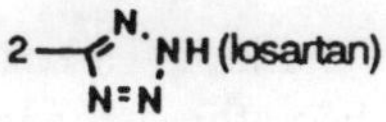 (losartan)	19	ED_{30} = 0.8	0.59	33
4–13	4-CH_3O-2-CN_4H[b]	580	ED_{30} = 1.75	>10	33
4–14	5-CH_3O-2-CN_4H[b]	120	ED_{30} = 6.03	>10	33
4–15	5-CN-2-CN_4H[b]	510	3	>30	33
4–16	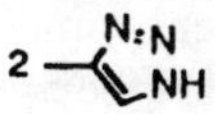	9,600	—	—	33
4–17	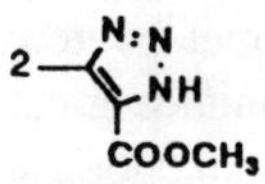	280	>1	>10	33
4–18	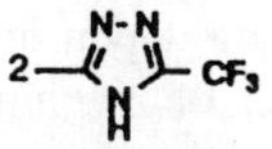	370	1	10	33
4–19	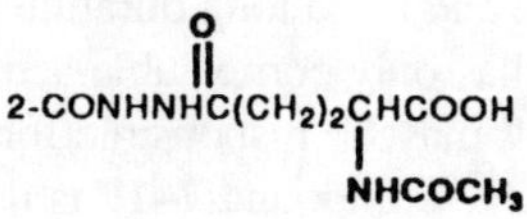	X[c]			35

[a] Dose in mg/kg that drops blood pressure >15 mm Hg in renal hypertensive rats.
[b] 5-Tetrazolyl.
[c] Comparable data not available.

affinity for the angiotensin II receptor and approximately one order of magnitude more potent as an antihypertensive agent in rats following intravenous administration, is a major metabolite of **4–12.**[4] Compound **4–12** is in clinical development under the name losartan, and it and its metabolite **7–15** are considered to be the prototypic AT_1 receptor antagonists.

Compounds with additional substituents on the acid-carrying phenyl ring show substantially lower affinity for the angiotensin II receptor than the unsubstituted analogs. One interesting analog is **4–19,** which was designed to be a substrate for γ-glutamyl hydrolase, an enzyme located in the kidney that should liberate the corresponding hydrazide in the kidney, thus providing renal selectivity. This compound lowers blood pressure when administered chronically to the hypertensive rat.

The effect of changing the imidazole substituents of losartan, or its 2′-carboxylic acid analog **3–5,** have been explored. As shown in Table 5 for the 2′-carboxybiphenyl series of the saturated 2-alkyl compounds, the 2-propyl analog has the highest affinity for angiotensin II receptors; however, the 2-butyl analog has the greatest potency when administered orally. The *trans*-2-butenyl analog, **5–5,** has the highest affinity for the angiotensin II receptor, and this compound is more potent *in vivo.* The 2-phenyl analog, **5–7,** is significantly less active than any of the 2-alkyl compounds. Alkylthio compounds have somewhat lower affinities for the angiotensin II receptor than the comparable alkyl analogs (**5–8** vs. **5–9**).

The effect of varying the 4-substituent of losartan and its 2′-carboxy analog, **3–5,** is shown in Table 6. The compounds with the highest affinity for angiotensin II receptors carry a 4-halo, 4-perfluoroalkyl, or carboxy group. With regard to position 5, a wide variety of functional groups allow for substantial affinity for angiotensin II receptors, as shown in Table 7. The more potent compounds can act as hydrogen bond acceptors. The most potent compounds are the 5-carboxylic acids, **7–2,** and the tetrazole, **7–19.** The latter is a major metabolite of losartan (**4–12**) in the rat and has approximately one order of magnitude greater affinity for angiotensin II receptors relative to losartan. However, unlike losartan, **7–19** is not a competitive antagonist of angiotensin II in rabbit aorta. Upon intravenous administration to renal hypertensive rats, **7–19** is an order of magnitude more potent than losartan in lowering blood pressure and has a long duration of action; however, following oral administration, **7–19** has only comparable activity, suggesting poor bioavailability. Nonetheless, **7–19** may be responsible for much of the efficacy when losartan is administered orally. Compound **7–19** is now a prototype for many potent orally active angiotensin II receptor antagonists.

Many of the compounds in Table 7 that have submicromolar affinity for angiotensin II receptors carry bulky 5-substituents. Compounds **7–16** and **7–22** are noteworthy in that they possess a functional group, similar to that found in the β-adrenoceptor antagonist, pindolol, attached to the imidazole via an amidomethyl link.[7] These compounds were designed to take advantage of the reduction in renin release produced by β-blockers in order to counteract the large increase in renin release produced by AT_1 receptor blockade. Both **7–22** and captopril have little effect on blood pressure in the normotensive dog; however, both compounds lower blood pressure in the high renin, sodium (with furosemide) depleted dog. In the pithed rat, **7–16** inhibits the pressor effect of angiotensin II, and at the same doses markedly inhibit isoproterenol-induced tachycardia, thus demonstrating the β-blocking activity of the compound.

A number of 2-butyl-4-chloro-5-hydroxymethyl (or 5-carboxy)-imidazoles with different N-benzyl sidechains have been studied (Table 8). The sidechains are essentially of the biphenyl type, as in **3–5,** or the 4-carboxybenzyl type, as in **2–1.** The most extensively studied is the 2-phenylbenzofuran series, which when combined with the 5-carboxyimidazole group, yields compounds (**8–2** and **8–14** to **8–16**) having high affinity for angiotensin II receptors. This series is discussed in detail later in this chapter. The corresponding indole and benzothiophene analogs are less active. The phenyl ring carrying the carboxyl group can be replaced by a heterocycle, as in **8–4** and **8–5.** In **8–6,** the two methoxyl groups *ortho* to the second phenyl ring restrict conformational freedom and markedly decrease affinity for angiotensin II receptors. Compounds **8–7** and **8–8** are analogs of **2–1.** As discussed above, the 5-carboxylic acid, **7–15,** is a major metabolite of losartan and may be responsible for the long duration and high potency of losartan *in vivo.* However, **7–15** has a low intravenous to oral potency ratio. Some representative analogs of **7–15** are shown in Table 9. The tetrazoles in some but not all examples are substantially more potent than the carboxylic acids. Thus, in the 2-butyl-4-pentafluoroethyl analogs, the 2′-carboxylic acid, **9–2,** and the 2′-tetrazole, **9–7,** show equivalent but modest affinity for angiotensin II receptors. However, in the closely related 2-propyl-4-pentafluoroethyl case, the 2′-tetrazole, **9–11,** shows 500-fold greater affinity for angiotensin II receptors than the corresponding carboxylic acid, **9–4.** Compound **9–11** (DuP 532)[8] has high affinity for angiotensin II receptors, shows noncompetitive antagonism of angiotensin II in rabbit aorta, and is well absorbed orally in the rat. It is noteworthy that the 4-methylsulfonyl compound, **9–8,** has an ED_{50} of 0.022 mg/kg, i.v., in inhibiting the pressor effect of angiotensin II in the rat,[9] while the 4-ethyl analog, **9–13,** has an ED_{50} of 0.027 mg/kg, i.v., and 0.03 mg/kg, p.o.,[10,11] despite the opposite electronic properties of the substituents.

Another important class of 4-carboxyimidazoles comprises those carrying an N-(3-bromo-2-(2′-tetrazolyl)phenylbenzofuranylmethyl group, as shown in Table 10. The structure–activity relationships reported in Table 10 represent potency in inhibiting the angiotensin II-induced contraction of rabbit aorta, and comparable data for losartan and its 5-carboxy metabolite, **7–15,** are also provided. The benzothiophene and indole analogs, **10–2** and **10–3,** while comparable in potency *in vitro* to the benzofuran, **10–1,** are inactive when dosed orally in the rat. The effect of various 3-benzofuran substituents on potency is correlated with their inductive electron withdrawing strength. The greatest oral activity is observed with the 3-chloro and 3-bromo analogs. In contrast to the biphenyl series, the 2′-tetrazolylphenyl and 2′-carboxyphenyl analogs have similar potencies as angiotensin II receptor antagonists. Although the 5-carboxylic acid analogs are the most potent in this series, other analogs, such as the 5-carboxamides, have comparable

TABLE 5
2-Substituted-1-(2′-carboxybiphenylylmethyl)-4-chloro-5-hydroxymethylimidazoles[33]

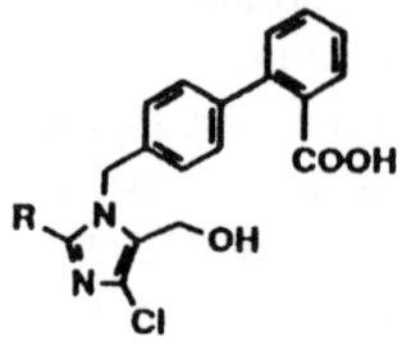

			In vivo activity[a]	
No.	**R**	**IC_{50} (n*M*)**	**i.v.**	**p.o.**
5–1	C_2H_5	1,700	10	100
5–2	n-C_3H_7	160	3	30
3–5	n-C_4H_9	230	3	ED_{30} = 11
5–3	n-C_5H_{11}	980	3	30
5–4	n-C_6H_{13}	1,300	10	>100
5–5	*trans* $CH{=}CHCH_2CH_3$	80	ED_{30} = 1	ED_{30} = 3.6
5–6	*cis* $CH{=}CHCH_2CH_3$	2,300	—	—
5–7	C_6H_5	24,000	10	100
5–8	n-C_3H_7S (4-H)	890	active	inactive
5–9	n-C_4H_9 (4-H)	270	active	inactive
5–10		1,800	—	—

[a] Dose in mg/kg that drop blood pressure >15 mm Hg in renal hypertensive rats.

potency.[12] Again, in contrast to the biphenyl series, the 2-ethyl and 2-propyl analogs are order of magnitude more potent as angiotensin II receptor antagonists than the 2-butyl compound. Detailed structure–activity relationship studies culminated in the selection of **10–11** (GR138950) for entry into clinical trials. GR138950 is a partially insurmountable angiotensin II receptor antagonist in rabbit aorta that is somewhat less bound to plasma protein than some of its close analogs. The related N-ethylcarboxamide undergoes metabolism to the carboxylic acid and primary carboxamide, with the primary carboxamide being metabolically stable. The trifluromethylsulfonamido group contributes to the oral absorption of this compound.[12]

B. IMIDAZOLE-5-ACRYLIC ACIDS

Imidazole-5-acrylic acids were designed to be conformationally restricted, extended analogs of CV 2947 (**1**) in order to better overlay an initial postulated bioactive model of angiotensin II. The development of the overlay hypothesis is discussed in detail in Chapter 3 of Volume II. The E-acrylic acid, **11–1** (Table 11), and the corresponding acetic acid, **1–6,** have comparable affinity for the

TABLE 6
4-Substituted-2-butyl-1-(2′-carboxy or tetrazolylbiphenylylmethyl)-5-hydroxymethylimidazoles[33]

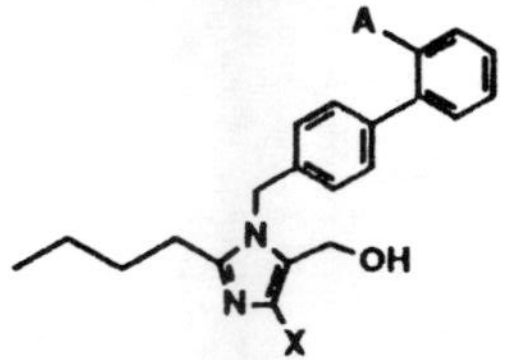

	A=COOH				A=Tetrazole			
			In vivo activity[a]				*In vivo* activity[a]	
X	No.	IC_{50}(n*M*)	i.v.	p.o.	No.	IC_{50}(n*M*)	i.v.	p.o.
Cl	3–5	230	ED_{30} = 3.7	ED_{30} = 11	4–12	19	ED_{30} = 0.80	ED_{30} = 0.59
Br	6–1	90	3	30	6–11	19	0.3	3
I	6–2	61	10	100	6–12	20	1	10
CF_3	6–3	62	3	30	6–13	12	ED_{30} = 0.53	>3
H	6–4	270	10	>100	6–14	29		
NO_2	6–5	260	10	>100				
CH_3 (2-Pr)	6–6	1,800	ED_{30} = 1.3	>30				
COOH	6–7	92	ED_{30} = 0.47	$ED_{30}\leq 10$	6–15 (2-Pr)	260		
CF_2CF_3	6–8	40	—	—	6–16	10		
$CF_2CF_2CF_3$	6–9	70	—	—	6–17	10		
	6–10	680	—	—				

[a] Dose in mg/kg that drops blood pressure >15 mm Hg in renal hypertensive rats.

TABLE 7
5-Substituted-2-butyl-1-(2′-carboxy or tetrazolylbiphenylylmethyl)-4-chloroimidazoles[34,7]

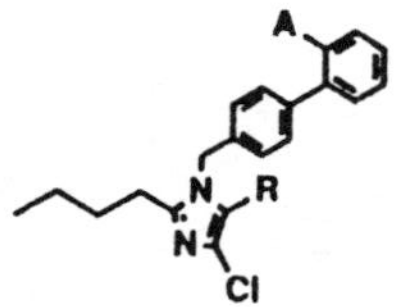

	A=COOH				A=Tetrazole			
			In vivo (mg/kg)[a]				*In vivo* (mg/kg)[a]	
R	No.	IC_{50} (n*M*)	i.v.	p.o.	No.	Ic_{50} (n*M*)	i.v.	p.o.
$—CH_2OH$	3–5	230	3	ED_{30} = 11	4–12	19	ED_{30} = 0.80	ED_{30} = 0.59
$—CH_2OCH_3$	7–1	99	0.47	≤10				
—COOH	7–2	92			7–19	1.3	ED_{30} = 0.038	ED_{30} = 0.66
$—CN_4H$	—	—			7–20	160		
$—CONH_2$	7–3	150						
$—COOCH_3$	7–4	340						
—CHO	7–5	940						
$—CON(CH_3)_2$	7–6	1,300						
$—CH_2COOH$	7–7	430						
$—CH(CH_3)COOH$	7–8	260						
$—(CH_2)_3COOH$	7–9	270						
$—CH{=}CHCH_2OH$	7–10	630						
$—(CH_2)_3F$	7–11	700						
$—CH_2NHCOOCH_3$	7–12	250			7–21	60		
$—CH_2NHCOO$-adamantyl	7–13	600						
t								
$—CH{=}CHCHOHCH_3$	7–14	2,100						
t								

TABLE 7 (continued)
5-Substituted-2-butyl-1-(2′-carboxy or tetrazolylbiphenylylmethyl)-4-chloroimidazoles[34,7]

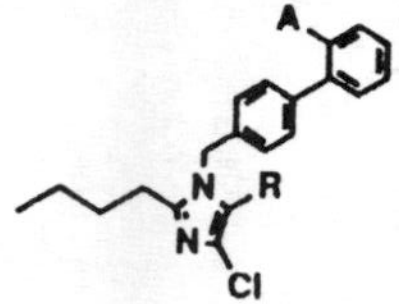

	A=COOH				A=Tetrazole			
			In vivo (mg/kg)[a]				*In vivo* (mg/kg)[a]	
R	No.	IC_{50} (n*M*)	i.v.	p.o.	No.	Ic_{50} (n*M*)	i.v.	p.o.
—CH=CHC$_6$H$_5$	7–15	1,600						
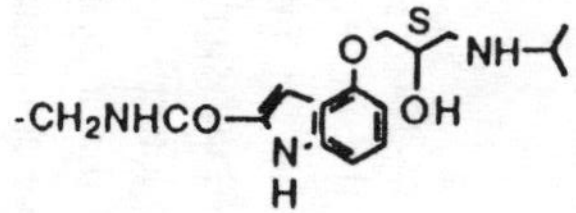	7–16	380			7–22	X[b]		
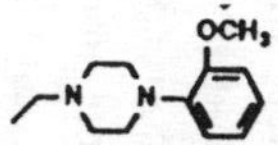	7–17	740			7–23	350		
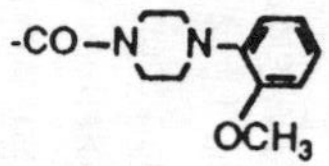					7–24	200		
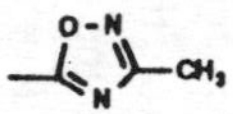					7–25[36]	X[6]		

[a] Dose which drops blood pressure >15 mm Hg in renal hypertensive rats.
[b] Comparable data not available.

TABLE 8
1-Substituted methyl-5-hydroxymethyl(or carboxy)imidazoles

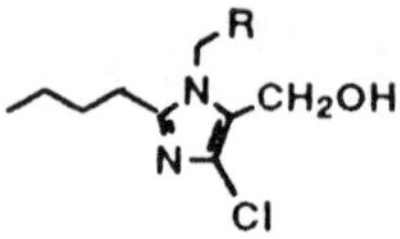

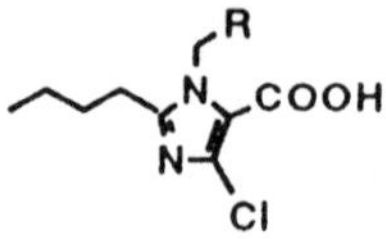

R	No.[Ref]	IC_{50}(n*M*)	No.[Ref]	IC_{50}(n*M*)
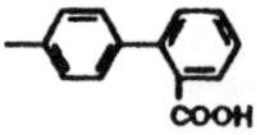	3–5	230	7–2	92
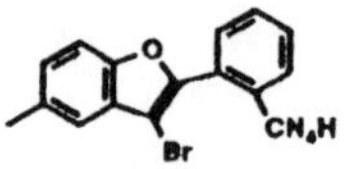	8–2[38]	pK_b = 9.1	8–14[38]	pK_b = 9.8
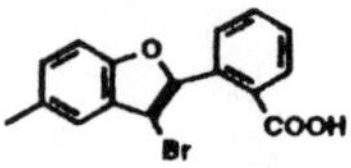	8–3[38]	pK_b = 8.6	8–15[38]	pK_b = 10.2
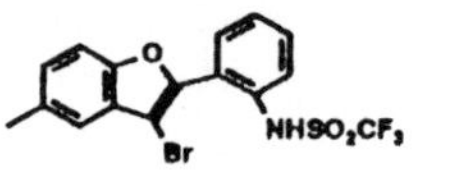	—	—	8–16[38]	pK_b = 7.8
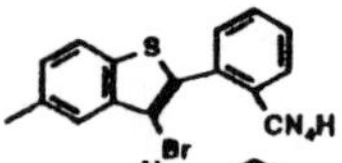	—	—	8–17[38]	pK_b = 8.7
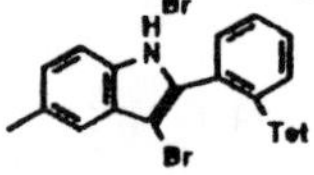	—	—	8–18[38]	pK_b = 9.0
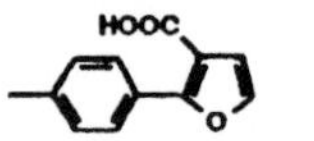	8–4[39]	4,200		
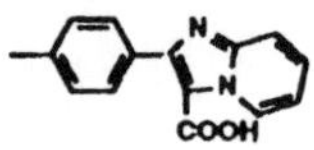	8–5[40]	X[a]		
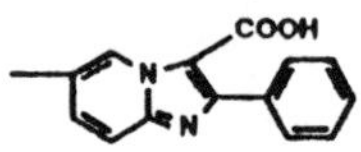	8–6[41]	X[a]		
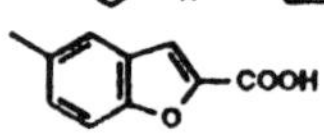	8–7[42]	X[a]		
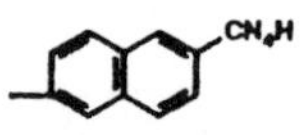	8–8[43]	600		

angiotensin II receptor; however, the *alpha* benzyl analog, **11–5,** which was prepared to make this portion of the molecule more closely mimic the Phe[8] of angiotensin II, has greater affinity for angiotensin II receptors than **11–1.** Both Z-acrylic acids, **11–2** and **11–6,** are significantly lower in affinity for angiotensin II receptors than their corresponding E isomers. Compounds in which the linkage between the phenyl ring and the acrylic acid has been shortened, lengthened, branched, or made *beta* all have comparable affinities for angiotensin II receptors

TABLE 8 (continued)
1-Substituted methyl-5-hydroxymethyl(or carboxy)imidazoles

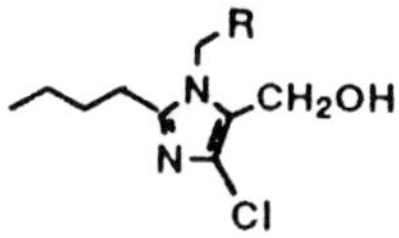

R	No.[Ref]	IC_{50}(n*M*)
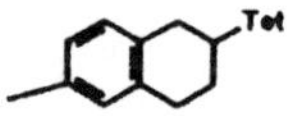	8–9[43]	5,400
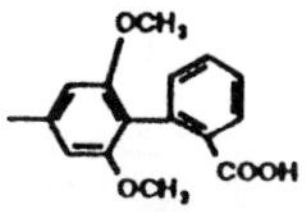	8–10[41]	1,900
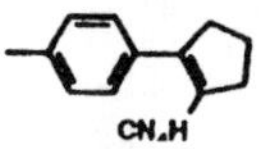	8–11[44]	3,500
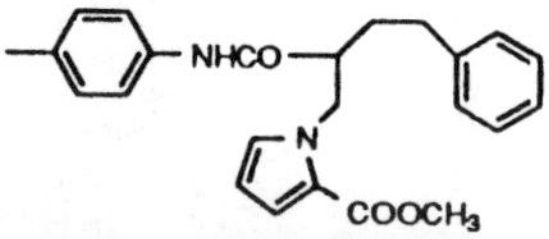	8–12[45]	X[a]
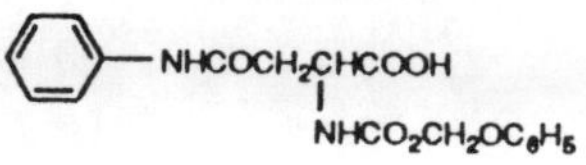	8–13[46]	X[a]

[a] Comparable data not available.

in radioligand binding studies but somewhat lower potency in inhibiting functional responses to angiotensin II.

Affinity for the angiotensin II receptor is sensitive to the nature of the 2-arylmethyl group of the acrylic acid. Both the 3,4-dimethoxybenzyl and the 3,4-dichlorobenzyl analogs, **12–4** and **12–7** (Table 12), have lower affinity for angiotensin II receptors, whereas, the 4-methoxybenzyl and 3,4-methylenedioxy analogs, **12–2** and **12–3,** have higher affinity. This observation suggested the substitution of small, electron-rich heterocyclic aryl groups. The 2-thienylmethyl analog, **12–8,** shows increased affinity for angiotensin II receptors compared to the benzyl analog. As in the benzyl series, the methoxythiophene analog, **12–10,** has higher affinity than **12–8.** Other heterocyclic analogs show comparable affinity. The increase in affinity for angiotensin II receptors resulting from replacing the phenyl ring of **12–1** with the thiophene ring in **12–8** is consistent with the increase in affinity observed when the Phe[8] phenyl ring of [Sar[1]]-angiotensin II (**12–14**) is replaced by a thiophene in [Sar[1], Thi[8]]-angiotensin II (**12–15**), thus supporting the overlay hypothesis used in the design of this class of angiotensin II receptor antagonists.

TABLE 9
2-and 4-Substituted-1-(2′-carboxy(or tetrazolyl)biphenylmethyl)-imidazole-5-carboxylic acids[39,47]

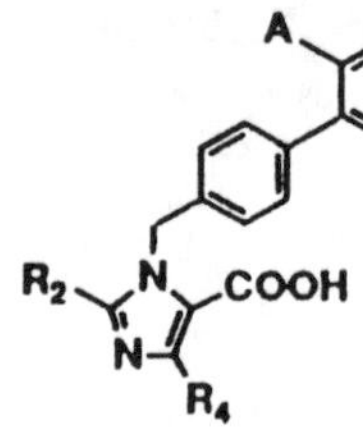

		A=COOH		A=Tetrazole	
R_2	**R_4**	**No.**	**IC_{50}(n*M*)**	**No.**	**IC_{50}(n*M*)**
n-Bu	Cl	7–2	92	7–19	1.3
n-Bu	CF_3	9–1	1,100	9–6	150
n-Bu	CF_2CF_3	9–2	1,600	9–7	1400
n-Bu	SO_2CH_3	—	—	9–8	X[a]
n-Pr	Cl	—	—	9–9	45
n-Pr	CF_3	9–3	1,100	9–10	1100
n-Pr	CF_2CF_3	9–4	1,600	9–11	3.1
n-Pr	COOH	9–5	6,700	9–12	76
n-Pr	CH_2CH_3	—	—	9–13	X[b]

[a] ED_{50} = 0.022 mg/kg, i.v., inhibition of AII pressor effect in the rat.
[b] ED_{30} = 0.027 mg/kg, i.v., 0.03 mg/kg, p.o., inhibition of AII pressor effect in the rat.

Further increases in potency are observed when the benzyl chlorine atom of **12–8** is replaced by hydrogen bond-accepting groups that could mimic the Tyr[4] hydroxyl group of angiotensin II. Thus the hydroxyl analogs, **13–1** and **13–2** (Table 13), and the 2-nitro analog, **13–3,** show higher affinity for angiotensin II receptors than **12–8.** However, the 4-carboxy analog, **13–6,** has more than two orders of magnitude higher affinity for angiotensin II receptors. The 2-chloro-4-carboxy analog, **13–7,** in which the orthogonal conformation of the phenyl and imidazole rings is favored, shows an additional increase in the ability to block functional responses to angiotensin II, in spite of having comparable affinity in radioligand binding studies. Radioligand binding affinity measurements may be distorted by the presence of bovine serum albumin in the assay mixture if the compounds are highly protein-bound.[6] It is noteworthy that **13–6** and its halogenated analogs show comparable affinity for angiotensin II receptors in radioligand binding assays and in functional studies compared to the peptide angiotensin II agonist **12–14** and antagonists **12–15** and **12–16.** The tetrazole analog, **13–9,** although potent *in vitro,* has a short duration of action *in vivo.* The 2′-carboxybiphenyl analog has sevenfold lower affinity for angiotensin II receptors than **13–6,** in contrast to the 4-chloro-5-hydroxymethylimidazole series, where it is fivefold more active than the 4-carboxy analog.

TABLE 10
2-Butyl-4-chloro-1-(2-phenylbenzofuran-5-ylmethyl(imidazole-5-carboxylic acids[38,48,49]

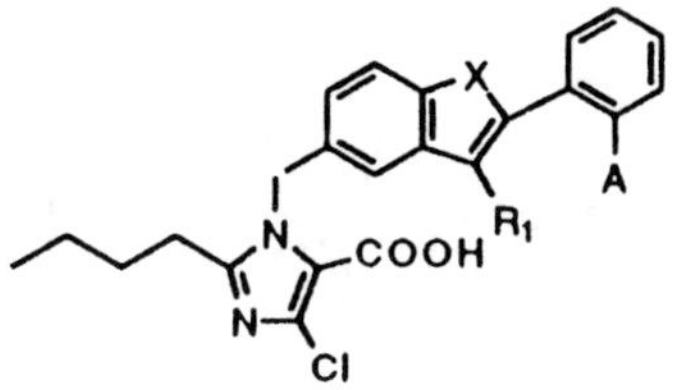

No.	X	A	R_1	pK_b[a]
10–1	O	CN_4H	Br	9.8
10–2	S	CN_4H	Br	8.7
10–3	NH	CN_4H	Br	9.0
10–4	O	CN_4H	Cl	10.5
10–5	O	CN_4H	CF_3	10.0
10–6	O	CN_4H	H	8.4
10–7	O	CN_4H	t-Bu	8.3
10–8	O	CN_4H	OCH_3	9.6
10–9	O	COOH	Br	10.2
10–10	O	$NHSO_2CF_3$	Br	7.8
4–12	losartan			8.4
7–15	losartan-5-COOH			10.6

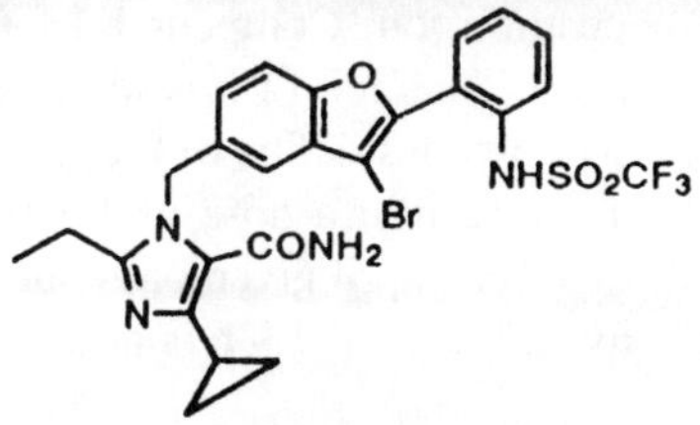

GR 138950[12,48]
10-11 $pK_b = 9.0$

[a] Inhibition of rat AII-induced contraction of rabbit aorta.

The dramatic increase in potency seen with the introduction of the 4-carboxy group suggests that this group is picking up a new binding site in the receptor and is possibly mimicking the Tyr[4] of angiotensin II. In this overlay model, as seen in Figure 4, the acrylic acid carboxyl and thienylmethyl groups then align with the corresponding elements of the Phe[8] of angiotensin II, and the 2-butyl lies in the hydrophobic region near Ile.[5] Also, the N-C-N imidazole region and the acrylic acid double bond may mimic amide bonds of the peptide.

The 4-carboxy analog, **13–6,** inhibits the pressor effect of angiotensin II in rats with an IC_{50} of 0.08 mg/kg following intravenous dosing.[13] Upon intraduodenal administration of 10 mg/kg, **13–6** reduces blood pressure in the renin-dependent

TABLE 11
(E)-2-Butyl-1-(2-chlorobenzyl)imidazole-5-acrylic acids[5,31]

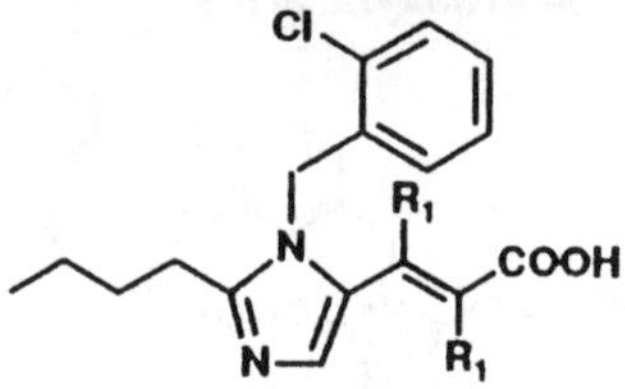

No.	R_1	R_2	IC_{50}(n*M*)	K_b(n*M*)
11–1	H	H	8,900	810
11–2	H (Z)	H	124,000	14,000
11–3	H	n-Bu	6,900	160
11–4	H	C_6H_5	20,100	600
11–5	H	$CH_2C_6H_5$	2,600	64
11–6	H (Z)	$CH_2C_6H_5$	38,000	8,700
11–7	H	$CH(CH_3)C_6H_5$	4,600	240
11–8	H	$CH_2CH_2C_6H_5$	2,900	420
11–9	$CH_2C_6H_5$	H	9,800	330
11–10	CH_3	$CH_2C_6H_5$	28,400	760

hypertensive rat by 25 mm Hg. A sustained infusion of **13–6,** at a rate of 25 μg/min in the hypertensive rat, normalizes blood pressure during 3 d of infusion, and this effect persisted 18 h postinfusion. Compound **13–6** also antagonizes the pressor effect of angiotensin I in dogs for 14 h when dosed orally at 10 mg/kg, and in dogs made hypertensive by constriction of the left renal artery, **13–6** lowered blood pressure for at least 12 h at a dose of 10 mg/kg orally.[14] The dog responses are similar to those of enalapril given orally at 1 mg/kg, but **13–6** has a more rapid onset of activity. Compound **13–6** is now undergoing clinical trials as SK&F 108566.

Data on other 5-substituted N-(4-carboxybenzyl)imidazoles designed in a manner similar to that used for SK&F 108566 are presented in Table 14.[15] While the 5-acetic acid analogs, **14–1** and **14–2,** are active in the high nanomolar range, compounds that incorporate other angiotensin II C-terminal mimetics are active in the low nanomolar range. The phenylalanine amide of **14–2** (i.e., **14–3**) is 23-fold more potent than the parent acid. Similarly, **14–4,** which contains the thienyl alanine moiety coupled to the imidazole by a methylene bridge, is 100-fold more potent than **14–1.** The α-propionylamidopropionic acid, **14–5,** is sevenfold more potent than **14–1.** These results reinforce the validity of the overlay hypothesis, which was used to design SK&F 108566.

C. NON-IMIDAZOLE 2′-TETRAZOLYL (OR CARBOXY) BIPHENYLYLMETHYL AII RECEPTOR ANTAGONIST DERIVATIVES

The 2′-tetrazolyl (or carboxy) biphenyl group imparts angiotensin II receptor antagonist activity when attached via an appropriate linker to a wide variety of

TABLE 12
3-(2-Butyl-1-(2-chlorobenzyl)imidazol-5-yl)-2-arylmethylacrylic acids[31]

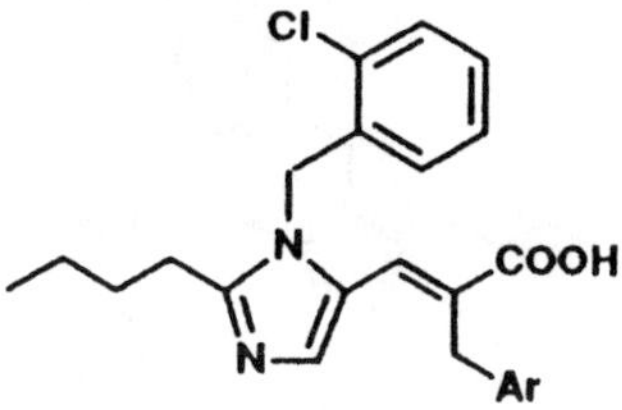

No.	Ar	IC_{50}(n*M*)	K_b(n*M*)
12–1	C_6H_5	2,600	64
12–2	4-$CH_3OC_6H_3$	810	270
12–3	3,4-$OCH_2OC_6H_3$	470	130
12–4	3,4-di-$CH_3OC_6H_3$	6,800	620
12–5	4-$NO_2C_6H_4$	400	310
12–6	4-$NH_2C_6H_4$	3,000	320
12–7	3,4-di-ClC_6H_3	25,600	13,700
12–8	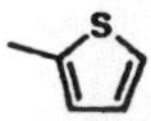	440	51
12–9	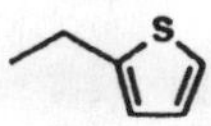	860	140
12–10	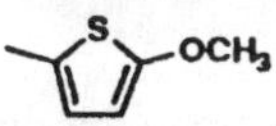	90	30
12–11		330	170
12–12	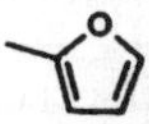	1,200	50
12–13	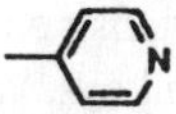	150	170
12–14	[Sar[1]]AII	1.3	
12–15	[Sar[1], Thi[8]]AII	0.11	2.5
12–16	[Sar[1], Ala[8]]AII	3.4	2.5

heterocyclic compounds, as well as some acyclic fragments. A survey of these compounds is shown in Table 15. In some cases, the heterocyclic ring contains a carboxylic acid group, but many potent angiotensin II receptor antagonists do not have the second acidic group, as in compound **15–14.** In many cases, an important function of the heterocyclic group is to allow hydrogen bonding with a component of the angiotensin II receptor site. The detailed structure–activity relationships of the major classes of angiotensin II receptor antagonists are discussed in the remainder of this chapter. The heterocyclic moiety of some of the more potent compounds has been used with novel biphenyl replacements to yield further interesting analogs. These will also be discussed below.

TABLE 13
3-(1-Arylmethyl-2-butylimidazol-5-yl)-2-(2-thienyl)methylacrylic acids[49]

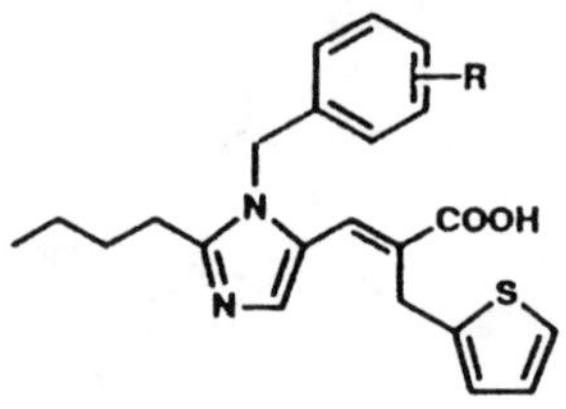

No.	R	IC_{50}(n*M*)	K_b(n*M*)
12–8	2-Cl	440	51
13–1	2-OH	90	34
13–2	4-OH-3-CH_3	21	33
13–3	2-NO_2	31	57
13–4	4-NO_2	620	50
13–5	2-COOH	6,000	2,250
13–6	4-COOH (**SK&F 108566**)	1.0	0.21
13–7	2-Cl-4-COOH	1.45	0.02
13–8	3-Cl-4-COOH	1.0	0.13
13–9	4-CN_4H	0.4	0.25
13–10	4-(2′-COOH-C_6H_4)	7	1.6

A noteworthy compound not discussed elsewhere in this chapter is **15–42,** also known as valsartan.[16] It is an N-acyl-N-biphenylylmethyl valine. This compound binds selectively to AT_1 receptors with high affinity (IC_{50} of 2.7 n*M*). The affinity for AT_2 receptors is extremely low (IC_{50} = 80,400 n*M*). Valsartan inhibits angiotensin II-induced contraction of rabbit aorta with a dissociation constant of 1.4 n*M* and shows antihypertensive effects in renal hypertensive rats at doses of 1, 3, and 10 mg/kg, p.o. In sodium-depleted marmoset monkeys, valsartan, at doses of 0.1 to 3 mg/kg, p.o., reduces blood pressure.

D. BENZIMIDAZOLES

1-Arylmethylbenzimidazoles with appropriate functional substituents can be potent angiotensin II receptor antagonists, as seen in Tables 16 and 17. The optimal 2-substituent for 1-(4-carboxybenzyl)benzimidazoles is n-hexyl (**16–3**), although most of the other compounds examined have an n-butyl substituent in this position. The 5,6-dichloro analog, **16–4,** and its unsubstituted analog, **16–2,** have similar potency. The tetrazole **16–5** is less potent than its corresponding carboxylic acid **16–2.** Substituents on the phenyl (**16–6** to **16–9**) have little effect on potency other than the 2-methoxy substitution, which decreases potency by tenfold. Although the 1-(2′-carboxybiphenyl) analog **16–10** has activity comparable to **16–2,** its corresponding tetrazole analog **16–11** is approximately 20-fold more potent. The

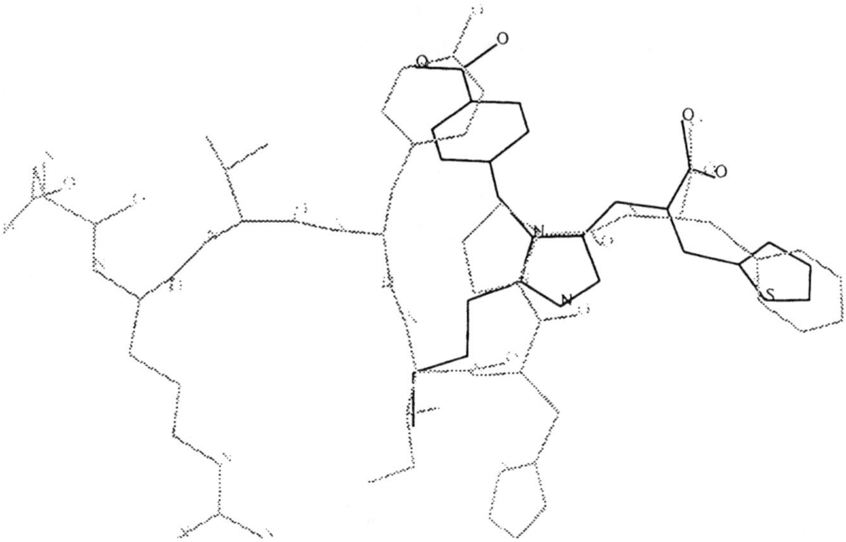

FIGURE 4. Overlay of SK&F 108566 (solid) and angiotensin Model 2 (dotted). In this overlay, the SK&F 108566 4-carboxyl phenyl is visualized as corresponding to the Tyr^4 phenol, the acrylic acid carboxyl, and thienylmethyl to the corresponding elements of Phe^8, and the 2-butyl to the Ile^5 sidechain. The N-C-N region and the acrylic acid double bond of SK&F 108566 may mimic amide bonds of angiotensin II.

low activity of the 2-unsubstituted analog **16–12** shows that an appropriate 2-alkyl substituent can markedly enhance affinity for angiotensin II receptors. A 7-carboxy group enhances potency markedly (**16–15** vs. **16–2**) and is present in the 2-ethoxy-1-(2′-tetrazol-5-yl)biphenylmethyl analog **16–17** and its prodrug **16–18.**

The most potent benzimidazoles reported to date are 1-biphenylmethyl-imidazoles with large 6-substitutents. Thus **17–1** and **17–2** (Table 17) have a benzimidazole at position 6, while **17–3** and **17–4** carry a 6-imidazotetrahydro pyridine. In the latter, the 2′-carboxybiphenyl and its corresponding tetrazole are equipotent. Compound **17–7** has both AT_1 and AT_2 receptor binding activity ($AT_1/AT_2=17$) and shows competitive antagonism of angiotensin II-induced contraction of rabbit aorta, with a pA_2 (i.e., $-\log K_D$) of 8.14. The related compound **17–8** has an AT_1/AT_2 receptor binding potency ratio of 37, but has a pA_2 of 7.1. Both compounds are short-acting in renal hypertensive rats following intravenous administration and are inactive following oral dosing.[17]

E. PYRAZOLES

1-Alkyl-3-butyl-5-carboxy-4-biarylmethylpyrazoles (Table 18) are among some of the most potent AT_1 receptor antagonists reported to date, with some compounds active in the subnanomolar range. Regioselectivity is critical for these compounds. For example, **18–2,** the 1-alkyl-5-carboxy isomer, is approximately

TABLE 14
2-Butyl-1-(4-carboxyphenylmethyl)-5-substituted imidazoles

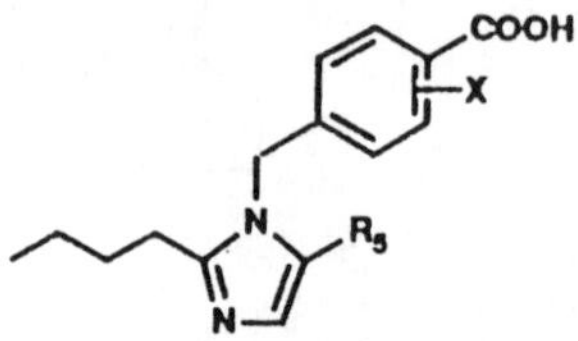

No.	X	R_5	$IC_{50}(nM)$	Ref.
13–6	H	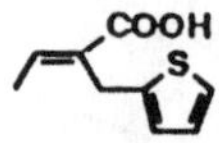	1	5
14–1	H	CH_2COOH	291	15
14–2	Benzo	CH_2COOH	980	15
14–3	Benzo	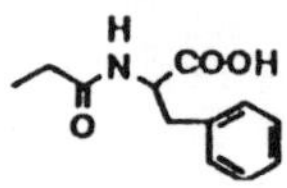	43	15
14–4	H	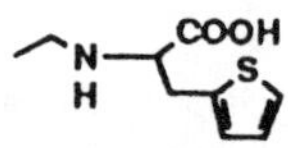	2.6	15
14–5	H	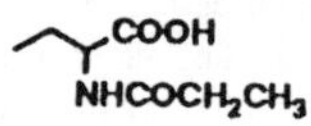	41	15

order of magnitude more potent as an angiotensin II receptor antagonist in the rabbit aorta than the unalkylated parent **18–1,** while the 2-alkyl-5-carboxy analog **18–3** is order of magnitude less potent than **18–1.** Many lower 1-alkyl- and 1-cycloalkyl-containing pyrazoles are potent angiotensin II receptor antagonists, but oral potency is critically dependent on the exact nature of the alkyl group. Thus, **18–10,** the cyclopropylmethyl analog, at a dose of 1 mg/kg, p.o., lowers blood pressure in the renal artery-ligated hypertensive rat for more than 24 h, whereas the isomeric equipotent *in vitro* cyclobutyl analog **18–11** is inactive following oral administration. It may be significant that the peak antihypertensive effect for **18–10** in the rat occurs 5 h following dosing. The arylalkyl analogs are also potent, with **18–14** completely inhibiting the pressor response to angiotensin II for more than 24 h when administered at a dose of 1 mg/kg. In this series, the nature of the acidic group on the biphenyl substituent has a profound effect on potency. Thus, the tetrazole **18–4** is nearly two orders of magnitude more potent than the corresponding carboxylic acid **18–5** or the trifluoromethyl sulfonylanilide **18–6.** No detailed activity data is available for **18–16** to **18–18,** which have been reported in the patent literature.

F. TRIAZOLES

The structure–activity relationships of some triazole angiotensin II receptor antagonists are shown in Table 19. The only 1,3,4-triazole with nanomolar potency

TABLE 15
2′-Carboxy (or 5-tetrazolyl)biphenylymethyl — nonimidazoles

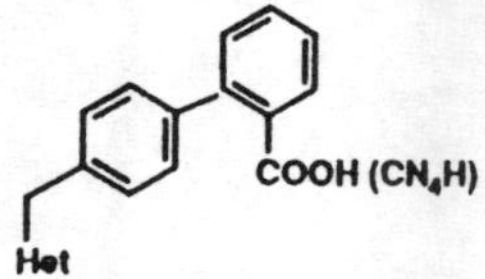

No.	Het	IC_{50}(n*M*) COOH	IC_{50}(n*M*) Tetrazole	Ref.	No.	Het	IC_{50}(n*M*) COOH	IC_{50}(n*M*) Tetrazole	Ref.
15–1	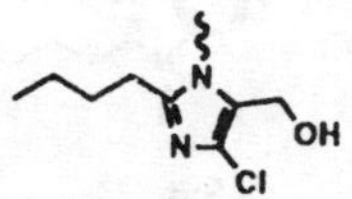	480	19	33	15–7	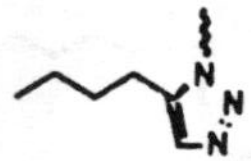	1,000	300	50
15–2	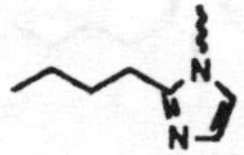	9,460	—	32	15–8	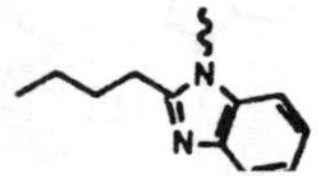	2,300	96	32
15–3	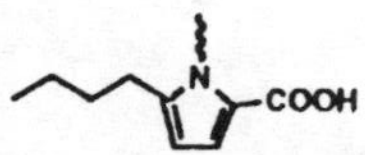	—	520	50	15–9	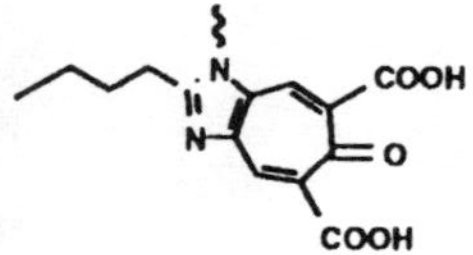	500	410	51
15–4	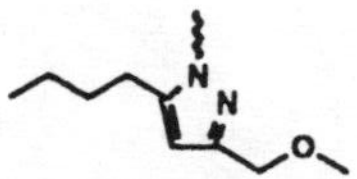	290	—	50	15–10	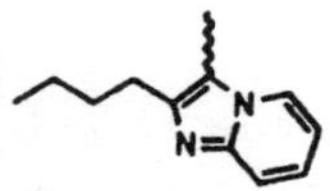	7,400	—	32
15–5	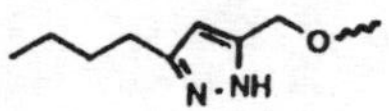	4,500	—	50	15–11	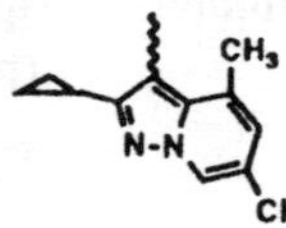	—	X[a]	52

TABLE 15 (continued)
2′-Carboxy (or 5-tetrazolyl)biphenylymethyl — nonimidazoles

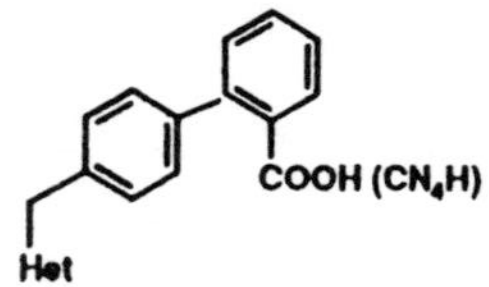

No.	Het	IC50(nM) COOH	IC50(nM) Tetrazole	Ref.	No.	Het	IC50(nM) COOH	IC50(nM) Tetrazole	Ref.
15–6	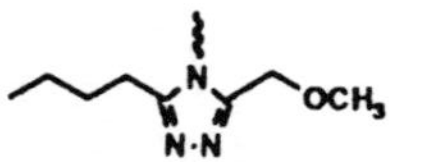	730	140	50	15–12	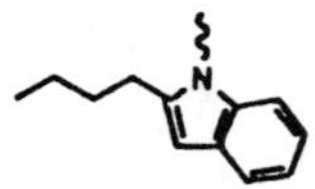	51,000	—	32
15–13	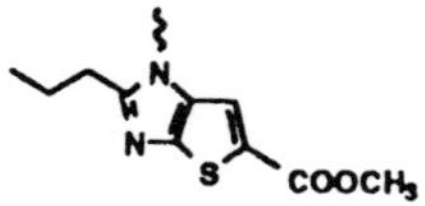	—	140		15–21	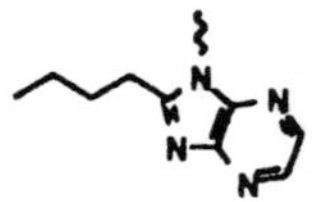	1,000	—	32
15–14	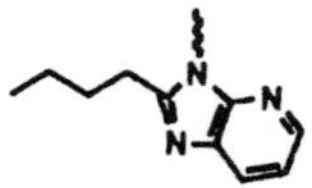	200	12	53	15–22	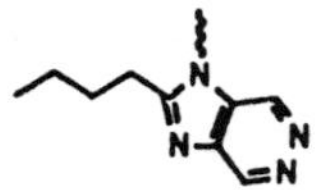	54% at $10^{-5}M$	—	32
15–15	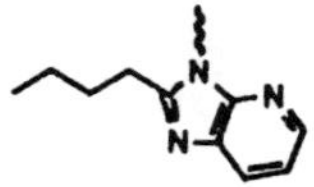	940	—	53	15–23	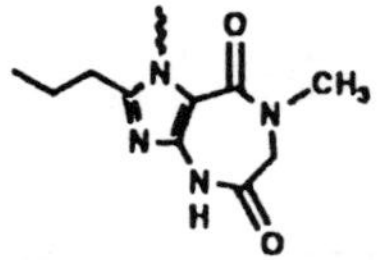	—	X[a]	54
15–16	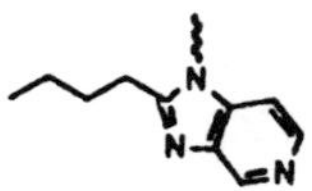	1500	—	53	15–24	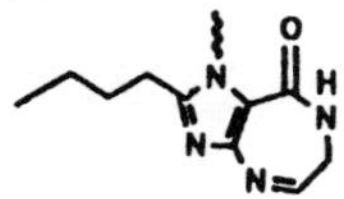	—	X[a]	54

TABLE 15 (continued)
2′-Carboxy (or 5-tetrazolyl)biphenylymethyl — nonimidazoles

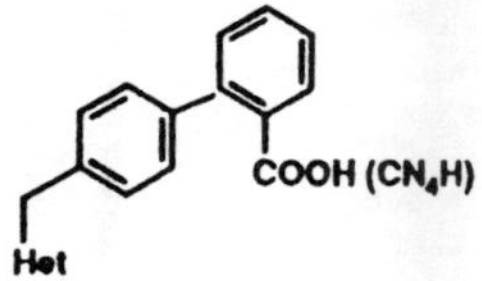

No.	Het	IC_{50}(n*M*) COOH	IC_{50}(n*M*) Tetrazole	Ref.	No.	Het	IC_{50}(n*M*) COOH	IC_{50}(n*M*) Tetrazole	Ref.
15–17	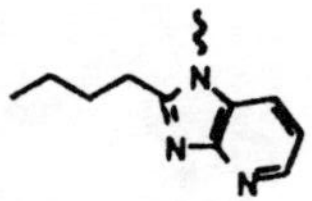	2,300	—	53	15–25	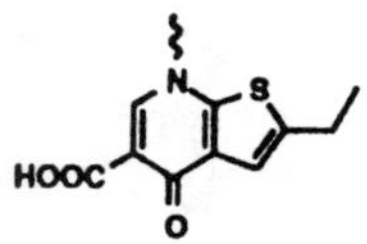	—	X[a]	55
15–18	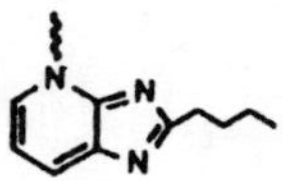	>10,000	—	53	15–26	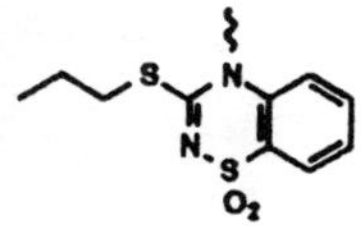	180	4.8 (k_i)	26
15–19	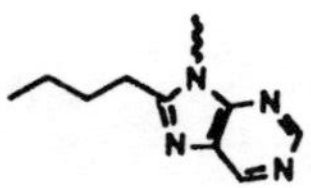	940	620	53	15–27	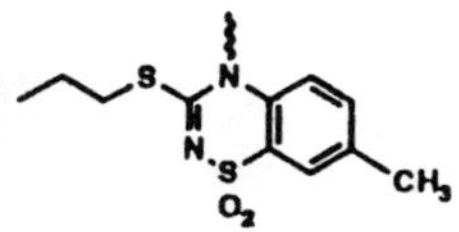	—	0.6 (Ki)	26,56
15–20	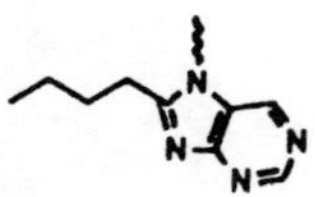	10,000	—	53	15–28	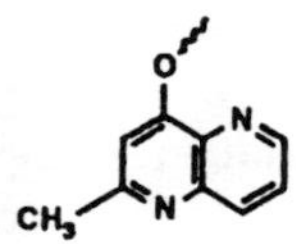	240	24	21,57

TABLE 15 (continued)
2′-Carboxy (or 5-tetrazolyl)biphenylymethyl — nonimidazoles

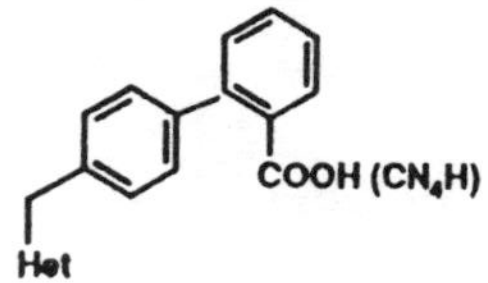

No.	Het	IC_{50}(n*M*) COOH	IC_{50}(n*M*) Tetrazole	Ref.	No.	Het	IC_{50}(n*M*) COOH	IC_{50}(n*M*) Tetrazole	Ref.
15–29	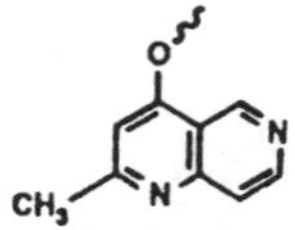	—	605	22	15–37	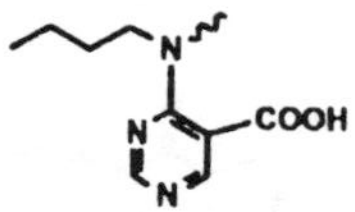	—	$pA_2 = 7.9$	24
15–30	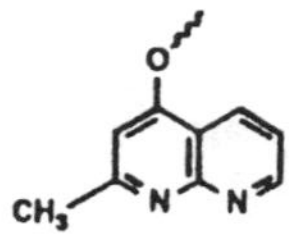	—	10,000	22	15–38	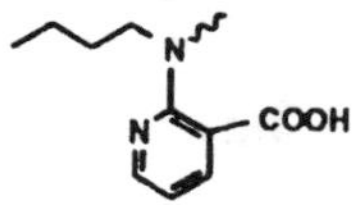	—	$pA_2 = 9.5$	24
15–31	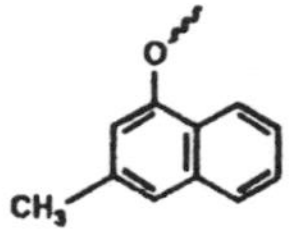	—	90,000	22	15–39	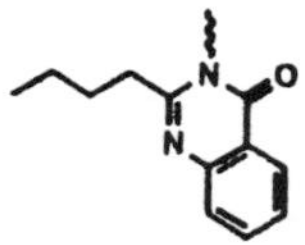	230	X[a]	58,59
15–32	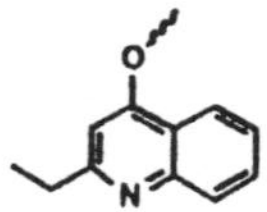	—	31	57	15–40	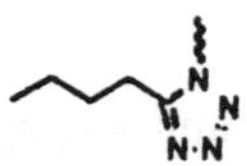	X[a]	—	60

TABLE 15 (continued)
2′-Carboxy (or 5-tetrazolyl)biphenylymethyl — nonimidazoles

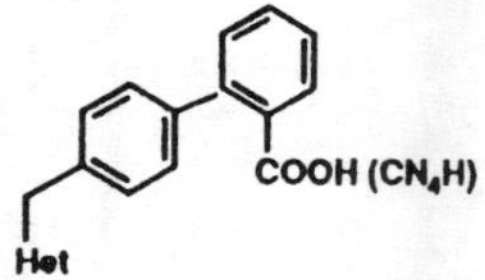

No.	Het	IC₅₀(nM)		Ref.	No.	Het	IC₅₀(nM)		Ref.
		COOH	Tetrazole				COOH	Tetrazole	
15–33	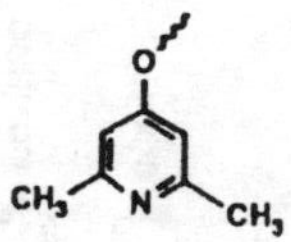	—	X[a]	61	15–41	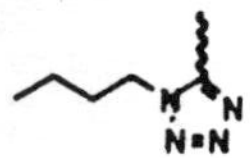	X[a]	—	60
15–34	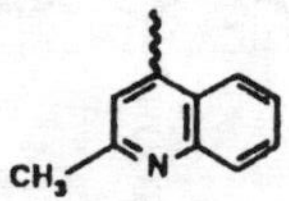	—	16	22	15–42	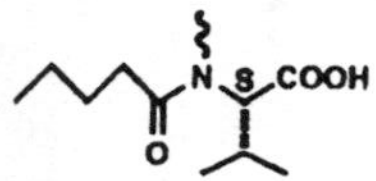	—	2.7	16
15–35	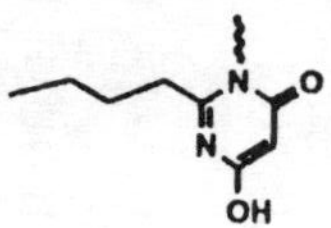	X[a]	—	62,63	15–43	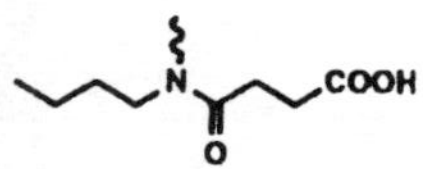	—	X[a]	64
15–36	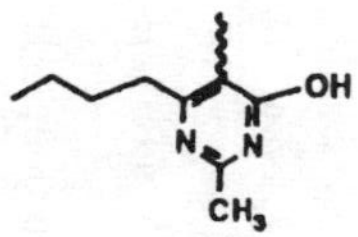	X[a]	—	56	15–44	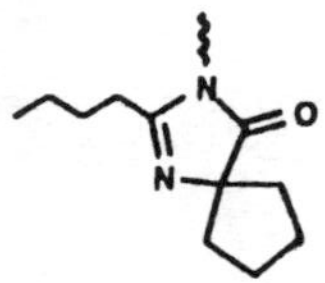	—	X[a]	65

[a] Comparable data not available.

TABLE 16
Benzmidazoles 1

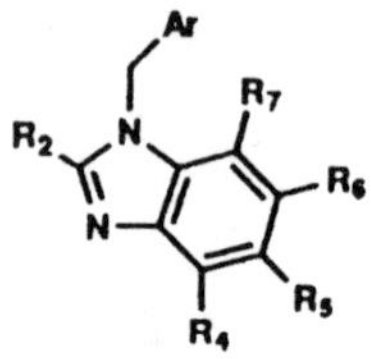

No.	R_2	R_4	R_5	R_6	R_7	Ar	IC_{50}(n*M*)	i.v. AII rat ED_{50}mg/kg	Ref.
16–1	n-Pr	H	H	H	H	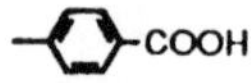	11,000		32
16–2	n-Bu	H	H	H	H	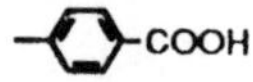	1,800	12.6	32
16–3	n-Hex	H	H	H	H	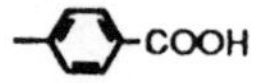	640	>20	32
16–4	n-Bu	H	Cl	Cl	H	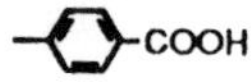	1,600		32
16–5	n-Bu	H	H	H	H	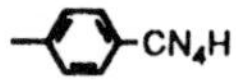	5,600		32
16–6	n-Bu	H	H	H	H	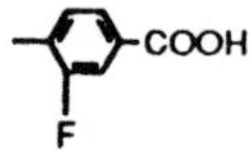	860		32
16–7	n-Bu	H	H	H	H	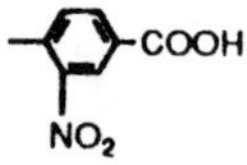	1,800		32
16–8	n-Bu	H	H	H	H	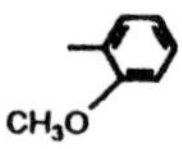	18,000		32
16–9	n-Bu	H	H	H	H	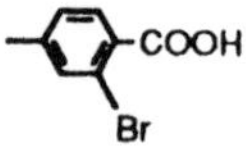	3,900	5.0	32
16–10	n-Bu	H	H	H	H	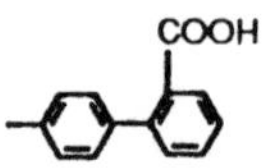	2,300	13.8	32
16–11	n-Bu	H	H	H	H	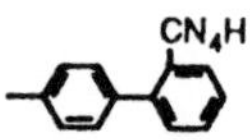	96	5.2	32
16–12	H	H	H	H	H	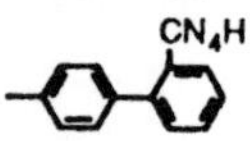	140,000		32
16–13	n-Bu	H	H	COOH	H	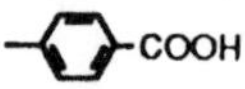	1,500		66
16–14	n-Bu	H	H	H	H	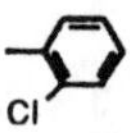	18,200		67
16–15	n-Bu	H	H	Cl	COOH	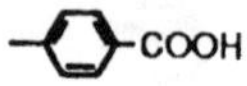	120		67
16–16	n-Bu	H	H	H	H		X[a]		68

TABLE 16 (continued)
Benzmidazoles 1

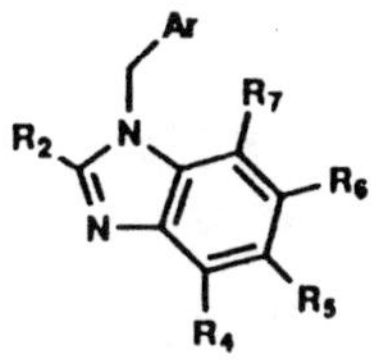

No.	R_2	R_4	R_5	R_6	R_7	Ar	IC_{50}(n*M*)	i.v. AII rat ED_{50}mg/kg	Ref.
16–17	EtO	H	H	H	COOH	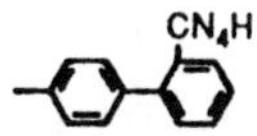	112		69
16–18	EtO	H	H	H	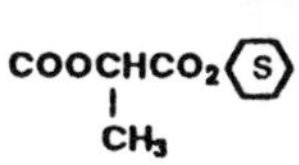	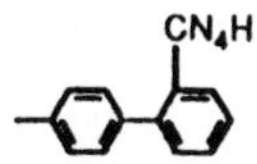	X[a]		69
(TCV-116, PRODRUG of 16-18)									
16–19	n-Pr	4,5,6,7-tetrahydro			COOEt	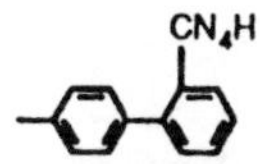	X[a]		70
16–20	n-Bu	H	H	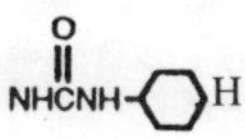	H	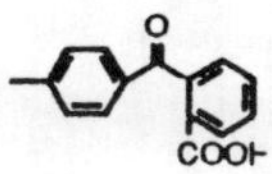	X[a]		71

[a] Comparable data not available.

(**19–5**) is a 4-(2′-(4-tetrazolyl)biphenylylmethyl compound with a carboxylic acid-containing 5-sidechain. In this class, the 2′-tetrazolylbiphenyl compounds are only slightly more potent than the corresponding 2′-carboxybiphenyl analogs. However, in the 1,2,4-triazoles, **19–9** and **19–10,** the tetrazole, is 17 times more potent than the carboxylic acid. In this series, a number of 3,5-dialkyl compounds are angiotensin II receptor antagonists in the 3 to 5 n*M* range. Furthermore, **19–9** is orally active. Compound **19–11** is the gamma glutamyl derivative of the 2′-carboxylic acid hydrazide. This compound has been designed to be a renal selective angiotensin II receptor antagonist, in that the renal enzyme, glutaminase, may cleave this compound to yield the corresponding hydrazide in the kidney. This compound can be further converted to **19–10,** which has modest AT_1 receptor antagonist activity. Replacement of the biphenyl ring system of **19–9** and **19–10** by a pyrrolyphenyl group yields **19–16** and **19–17,** which have lower affinity for angiotensin II receptors.

G. IMIDAZOPYRIDINES

Inspection of the structure–activity relationships of the imidazole angiotensin II receptor antagonists indicates that affinity for angiotensin II receptors is enhanced by the presence of groups capable of hydrogen-bond interactions at the

TABLE 17
1-Biphenylmethyl-6-Functionalized Imidazoles

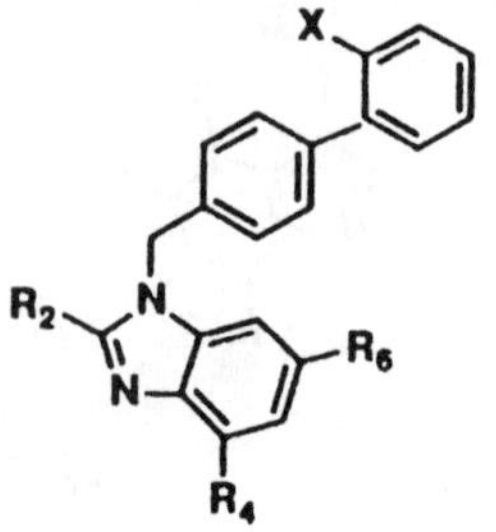

No.	X	R_2	R_4	R_6	IC_{50}(n*M*)	Ref.
17–1	COOH	Pr	Me	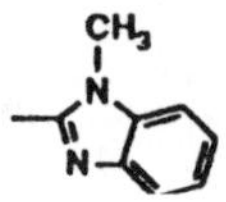	3.7	72
17–2	CN_4H	Pr	Me		14	72
17–3	COOH	Pr	Me	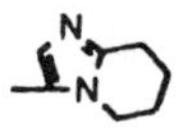	1.2	72
17–4	CN_4H	Pr	Me		1.7	72
17–5	CN_4H	Pr	Me	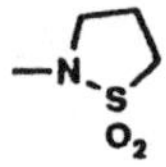	1.2	72
17–6	COOH	Bu	Me		3.5	72
17–7	COOH	Bu	H		29	17
17–8	COOH (3,5diCl)	Bu	H		20	17
17–9	COOH	Bu	H	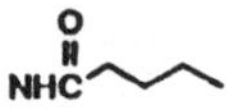	X[a]	73
17–10	COOH	Bu	H	CH_2OH	X[a]	74
17–11	COOH	Pr	H	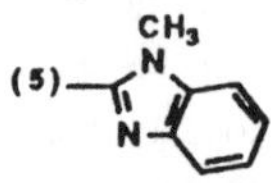	X[a]	75

[a] Comparable data not available.

imidazole-5 position. This observation suggests that fusing the imidazole to another heterocyclic ring with a heteroatom corresponding to a substituent in the imidazole-5-position would result in a compound with high affinity. Indeed, heterocyclic imidazole replacements designed using this rationale yielded potent angiotensin II receptor antagonists. As predicted, of the imidazolopyridine isomers shown in Table 20, **20–5,** the imidazo[4,5-b]pyridine with the pyridine nitrogen in the 5-position of the fused imidazole, has the highest affinity for angiotensin II receptors. The tetrazole analog **20–6** has 20-fold higher affinity than the carboxylic acid, an effect similar to that seen in the imidazole series.

TABLE 18
Pyrazoles

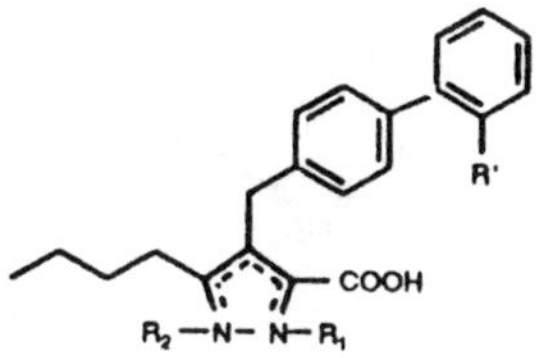

No.	R_1	R_2	R′	IC_{50}(n*M*)	pK_B	Ref.
18–1	H	—	CN_4H	2.0	8.8	76,77,78
18–2	CF_3CH_2	—	CN_4H		9.8	77
18–3	—	CF_3CH_2	CN_4H		7.7	77
18–4	Et	—	CN_4H		9.8	77
18–5	Et	—	COOH		8.0	77
18–6	Et	—	$NHSO_2CF_3$		8.0	77
18–7	Pr	—	CN_4H		10.0	77
18–8	i-Bu	—	CN_4H		11.0	77
18–9	t-$BuCH_2$	—	CN_4H		10.6	77
18–10		—	CN_4H		10.5	77
18–11	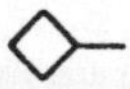	—	CN_4H		10.9	77
18–12	C_6H_5	—	CN_4H	2.0		78
18–13	2,6-di-ClC_6H_5	—	CN_4H	0.18		78
18–14	$C_6H_5CH_2$	—	CN_4H	0.20		78
18–15	$C_6H_5CH_2CH_2$	—	CN_4H	0.24		78

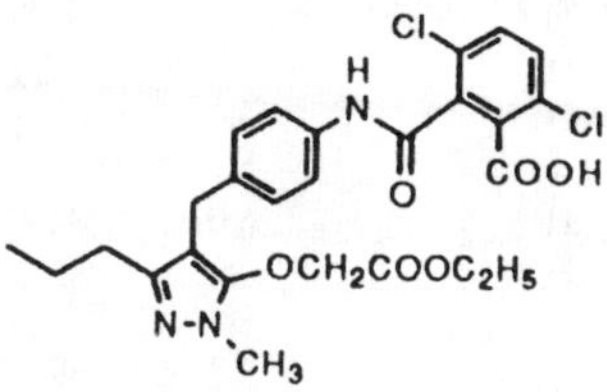

18–16[79]

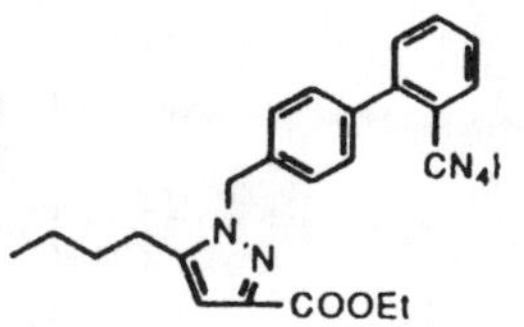

18–17[80]

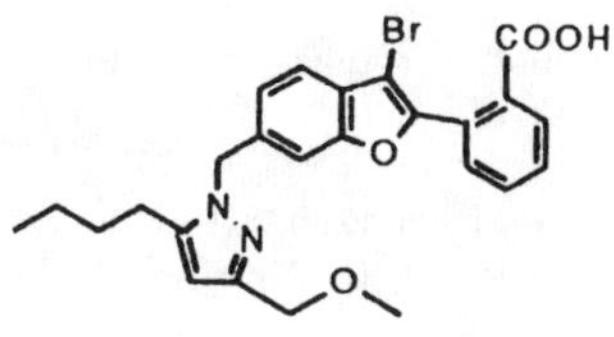

18–18[81]

As shown in Table 21, modification of **20–6** by optimizing the 2-alkyl substituent and adding appropriate substituents to the pyridine ring gave compounds with subnanomolar affinity for angiotensin II receptors. Both the 5-methyl analog **21–2** and the 7-methyl analog **21–4**[18,19] have higher affinity than the unsubstituted analog **21–1.** The effect of methyl substitution is additive in that the 5,7-dimethyl compound **21–5** is more active than either of the monomethyl analogs. Based on affinity for angiotensin II receptors, the 2-ethyl derivative **21–6** and its propyl congener **21–5** have equal activity, and both compounds have higher affinity than the butyl derivative. However, **21–6** (L 158809) is the most potent compound of this series *in vivo:* i.v. ED_{50} in inhibiting the pressor response to angiotensin II of 0.048 mg/kg vs. 0.087 mg/kg for **21–5.** Substituents in the 6-position (i.e., **21–3, 21–8,** and **21–9**) decrease affinity for angiotensin II receptors.

TABLE 19
Triazoles

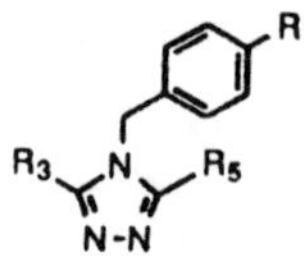

No.	R_3	R_5	R	IC_{50}(n*M*)	Ref.
19–1	Bu	Bu	2-COOH - C_6H_4	305	82
19–2	Bu	Bu	2-CN_4H - C_6H_4	110	82
19–3	Bu	CH_2OCH_3	2-COOH - C_6H_4	730	83
19–4	Bu	CH_2OCH_3	2-CN_4 - C_6H_4	140	83
19–5	Bu	S-CH_2–2-COOH-C_6H_4	2-CN_4 - C_6H_4	1.5	84
19–6	Bu	=O(N_4Bu)	2-COOH - C_6H_4	101	85
19–7	Bu	S-4Cl-C_6H_4	OCH(COOH) - C_6H_5	X	86
19–8	Bu	S-CH_2COOCH_3	NHCO(2-COO-C_6H_4)	X	87

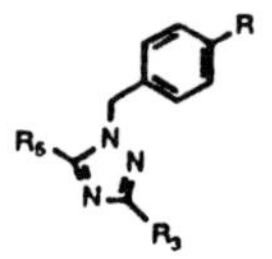

No.	R_3	R_5	R	IC_{50}(n*M*)	Ref.
19–9	Bu	Bu	2-CN_4H - C_6H_4	5.6	82
19–10	Bu	Bu	2-COOH - C_6H_4	95	82
19–11	Bu	Bu	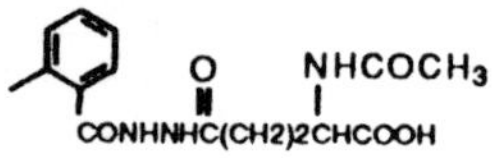	X	82,88
19–12	t-Amyl	Bu	2-CN_4H - C_6H_4	4.0	82
19–13		Bu	2-CN_4H - C_6H_4	4.4	82
19–14	1-adamantyl	Bu	2-CN_4H - C_6H_4	5.0	82
19–15	$C_6H_5CH_2CH_2$	Bu	2-CN_4H - C_6H_4	3.2	82
19–16	Bu	Bu	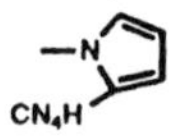	33	89
19–17	Bu	Bu	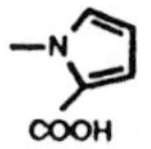	270	89
19–18			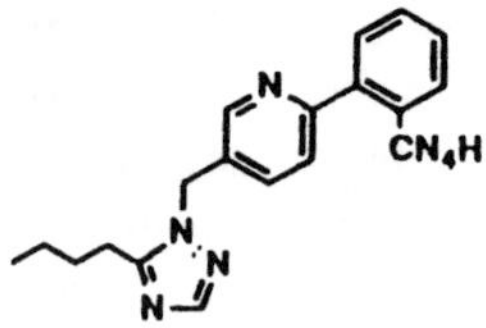	2.8	90

TABLE 20
2-Butyl-1-(2′-carboxybiphenylylmethyl)imidazopyridines[53]

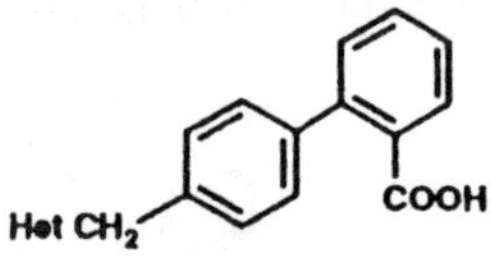

No.	Het	IC_{50}(n*M*)
20–1	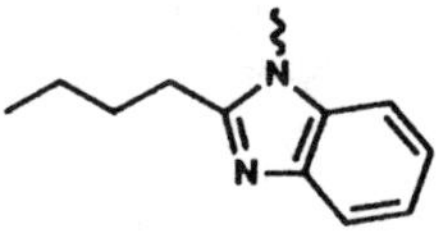	650
20–2	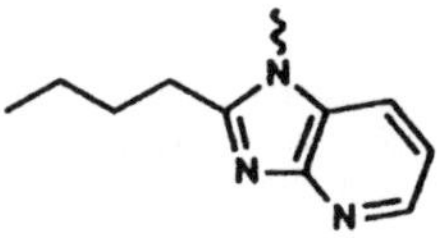	2,300
20–3	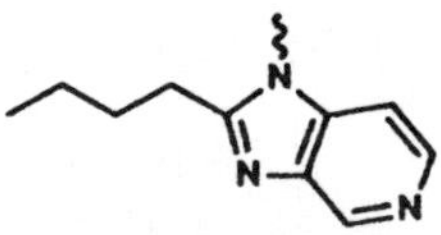	1,500
20–4	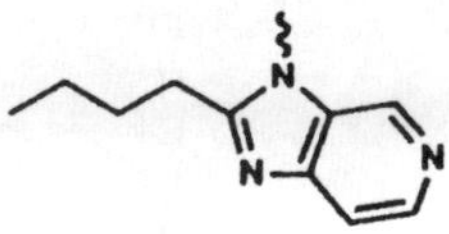	940
20–5	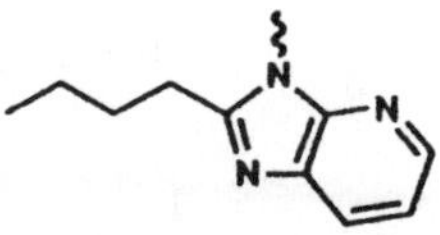	200
20–6	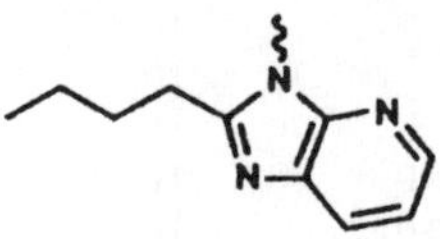 (2'-tetrazole)	10

Of the 7-substituents studied to date (**21–6** and **21–11** to **21–14**), none is superior to the methyl. Substitution at the 5-position maintains high affinity for angiotensin II receptors with a variety of functional groups. The secondary amine **21–19** has higher affinity than **21–6** and shows a duration of action of more than 24 h in inhibiting the pressor response to angiotensin II when administered at a dose of 0.3 mg/kg, i.v., in the rat. Even the biotin amide **21–22** has nanomolar affinity for angiotensin II receptors. Another interesting compound is the 5-carboxylic acid, **21–26.** In the imidazopyridine series, this substitution is not markedly more active than the hydroxymethyl substitution **21–17,** in contrast to the situation for the imidazole series.

A number of imidazo[4,5-b] pyridines with novel substituents are shown in Table 22. Some of these were reported in the patent literature, and little

TABLE 21
2-Alkyl-3-(2′-(5-tetrazolyl)biphenylylmethyl) imidazo[4,5-b]pyridines[18,19]

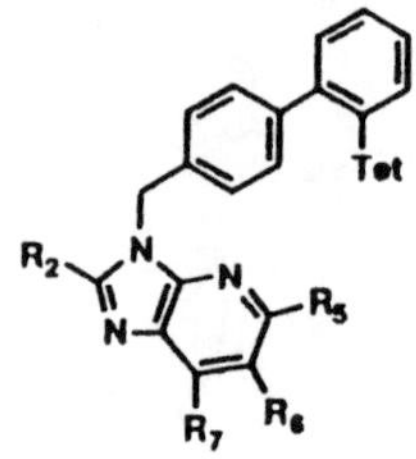

No.	R_2	R_5	R_6	R_7	IC_{50}(n*M*)
21–1	Pr	H	H	H	8
21–2	Pr	CH_3	H	H	2
21–3	Pr	H	CH_3	H	22
21–4	Pr	H	H	CH_3	1
21–5	Pr	CH_3	H	CH_3	0.3
21–6	Et	CH_3	H	CH_3	0.3
21–7	Bu	CH_3	H	CH_3	0.5
21–8	Et	CH_3	NH_2	CH_3	4
21–9	Et	CH_3	Cl	CH_3	3
21–10	Et	CH_3	H	H	2
21–11	Et	CH_3	H	Et	2
21–12	Et	CH_3	H	CH_2OH	5
21–13	Et	CH_3	H	$NHCH_3$	12
21–14	Et	CH_3	H	$N(CH_3)_2$	270
21–15	Et	$CH(CH_3)_2$	H	CH_3	0.5
21–16	Et	$CH_2C_6H_5$	H	CH_3	0.2
21–17	Et	CH_2OH	H	CH_3	0.7
21–18	Et	NH_2	H	CH_3	0.5
21–19	Et	$NHCH_3$	H	CH_3	0.2
21–20	Et	$NH(CH_2)_5CH_3$	H	CH_3	0.5
21–21	Et	$NH(CH_2)_6NH_2$	H	CH_3	0.3
21–22	Et	$NH(CH_2)_6NHB$[a]	H	CH_3	0.92
21–23	Et	OH	H	CH_3	2
21–24	Et	Br	H	CH_3	4
21–25	Et	CN	H	CH_3	7
21–26[91]	Et	COOH	H	CH_3	0.5

[a] B = Biotin.

quantitative biological data are available. However, a number of others with novel 2-substituents were reported with more detailed biological data.[20] The more interesting 2-substituents include cyclopropyl, propynyl, and ethoxy. Compound **22–20,** the 2-ethoxy analog, has pA_2 in rabbit aorta of 10.58 and shifts the AII dose-response curve twofold at a dose of 0.0084 mg/kg in the anesthetized rat. The corresponding 2-propyl analog has similar *in vitro* potency but is three times less potent *in vivo.*

TABLE 22
3-(4-substitutedbenzyl)imidazo[4,5-b]pyridines

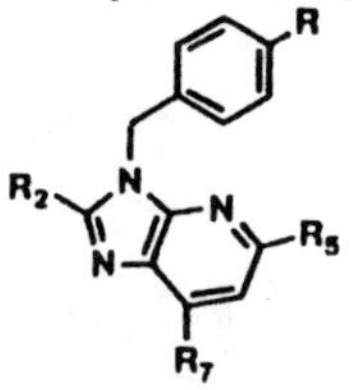

No.	R_2	R_5	R_7	R	pA_2[a]	i.v. C (mg/kg)[b]	Ref.
22–1	Et	COOH	CH_3	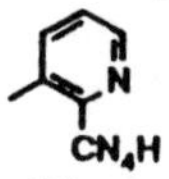	X[c]		92
22–2	n-Pr	CH_3	CH_3	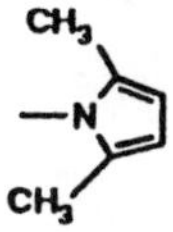	X[c]		93
22–3	n-Pr	CH_3	CH_3	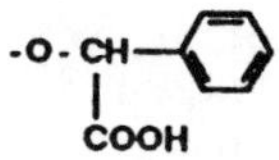	X[c]		94
22–4	Et	CH_3	CH_3	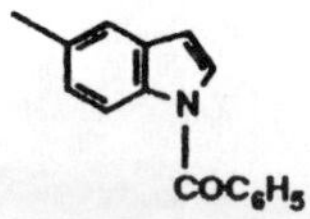	X[c]		95
				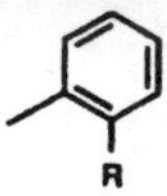			
22–5	n-Bu	CH_3	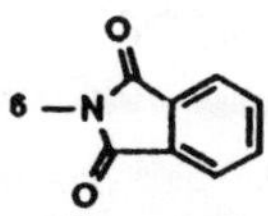	COOH	X[c]		96
22–6	Et	H	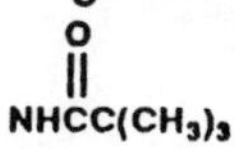	CN_4H	X[c]		96
22–7	n-Bu	H	H	CN_4H	8.11	0.74	20
22–8	n-Pr	H	H	CN_4H	9.08	0.15	20
22–9	n-Pr	H	CH_3	CN_4H	10.33	0.029	20
22–10	n-Pr	CH_3	CH_3	CN_4H	10.62	0.040	20
22–11	n-Pr	CH_3	CH_3	COOH	8.94	0.080	20
22–12	Et	H	CH_3	CN_4H	10.04	0.019	20
22–13	Et	H	CH_3	COOH	8.50	0.29	20
22–14	cyclo-Pr	H	CH_3	CN_4H	10.64	0.022	20
22–15	cyclo-Pr	H	CH_3	COOH	8.67	0.075	20
22–16	cyclo-Bu	H	CH_3	COOH	8.31	0.25	20
22–17	$CH_3C{\equiv}C$—	H	CH_3	CN_4H	9.88	0.040	20
22–18	EtS	H	CH_3	CN_4H	9.74	0.04	20
20–19	MeO	H	CH_3	CN_4H	9.88	0.016	20
22–20	EtO	H	CH_3	CN_4H	10.58	0.0084	20
22–21	n-Pro	H	CH_3	CN_4H	9.43	0.034	20

[a] Rabbit aorta.

[b] Dose that shifts pressor response of AII twofold to the right.

[c] Comparable data not available.

TABLE 23
Purines[53]

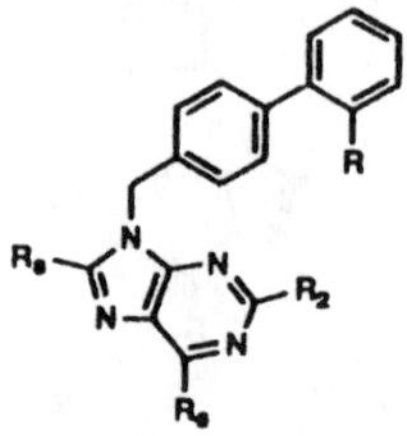

No.	R_2	R_6	R_8	R	IC_{50}(n*M*)
23–1	H	H	Bu	CN_4H	620
23–2	H	H	Pr	CN_4H	180
23–3	H	Cl	Pr	COOH	450
23–4	H	Cl	Pr	CN_4H	21
23–5	H	CH_3S	Pr	CN_4H	40
23–6	H	CH_3NH	Pr	CN_4H	23
23–7	H	CH_3	Pr	CN_4H	17
23–8	Cl	CH_3	Pr	CN_4H	28
23–9	CH_3S	CH_3	Pr	CN_4H	8
23–10	$(CH_3)_2N$	CH_3	Pr	CN_4H	3
23–11	CH_3NH	CH_3	Pr	CN_4H	0.9
23–12	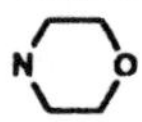	CH_3	Pr	CN_4H	0.7

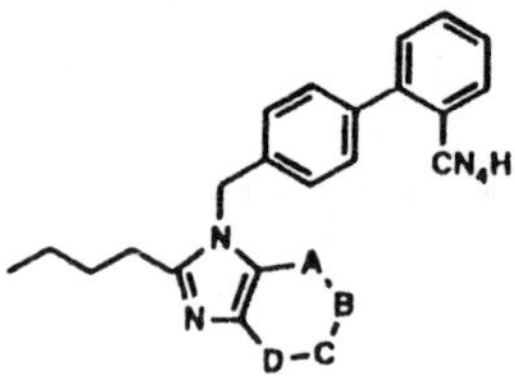

No.	A	B	C	D	IC_{50}(n*M*)
23–13[97]	C=O	NH	C	N	66
23–14[98]	C=O	NCH_3	C=O	NCH_3	280
23–15[99]	C=O	NH	C=O	NH	X[a]
23–16[53]	CH	CH (2-Propyl)	CH	CCH_3	12

[a] Comparable data not available.

H. PURINES

Purines were conceived as benzimidazole analogs with additional heteroatoms that could provide added interaction with angiotensin II receptors and thus improve pharmacological properties. Although **23–7** (Table 23) and its benzimidazole analog, **23–16,** have comparable affinity for angiotensin II receptors, electron-supplying substituents in the two position of the purine enhance activity.

Some of the more active compounds in this series are orally active. Thus, **23–10,** when administered at a dose of 1 mg/kg, either orally or intravenously, inhibits the pressor effect of angiotensin II in the rat for more than 6 h. The oxopurines **23–13** and **23–14** have modest affinity for angiotensin II receptors, albeit less than that of the similar benzimidazole **23–16.**

I. PYRIDINES

The 2-(2′-tetrazolylbiphenylmethylamino)-3-carboxypyridines are potent AT_1 receptor antagonists. The N-propyl analog **24–2** (Table 24) is more potent in inhibiting the contractile effect of angiotensin II in the rabbit aorta than either the corresponding N-ethyl or N-butyl analogs **24–1** and **24–3.** The 6-methyl group in **24–4** reduces antagonist activity at AT_1 receptors by more than one order of magnitude, possibly because it may interfere with the H-bonding activity of the pyridine nitrogen. Compound **24–2** has an oral bioavailability of 97% in the rat, with a $t_{1/2}$ of 11.4 h. Following oral administration at 0.3 mg/kg to renal hypertensive rats, **24–2** lowered mean arterial blood pressure by 32 to 35% for at least 24 h.

The related 4-aminopyridine analogs also show high affinity for angiotensin II receptors. It is remarkable that **24–5** has only the one acidic group, which is in the biphenyl sidechain. The compound inhibits the pressor effect of angiotensin II with an ED_{50} of 0.13 mg/kg, i.v., and 1 mg/kg, p.o., and is effective in this test for >7 h, suggesting acceptable bioavailability.

Piperidin-2-one biphenyltetrazoles with two small alkyl substituents at the 4-position have reasonable affinity for both AT_1 and AT_2 receptors and are also potent in inhibiting angiotensin II-induced contraction in the rabbit aorta. The radioligand binding studies were carried out using bovine adrenal cortex membranes, which as seen with losartan, yield substantially higher IC_{50} values than those obtained using rat adrenal cortex membranes (420 vs. 19 n*M*). The compounds with the highest affinity for angiotensin II receptors are disubstituted in the 4-position of the piperidine ring, with the 4,4-diethyl compound, **24–10,** and the spirocyclohexyl analog, **24–12,** being the most active. The 4-ethyl-4-methyl analog **24–9** is the most potent antagonist of angiotensin II-induced contraction in the rabbit aorta, where the compound exhibits insurmountable antagonism. This compound lowers blood pressure in the spontaneously hypertensive rat with a duration of action of 24 h, similar to that of losartan.

The structure–activity relationships of biphenylylmethoxy pyridines are discussed in a separate section of this chapter with other similar compounds (Table 26).

J. QUINOLINES

Biphenylylmethyl groups have been placed onto quinolines in a variety of ways. A number of compounds in which the linkage is at the quinoline 4-position are shown in Table 25. One of those compounds with the highest affinity for angiotensin II receptors is **25–2,** in which the link is an oxymethylene. The

TABLE 24
Pyridines[100]

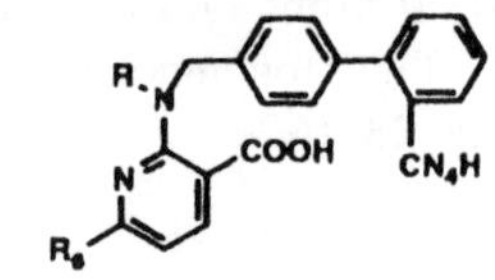

No.	R	R_6	Ki (n*M*)	pA_2	Ref.
24–1	CH_2CH_3	H	—	8.76	24
24–2	$CH_2CH_2CH_3$	H	0.76	10.4	24
24–3	$CH_2CH_2CH_2CH_3$	H	—	9.5	24
24–4	$CH_2CH_2CH_2CH_3$	CH_3	122	8.15	24
	Losartan		14.6	8.5	24

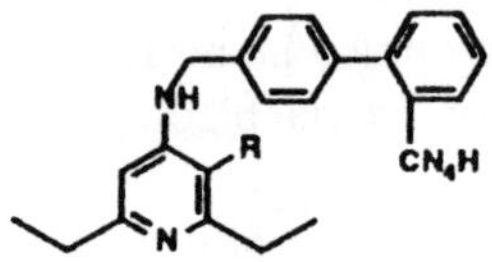

No.	R	IC_{50}(n*M*)	Ref.
24–5	I	4	21
24–6	$COOCH_3$	8	21

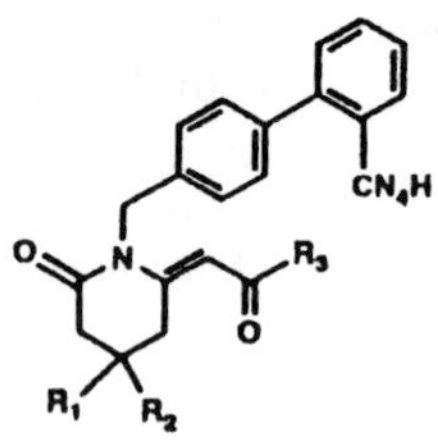

No.	R_1	R_2	R_3	Bovine adrenal IC_{50}(n*M*)	Rabbit aorta pA_2	Ref.
24–7	CH_3	H	OC_2H_5	200	6.9	100
24–8	CH_3	CH_3	OC_2H_5	5,800	7.4	100
24–9	CH_3	C_2H_5	OC_2H_5	250	9.0	100
24–10	C_2H_5	C_2H_5	OC_2H_5	20	8.1	100
24–11	C_2H_5	n-C_3H_7	OC_2H_5	90	7.9	100
24–12	—$(CH_2)_5$—		OC_2H_5	40	8.3	100
24–13	C_2H_5	C_2H_5	OCH_3	210	8.4	100
24–14	C_2H_5	C_2H_5	CH_3	3,100	6.8	100
	Losartan			420	8.8	100

TABLE 25
Quinolines

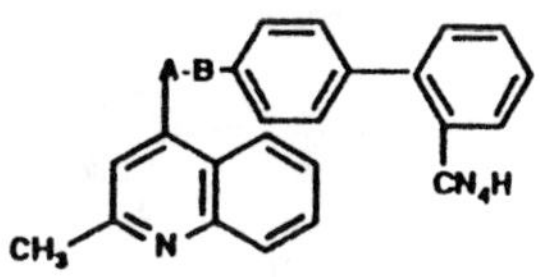

No.	A - B	IC_{50}(n*M*)	Ref.
25–1	CH_2	17	101
25–2	OCH_2	16	22
25–3	$OCH(CH_3)$	40	22
25–4	SCH_2	37	22
25–5	CH=CH (2/1 E to Z)	1,300	22
25–6	CH_2CH_2	270	22

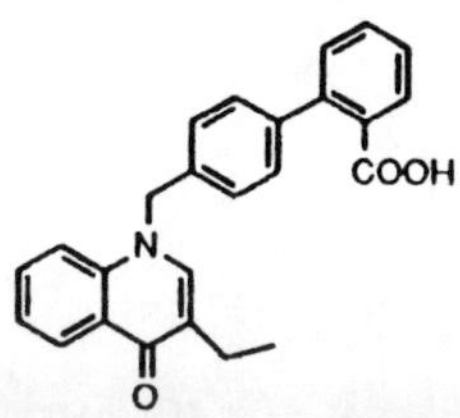

25–7[22]
IC_{50} 1400 n*M*

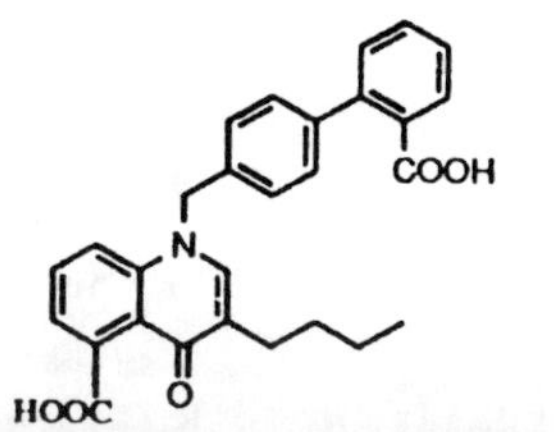

25–8[102]
IC_{50} 92n*M*

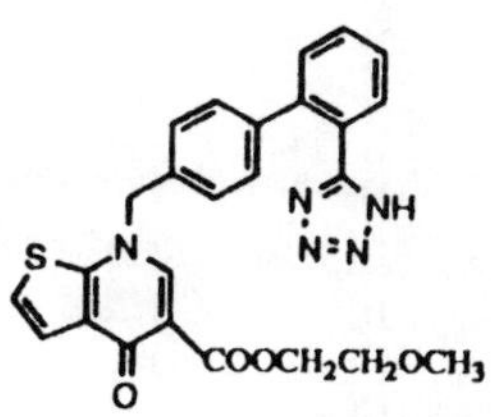

25–9[103]
IC_{50} 10 n*M*

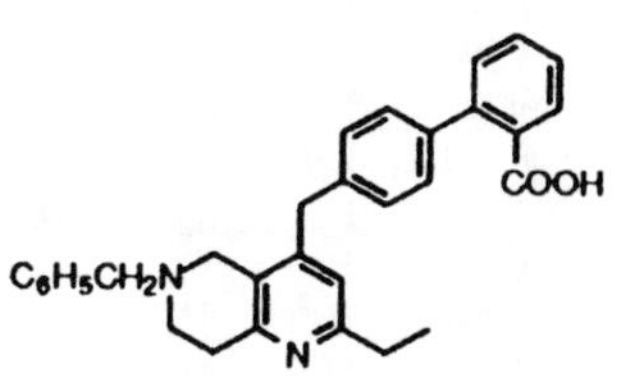

25–10[104]
X[a]

[a] Comparable data not available.

structure–activity relationships of this series of compounds are presented in Table 26. Compounds **25–7** to **25–9** are 4-quinolines with the biphenylylmethyl group attached to the nitrogen. The patent describing **25–10** did not disclose detailed biological data.

K. BIPHENYLYLMETHOXY HETEROCYCLES

The biphenylylmethoxy group has been studied using pyridine, quinoline, and napthyridine as the basic heterocycle (Table 26). At least in the quinoline series, the 2′-tetrazolylbiphenyl compounds are approximately 1 order of magnitude higher in affinity for angiotensin II receptors than the corresponding carboxylic

TABLE 26
Biphenylylmethoxy Heterocycles
(Pyridines, Quinolines, Napthyridines)

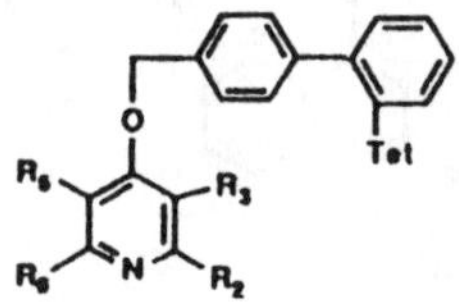

No.	R_2	R_3	R_5	R_6	IC_{50}(n*M*)	Ref.
26–1	H	H	H	H	7	105
26–2	CH_3	CH_3	H	C_2H_5	300	105
26–3	—$(CH_2)_3$—		H	C_2H_5	7	21
26–4	—$(CH_2)_4$—		H	C_2H_5	5	21
26–5	CH_3	$COOCH_3$	H	C_2H_5	5	21
26–6	CH_3	$COOCH_3$	H	C_2H_5	5	21

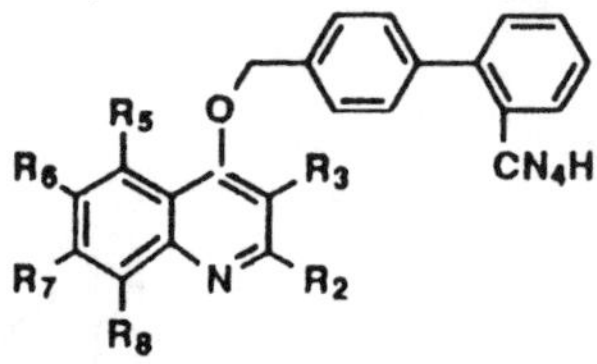

No.	R_2	R_3	R_5	R_6	R_7	R_8	IC_{50}(n*M*)	Ref.
26–7	H	H	H	H	H	H	6,300	22
26–8	CH_3	H	H	H	H	H	16	22
26–9	C_2H_5	H	H	H	H	H	31	22
26–10	n-Pr	H	H	H	H	H	60	22
26–11	CH_3	CH_3	H	H	H	H	4,600	22
26–12	C_2H_5	H	CH_3	H	H	H	13	22
26–13	C_2H_5	H	Cl	H	H	H	120	22
26–14	C_2H_5	H	H	Cl	H	H	1,200	22
26–15	C_2H_5	H	H	H	Cl	H	160	22
26–16	C_2H_5	H	H	H	H	Cl	140	22
26–17	C_2H_5	H	H	OCH_3	H	H	320	22
26–18	C_2H_5	H	H	H	OCH_3	H	220	22
26–19	C_2H_5	H	H	H	H	OCH_3	960	22
26–20	C_2H_5	H	CN	H	H	H	60	22
26–21	C_2H_5	H	H	OCH_2CH_2F	H	H	7	22
26–22	C_2H_5	H	H	OCH_2CF_3	H	H	26	22
26–23	C_2H_5	H	H	$OCH_2CH_2NH_2$	H	H	X[a]	106

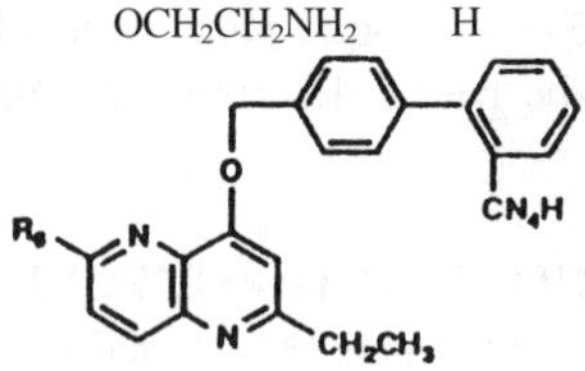

No.	R_6	IC_{50}(n*M*)	Ref.
26–24	H	24	21
26–25	$OCH(CH_3)_2$	8	21,105,23
26–26	$OCH_2CH_2OCONHC_2H_5$	18	107

[a] Comparable data not available.

acids. One compound that stands out in the pyridine series is **26–4,** which has a pA_2 of 10.3 in the rabbit aorta and, when dosed orally at 3 mg/kg, lowers blood pressure in the renal hypertensive rat for >7 h. This compound, referred to as D 6888, is in clinical studies.[21]

In the quinoline series, a small 2-alkyl group is necessary for high affinity for angiotensin II receptors. Electron-donating or electron-attracting substituents on the benzo ring do not improve potency. Compound **26–9** has an ED_{50} of 1 mg/kg, i.v., for antagonizing the pressor effect of angiotensin II in the rat, and at a dose of 10 mg/kg, i.v., has a duration of action >6 h. When dosed orally at 5 mg/kg to renal hypertensive rats, **26–9** normalizes blood pressure within 2 h of dosing, and an antihypertensive effect was still evident 24 h after dosing.[22]

In the napthyridine series, optimal activity is found in the 6-isopropyloxy-2-ethyl analog, **26–25.** This compound has a pA_2 of 8.3 in the rabbit aorta, which is similar to the best compounds in the quinoline series.[23]

L. PYRIMIDINES AND QUINAZOLONES

4-(2′-(Tetrazolyl)biphenylylmethylamino)pyrimidines analogous to those reported in the pyridine series are also active (Table 27). The most potent compounds, **27–4** and **27–5,** have a 3- or 4-carbon straight alkyl chain on the exocylic nitrogen, and either a hydrogen or a methyl group at the 2-position of the pyrimidine ring. The 5-carboxy substituent is required for high affinity. Although **27–4** lowers blood pressure when administered to renal artery-ligated hypertensive rats at a dose of 0.3 to 1 mg/kg, i.v., the compound produces a poor antihypertensive response when dosed orally at 1 to 10 mg/kg. The compound has a bioavailability of only 5.8% in fasted rats.[24]

Some dihydropyrimidines also have nanomolar affinity for angiotensin II receptors.[25] The more active compounds are 2-methyl-4-chloro-1,2-dihydropyrimidines and 2-chloro-4,4-dimethyl-1,4-dihydropyrimidines with a 2′-tetrazole in the biphenyl moiety. The 3-carboxylic acid **27–11** is more active than its corresponding ester **27–10.** Although **27–11** is fairly potent in antagonizing the pressor effects of angiotensin II when administered intravenously to the rat, the compound is not active when administered orally. The corresponding ester **27–10,** although less potent *in vitro* and when administered intravenously to the rat, is active when given orally, although it is only approximately one-third as potent as losartan. The tetrazole 1,4-dihydropyrimidine, **27–16,** although having high affinity for angiotensin II receptors, shows only modest antagonism of angiotensin II in the rabbit aorta and does not antagonize the pressor effects of angiotensin II in rats following oral administration.

Quinazolones with appropriate pharmacophores may also have potent angiotensin II receptor affinity (Table 28). Compound **28–5** has the ubiquitous biphenyltetrazole on the nitrogen adjacent to the hydrogen-bond-accepting 4-carbonyl and also a hydrogen-bond- accepting dialkyl urea on the benzo ring. It has subnanomolar angiotensin II receptor affinity and a good duration of action in the rat.

TABLE 27
Pyrimidine

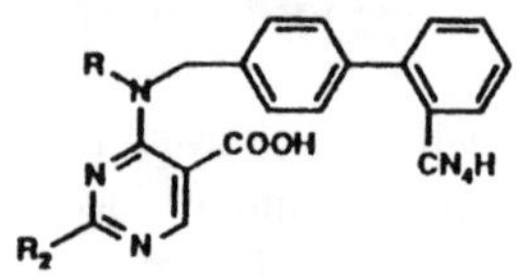

No.	R	R_2	Ki (n*M*) (rat liver)	pA_2 (rabbit aorta)	Ref.
27–1	H	n-Bu	198	7.94	24
27–2	CH_3	n-Bu	—	7.15	24
27–3	n-Bu	CH_3	15.3	8.81	24
27–4	n-Bu	H	3.13	9.93	24
27–5	n-Pr	CH_3	2.78	9.60	24
27–6	H	H	—	6.56	24
	Losartan		14.6	8.5	24

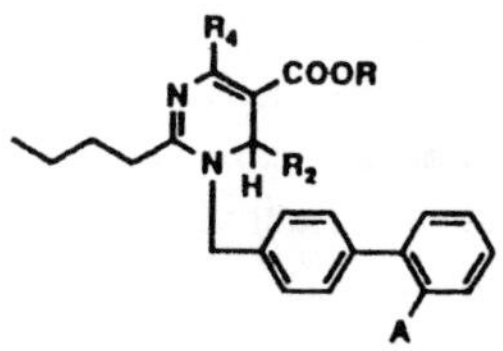

No.	R	R_2	R_4	A	Rat adrenal cortex binding Ki (n*M*)	Rabbit aorta K_b(n*M*)	Ref.
27–7	Et	CH_3	CH_3	COOH	990	660	25
27–8	Et	CH_3	C_6H_5	COOH	850	630	25
27–9	Et	CH_3	Cl	COOH	41	70	25
27–10	Et	CH_3	Cl	CN_4K	8.3	5.3	25
27–11	H	CH_3	Cl	CN_4H	1	0.45	25
27–12	Et	C_6H_5	Cl	COOH	1,430	1,000	25
27–13	Et	═0	H	COO Na	82	610	25
27–14	Et	═0	CH_3	COO Na	150	95	25

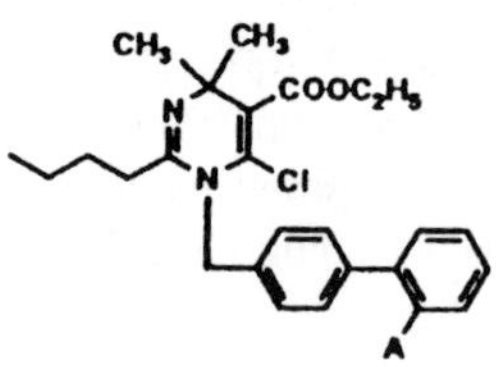

No.	A	Rat adrenal cortex binding Ki(n*M*)	Rabbit aorta K_b (n*M*)	Ref.
27–15	A═COONa	58	23	25
27–16	A═CN_4H	1.1	24	25

TABLE 27 (continued)
Pyrimidine

Reported in patents with no comparable data

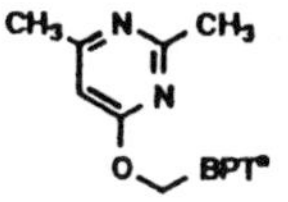
27-17[61]

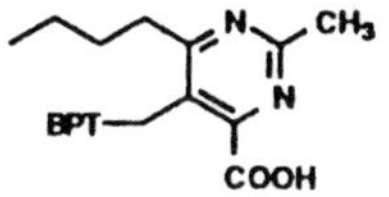
27-18[108]

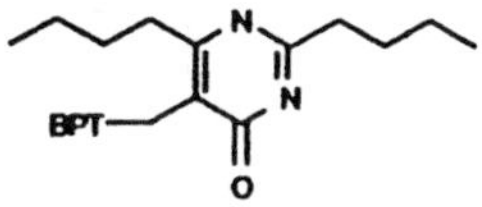
27-19[109]

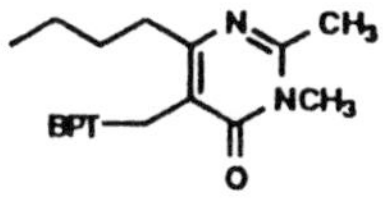
27-20[56]

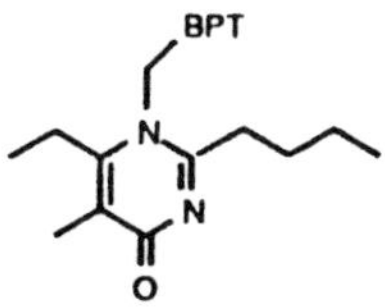
27-21[62]

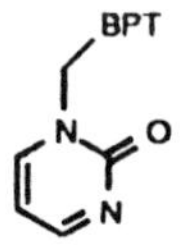
27-22[110]

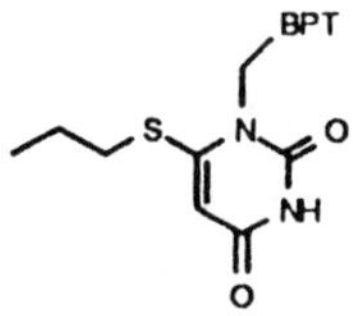
27-23[111]
IC_{50} = 20 n*M*

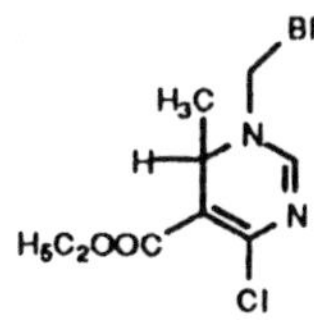
27-24[112]

[a]BPT =
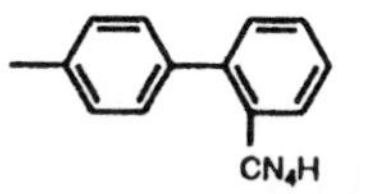

[b]BPC =
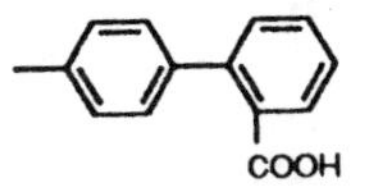

M. BENZOTHIADIAZINE DIOXIDES

3-Butylbenzothiadiazine dioxides with a 2′-acidic biphenylylmethyl group on either the 2- or 4-nitrogen can show high affinity for angiotensin II receptors (Table 29).[26] A possible drawback to their utility is their susceptibility to hydrolysis at pH extremes. However, the corresponding 3-propylthio analogs are considerably more resistant to hydrolysis and retain high affinity for angiotensin II receptors. In this series, the 4-(2′-tetrazolylbiphenylyl)methyl compounds **29–6** and **29–7** showed the highest affinity for angiotensin II receptors, but only **29–7,** the 5-carboxylic acid, exhibited acceptable potency in the functional assays. However, **29–9,** the corresponding pyridothiadiazine dioxide with a 3-ethylthio group, does have subnanomolar potency in both the radioligand binding and functional assays.

N. THIENOIMIDAZOLES

A few thienoimidazoles have been reported without biological data. Their structures are shown in Table 30.

III. ANGIOTENSIN AT_2 RECEPTOR ANTAGONISTS

The 1-benzyl-5-acyl-tetrahydroimidazo [4,5-c] pyridine-6-carboxylic acids were chosen for study as possible angiotensin II receptor antagonists because they include in a compact form structural elements of the C-terminal tetra- or pentapeptide sequence of angiotensin II, which was known to be the minimal binding

TABLE 28
Quinazolones

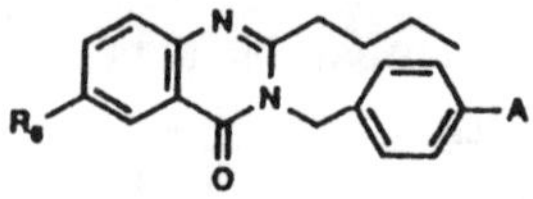

No.	R_6	A	IC_{50}(n*M*)	Ref.
28–1	H	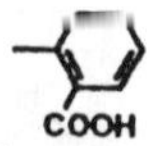	230	28,58
28–2	H	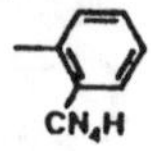	X[a]	28,59
28–3	H	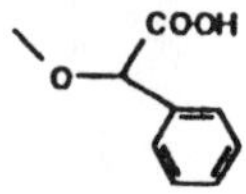	X[a]	113
28–4	NH_2	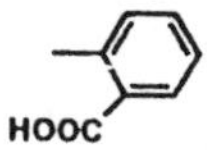	33	28
28–5		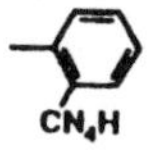	0.1	28
28–6		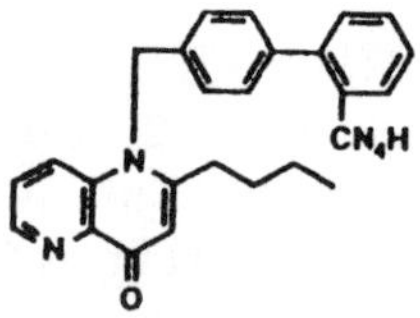	X[a]	114

[a] Comparable data not available.

and functional fragment.[27] The data in Table 31 were obtained using rat adrenal membranes in the presence of 5 m*M* dithiothreitol, which inhibits binding to AT_1 receptors but not to AT_2 receptors. 5-Arylalkyl amides with two atoms between the aryl group and N-5 have affinity for AT_2 receptors, with the best unsubstituted analog being the diphenylacetyl derivative **31–7.** Substituents on the diphenylacetyl phenyls do not markedly improve affinity (**31–8** to **31–12**). In contrast, substituents on the N-benzylphenyl group can improve affinity for AT_2 receptors markedly. The compounds with the highest affinity for AT_2 receptors include **31–15, 31–19,** and **31–20,** which contain in the acyl group a 3-methyl-4-methoxy, amino, or dimethylamino substituent. These compounds have low affinity for the AT_1 receptor and have become the prototypic nonpeptide AT_2 receptor antagonists. Compounds with alicyclic N-1 substituents (**31–25** to **31–27**), with more than one carbon atom between the aryl and the imidazole nitrogen (**31–28** and **31–29**), and the diphenymethyl analog all have affinities for the AT_2 receptor

TABLE 29
Benzothiadiazene Dioxides[26]

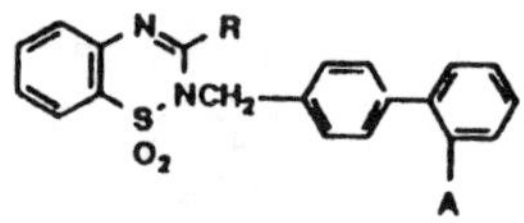

No.	R	A	Ki(n*M*)	K_b(n*M*)
29–1	$CH_3CH_2CH_2CH_2$—	COOH	230	160
29–2	$CH_3CH_2CH_2S$—	COOH	4,900	1,200
29–3	$CH_3CH_2CH_2S$—	CN_4H	57	640

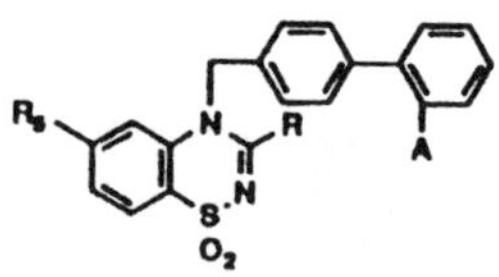

No.	R_5	R	A		
29–4	H	$CH_3CH_2CH_2CH_2$—	COOH	550	130
29–5	H	$CH_3CH_2CH_2S$—	COOH	180	20
29–6	H	$CH_3CH_2CH_2S$—	CN_4H	4.8	93
29–7	COOH	$CH_3CH_2CH_2S$—	CN_4H	8.5	0.08

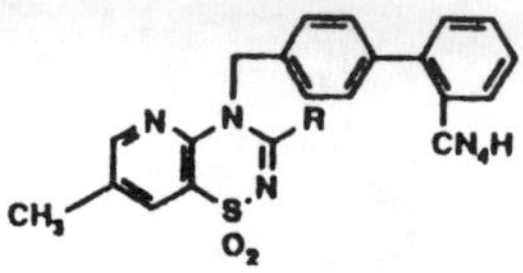

No.	R		
29–8	$CH_3CH_2CH_2S$—	0.6	11
29–9	CH_2CH_2S—	0.8	0.6
	Losartan	5.7	2.6

comparable to that of **31–7,** the benzyl analog. Compounds **31–22** to **31–24,** which are not 6-carboxylic acids, have much lower affinity for AT_2 receptors than their carboxy analog **31–15.** However, those analogs with substituents on the 6-position that could act as H-bond acceptors are approximately tenfold more potent than the 6-unsubstituted compound. This pattern of structure–activity relationships is similar to that observed previously for the 5-substituent of the imidazole AT_1 receptor antagonists.

Another class of potent AT_2 receptor antagonists are the 2-carboxypiperazines with a diphenylacetyl and/or diphenylcarbamyl on the nitrogens[28,29] in Table 32. Compounds **32–1** to **32–3** show comparable (i.e., 1 n*M*) affinity for AT_2 receptors. Even the *trans*-2,5-dimethyl analog without the carboxy group (i.e., **32–4**) has high affinity for the AT_2 receptor. Compound **32–3** has 20% oral bioavailability,

TABLE 30
Thienoimidazoles

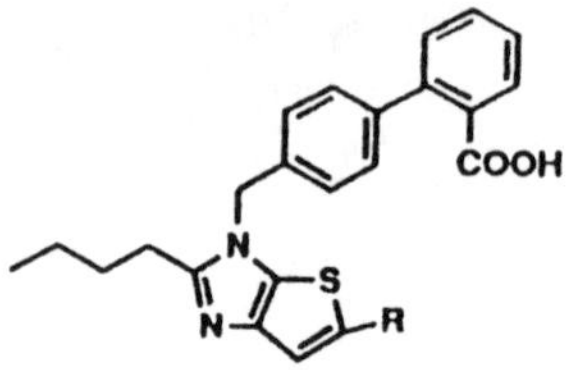

No.	R	IC_{50}(n*M*)	Ref.
30–1	H	X[a]	115,116
30–2	COOH	X[a]	117

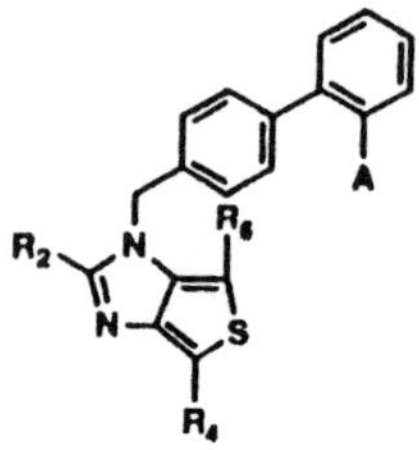

No.	R_2	R_4	R_6	A	IC_{50}(n*M*)	Ref.
30–3	n-C_4H_9	H	H	COOH	X[a]	118
30–4	n-C_3H_7S	CH_3	COOH	CN_4	X[a]	119
30–5	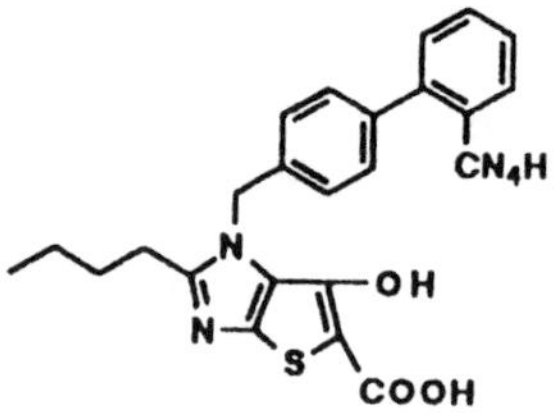				X[a]	

[a] Comparable data not available.

with a >6 h duration of action. The compound is highly selective for AT_2 receptors (i.e., >20,000-fold selectivity). Compound **32–5** was reported to be a combination angiotensin II receptor antagonist and calcium channel blocker. The diphenylmethyl on the piperazine is similar to the diphenylacetylpiperazines, but no data has been reported showing that this compound has affinity for AT_2 receptors.

IV. SUMMARY AND CONCLUSIONS

Potent AT_1 receptor antagonists of many types are now known. Most contain the 2′-tetrazolylbiphenylylmethyl group as the key pharmacophore. The tetrazole

TABLE 31
4,5,6,7-Tetrahydroimidazo[4,5-C]pyridine AT-2 Antagonists[27]

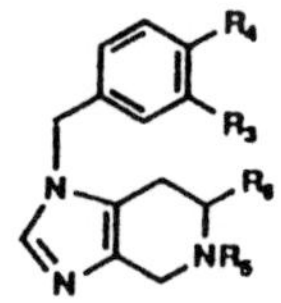

No.	R_5	R_3	R_4	R_6	IC_{50}(n*M*)
31–1	$CONHC_6H_5$	H	H	COOH	2,100
31–2	$CON(CH_3)C_6H_5$	H	H	COOH	1,200
31–3	COC_6H_5	H	H	COOH	12% at 10 μ*M*
31–4	$COCOC_6H_5$	H	H	COOH	2,400
31–5	$COCH_2C_6H_5$	H	H	COOH	1,400
31–6	$COCH_2CH_2C_6H_5$	H	H	COOH	13,000
31–7	$COCH(C_6H_5)_2$	H	H	COOH	1,000
31–8	$COCH(4\text{-}MeOC_6H_4)_2$[a]	H	H	COOH	610
31–9	$COCH(4\text{-}NH_2C_6H_4)_2$[a]	H	H	COOH	8,400
31–10	$COCH(4\text{-}CH_3C_6H_4)_2$[a]	H	H	COOH	1,300
31–11	$COCH(4\text{-}ClC_6H_4)_2$[a]	H	H	COOH	6,800
31–12	$COCH(4\text{-}NO_2C_6H_4)_2$[a]	H	H	COOH	13,000
31–13	$COCH(C_6H_5)_2$[a]	CH_3	H	COOH	150
31–14	$COCH(C_6H_5)_2$[a]	H	OCH_3	COOH	290
31–15	$COCH(C_6H_5)_2$[a]	CH_3	OCH_3	COOH	70
31–16	$COCH(4\text{-}CH_3OC_6H_4)_2$[a]	CH_3	OCH_3	COOH	2,000
31–17	$COCH(4\text{-}CH_3C_6H_4)_2$[a]	CH_3	OCH_3	COOH	510
31–18	$CON(CH_3)C_6H_5)$	H	H	COOH	140
31–19	$COCH(C_6H_5)_2$[a]	CH_3	NH_2	COOH	66
31–20	$COCH(C_6H_5)_2$[a]	CH_3	$N(CH_3)_2$	COOH	34
31–21	$COCH(C_6H_5)_2$[a]	NO_2	OCH_3	COOH	51
31–22	$COCH(C_6H_5)_2$	CH_3	OCH_3	H	4,400
31–23	$COCH(C_6H_5)_2$	CH_3	OCH_3	CH_2OH	370
31–24	$COCH(C_6H_5)_2$	CH_3	$N(CH_3)_2$	$COOCH_3$	350

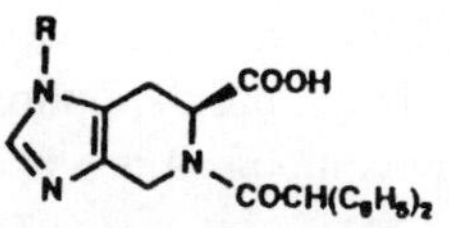

No.	R	IC_{50}(n*M*)
31–25	$CH_2C_6H_{11}$	340
31–26	$CH_2CH_2C_6H_{11}$	240
31–27	CH_2CH_2(1-adamantyl)	130
31–28	$CH_2CH_2C_6H_5$	240
31–29	$(CH_2)_3C_6H_5$	700
31–30	$CH(C_6H_5)_2$	110

[a] S-enantiomer.

TABLE 32
Other AT-2 Antagonists

No.	Structure	IC_{50}(n*M*)	Ref.
32–1	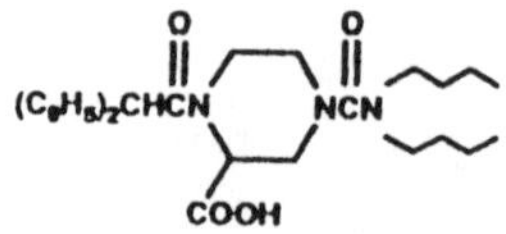	1.0	28
32–2	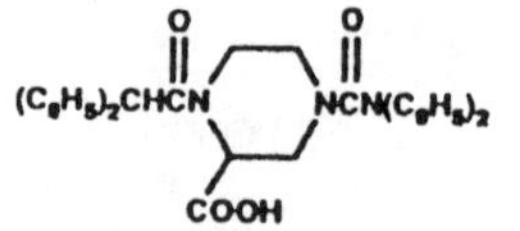	1.5	28
32–3	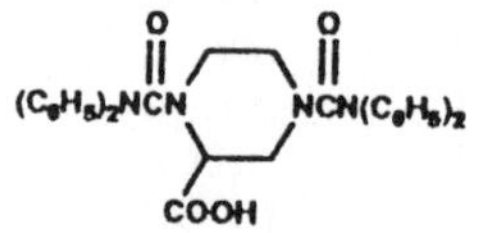	1.7 (AT-1 30,000 n*M*)	28
32–4	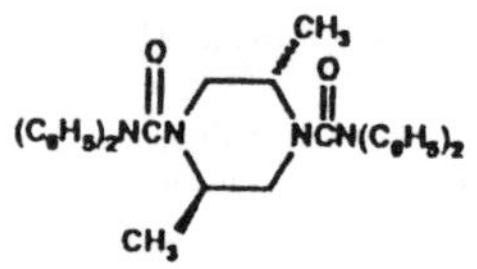	10	28
32–5	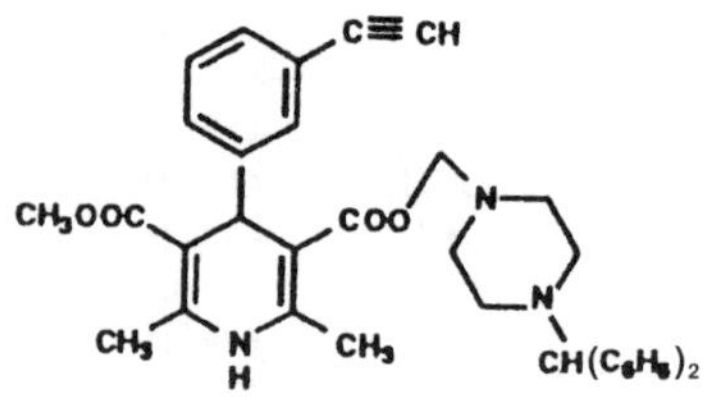	Equipotent to saralasin at $10^{-7}M$	120

may be replaced by other acidic groups, such as carboxyl and trifluoromethylsulfonylamido, with variable effects on affinity, depending on the remainder of the molecule. Another key pharmacophore is an alkyl group in the vicinity of the biphenyl. Most often this is n-butyl, but the optimal chain length varies in individual cases. A third important component is a hydrogen bond acceptor near the biphenyl substituent. In many cases, this is a carboxylic acid, but basic groups, such as the pyrido nitrogen of the imidazo[4,5-b]pyridines, are also efficacious.

A second class of potent AT_1 receptor antagonists contains as its key pharmacophore an α-thienylmethylacrylic acid. In this series, a 4-carboxybenzyl and an adjacent alkyl group are also necessary. In both classes, it is possible to relate the structures of the nonpeptide angiotensin II receptor antagonists to that of angiotensin II, and this proved to be a useful tool for drug design.

A necessary property of any therapeutically useful angiotensin II receptor antagonist is oral activity and long duration of action. In most cases, this has been achieved by empirical manipulation, with small changes in structure providing significant changes in these properties. In some cases, long duration of action is associated with insurmountable antagonism observed in *in vitro* functional assays, possibly resulting from a slow off-rate from the receptor.

The area of AT_2 receptor antagonists is much more limited. No clear therapeutic goal currently exists for this class, in spite of the fact that selective antagonists of AT_2 receptors have been available for several years. With the advent of orally active piperazine AT_2 receptor antagonists, research in this field will accelerate.

V. ADDENDUM

Since the completion of this manuscript a significant number of new papers have appeared, some detailing areas discussed in this chapter and others highlighting previously undisclosed work. Many have been published in the *Journal of Medicinal Chemistry,* but also as a Symposium-in-print in *Bioorganic and Medicinal Chemistry Letters* (4, 1–222, 1994) containing 36 contributions. Among the significant new developments are the balanced AT_1/AT_2 receptor antagonists. For L-162,620 (1) the AT_1 and AT_2 binding IC_{50} values are 0.33 and 0.94 n*M*.[121,122] Another significant development is the closely related AT_1 selective non-tetrazole L-159,282 (2) which has AT_1 and AT_2 binding IC_{50} values of 0.3 and 3000 nM respectively.[123] A third is SB 203220 (3), a napthyl analog of SK&F 108566, which exhibits insurmountable rather than competitive AT_1 antagonism, and possibly as a consequence has a longer duration of action.[124] Finally, SR 47436 (4), an interesting spiro dihydroimidazol-4-one diphenyltetrazole which exhibits competitive AT_1 antagonist activity with a binding IC_{50} of 1.3 nM.[125]

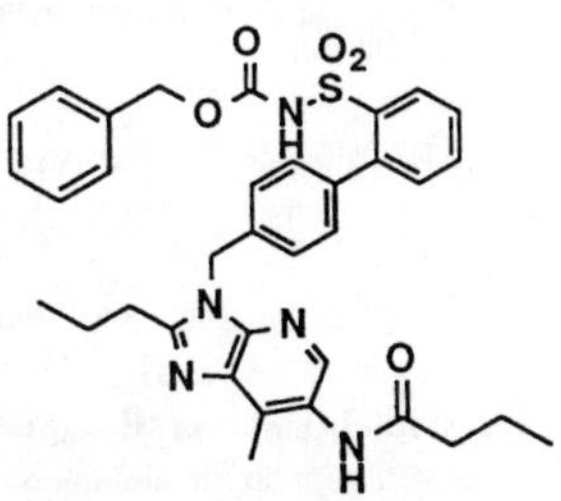

1

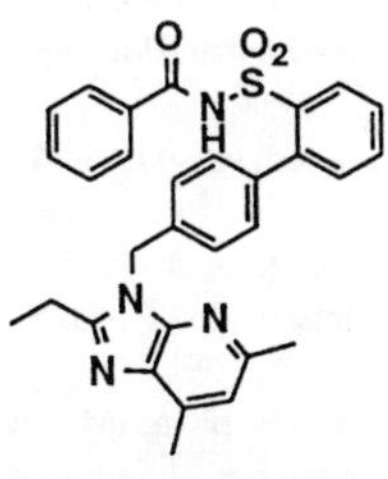

2

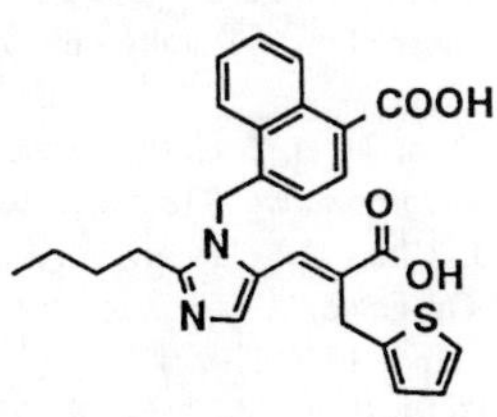

3

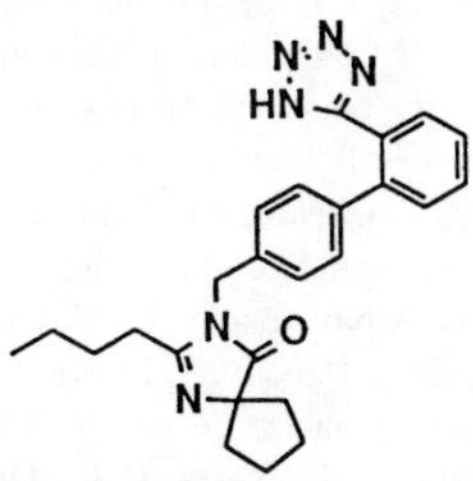

4

REFERENCES

1. **Furukawa, Y., Kishimoto, S., and Nishikawa, K.,** U.S. Patent 4,355,040, 1982.
2. **Furukawa, Y., Kishimoto, S., and Nishikawa, K.,** U.S. Patent 3,340,598, 1982.
3. **Furukawa, Y. and Nishikawa, K.,** U. S. Patent 4,582,847, 1986.
4. **Duncia, J. V., Carini, D. J., Chiu, A. T., Johnson, A. L., Price, W. A., Wong, P. C., Wexler, R. R., and Timmermans, P. B. M. W. M.,** The discovery of DuP 753, a potent, orally active nonpeptide angiotensin II receptor antagonist, *Med. Res. Rev.,* 12, 149, 1992.
5. **Weinstock, J., Keenan, R. M., Samanen, J., Hempel, J., Finkelstein, J. A., Franz, R. G., Gaitanopoulos, D. E., Girard, G. R., Gleason, J. G., Hill, D. T., Morgan, T. M., Peishoff, C. E., Aiyar, N., Brooks, D. P., Fredrickson, T. A., Ohlstein, E. H., Ruffolo, R. R. J., Stack, E. J., Sulpizio, A. C., Weidley, E. F., and Edwards, R. M.,** 1-(Carboxybenzyl)imidazole-5-acrylic acids: potent and selective angiotensin II receptor antagonists, *J. Med. Chem.,* 34, 1516, 1991.
6. **Chiu, A. T., Carini, D. J., Duncia, J. V., Leung, K. H., McCall, D. E., Price, W. A., Wong, P. C., Smith, R. D., Wexler, R. R., and Timmermans, P. B. M. W. M.,** DuP 532: a second generation of nonpeptide angiotensin II receptor antagonists, *Biochem. Biophys. Res. Commun.,* 177, 209, 1991.
7. **Carini, D. J. and Duncia, J. J. V.,** U.S. Patent 4,916,129, 1990.
8. **Wong, P. C., Hart, S. D., Chiu, A. T., Herblin, W. F., Carini, D. J., Smith, R. D., Wexler, R. R., and Timmermans, P. B. M. W. M.,** Pharmacology of DuP 532, a selective and noncompetitive AT_1 receptor antagonist, *J. Pharmacol. Exp. Ther.,* 259, 861, 1991.
9. **Caille, J. C., Corbier, A., Fortin, M., Hamon, G., Jouguey, S., and Levert, J. P.,** European Patent Appl. 465,368, 1992.
10. **Carini, D. J.,** World Patent Appl. 9,200,977, 1992.
11. **Wexler, R.,** presented at Royal Society of Chemistry, Fine Chemicals and Medicinals Group, Glasgow, October 15, 1992.
12. **Middlemiss, D.,** presented at Royal Society of Chemistry, Fine Chemicals and Medicinals Group, Glasgow, October 15, 1992.
13. **Edwards, R. M., Aiyar, N., Ohlstein, E. H., Weidley, E. F., Griffin, E., Ezekiel, M., Keenan, R. M., and Ruffolo, R. R., and Weinstock, J.,** Pharmacological characterization of the nonpeptide angiotensin II receptor antagonist SK&F 108566, *J. Pharmacol. Exp. Ther.,* 260, 175, 1992.
14. **Brooks, D. P., Fredrickson, T. A., Weinstock, J., Ruffolo, R. R., Edwards, R. M., and Gellai, M.,** Antihypertensive activity of the non-peptide angiotensin II receptor antagonist SK&F 108566, in rats and dogs, *Arch. Pharmacol.,* 345, 673, 1992.
15. **Weinstock, J.,** presented at Royal Society of Chemistry, Fine Chemicals and Medicinals Group, Glasgow, October 15, 1992.
16. **Criscione, L., DeGasparo, M., Buhlmayer, P., Whitebread, S., Ramjoue, H. P., and Wood, J.,** Pharmacological profile of CGP 48933, a novel, nonpeptide antagonist of AT_1 angiotensin II receptor, *J. Hypertension,* 10(Suppl. 4), 196, 1992; Buhlmayer, P., Fuset, P., Criscione, L., de Gasparo, M., Whitebread, S., Schmidlin, T., Lattman, R., and Wood, J., Valsartan, a potent, orally active angiotensin II antagonist developed from the structurally new amino acid series, *Bioorg. Med. Chem.* Lett., 4, 29, 1994.
17. **Zhang, J. S., Entzeroth, M., Wienen, W., and Van Meel, J. C. A.,** Characterization of two novel nonpeptide angiotensin II antagonists, *Eur. J. Pharmacol.,* 218, 35, 1992.
18. **Mantlo, N. B., Chakravarty, P. K., Ondeyka, D., Chen, A., Camara, V. J., Chang, R. S. L., Lotti, V. J., Siegl, P. K. S., Patchett, A. A., and Greenlee, W. J.,** presented at 201st American Chemical Society National Meeting, Atlanta, GA, April 14–19, 1991.
19. **Mantlo, N. B., Chakravarty, P. K., Ondeyka, D. L., Lotti, V. J., Faust, A. K., Chen, T.-B. Schorn, T. W., Sweet, C. S., Emmert, S. E., Patchett, A. A., and Greenlee, W. J.,** Potent, orally active imidazo[4,5-b]pyridine-based angiotensin II receptor antagonists, *J. Med. Chem.,* 34, 2919, 1991.

20. **Miyake, K., Matsukura, M., Yoneda, N., Hiroshima, O., Mori, N., Ishihara, H., Musha, T., Matsuoka, T., Hamano, S., Minami, N.,** European Patent Appl. 420, 237, 1991.
21. **Edwards, M. P.,** presented at Royal Society of Chemistry, Fine Chemicals and Medicinals Group, Glasgow, October 15, 1992.
22. **Bradbury, R. H., Allott, C. P., Dennis, M., Fisher, E., Major, J. S., Masek, B. B., Oldham, A. A., Pearce, R. J., Rankine, N., Revill, J. M., Roberts, D. A., and Russell, S. T.,** New nonpeptide angiotensin II receptor antagonists. II. Synthesis, biological properties and structure–activity relationships of 2-Alkyl-4-(biphenylylmethoxy)quinoline derivatives, *J. Med. Chem.,* 35, 4027, 1992.
23. **Oldham, A. A., Allott, C. P., Major, J. S., Smith, C. F. C., Luke, R. W. A., Roberts, D. A., Bradbury, R. H., and Pearce, R. J.,** Substituted 4-benzyloxyl-naphthyridine derivative: a novel, potent and orally-effective angiotensin II receptor antagonist., *Br. J. Pharmacol.,* 107 (Proc. Suppl.), 139, 1992.
24. **De, B., Winn, M., Zydowsky, T. M., Kerkman, D. J., DeBernandis, J. F., Le, J., Buckner, S., Warner, R., Brune, M., Hancock, A., Opgenorth, T., and Marsh, K.,** Discovery of a novel class of orally active, non-peptide angiotensin II antagonists, *J. Med. Chem.,* 35, 3714, 1992.
25. **Atwal, K. S., Ahmed, S. Z., Bird, J. E., Delaney, C. L., Dickinson, K. E., Ferrara, F. N., Hedberg, A., Miller, A. V., Moreland, S., O'Reilly, B. C., Schaeffer, T. R., Waldron, T. L., and Weller, H. N.,** Dihydropyrimidine angiotensin II receptor antagonists, *J. Med. Chem.,* 35, 4751, 1992.
26. **Weller, H. N., Miller, A. V., Moquin, R. V., Dickinson, K. E., Hedberg, S. A., Moreland, S., Cohen, R. B., Delaney, C. L., Skwish, S., and Williams, S.,** Benzothiadiazine dioxides: a new class of potent angiotensin-II (AT_1) receptor antagonists, *Bioorg. Med. Chem. Lett.,* 2, 1115, 1992.
27. **Blankley, C. J., Hodges, J. C., Klutchko, S. R., Himmelsbach, R. J., Chucholowski, A., Essenburg, A. D., and Cohen, D. M.,** Synthesis and structure–activity relationships of a novel series of nonpeptide angiotensin II receptor binding inhibitors specific for the AT_2 subtype, *J. Med. Chem.,* 34, 3248, 1991.
28. **Greenlee, W. J.,** presented at Royal Society of Chemistry, Fine Chemicals and Medicinals Group, Glasgow, October 15, 1992.
29. **Wu, M. T., Ikeler, T. J., Ashton, W. T., Chang, R. S. L., Lotti, V. J., and Greenlee, W. J.,** presented at the 205th American Chemical Society National Meeting, Denver, CO, March 28–April 2, 1993.
30. **Duncia, J. V., Chiu, A. T., Carini, D. J., Gregory, G. B., Johnson, A. L., Price, W. A., Wells, G. J., Wong, P. C., Calabrese, J. C., and Timmermans, P. B. M. W. M.,** The discovery of potent nonpeptide angiotensin II receptor antagonists: a new class of potent antihypertensives, *J. Med. Chem.,* 33, 1312, 1990.
31. **Keenan, R. M., Weinstock, J., Finkelstein, J. A., Franz, R. G., Gaitanopoulos, D. E., Girard, G. R., Hill, D. T., Morgan, T., Samanen, J. M., Hempel, J., Eggleston, D. S., Aiyar, N., Griffin, E., Ohlstein, E. H., Stack, E. J., Weidley, E. F., and Edwards, R.,** Imidazole-5-acrylic acids: potent nonpeptide angiotensin II receptor antagonists designed using a novel peptide pharmacophore model, *J. Med. Chem.,* 35, 3858, 1992.
32. **Thomas, A. P., Allott, C. P., Gibson, K. H., Major, J. S., Masek, B. B., Oldham, A. A., Ratcliffe, A. H., Roberts, D. A., Russell, S. T., and Thomason, D. A.,** New nonpeptide angiotensin II receptor antagonists. I. Synthesis, biological properties, and structure–activity relationships of 2-alkyl benzimidazole derivatives, *J. Med. Chem.,* 35, 877, 1992.
33. **Carini, D. J., Duncia, J. V., Aldrich, P. E., Chiu, A. T., Johnson, A. L., Pierce, M. E., Price, W. A., Santella, J. B. I., Wells, G. J., Wexler, R. R., Wong, P. C., Yoo, S.-E., and Timmermans, P. B. M. W. M.,** Nonpeptide angiotensin II receptor antagonists: the discovery of a series of N-(biphenylylmethyl)imidazoles as potent, orally active antihypertensives, *J. Med. Chem.,* 34, 2525, 1991.
34. **Carini, D. J. and Dunica, J. J. V.,** European Patent 0,253,310, 1988.

35. **Manning, R. E. and Reitz, D. B.,** World Patent 92/02257, 1992.
36. **Binder, D., Weinberger, J., and Koch, A.,** European Patent Appl. 497,121, 1992.
37. **Wong, P. C., Chiu, A. T., Price, W. A., Thoolen, M. J. M. C., Carini, D. J., Johnson, A. L., Taber, R. I., and Timmermans, P. B. M. W. M.,** Nonpeptide angiotensin II receptor antagonists. I. Pharmacological characterization of 2-n-butyl-1-(2-chlorobenzyl)imidazole-5-acetic acid, sodium salt (S8307), J. *Pharmacol. Exp. Ther.,* 247, 1, 1988.
38. **Middlemiss, D., Drew, G. M., Ross, B. C., Robertson, M. J., Scopes, D. I. C., Dowle, M. D., Akers, J., Cardwell, K., Clark, K., Coote, S., Eldred, C. D. Hamblett, J., Hilditch, A., Hirst, G. C., Jack, T., Montana, J., Panchal, T. A., Paton, J. M. S., Shah, P., Stuart, G., and Travers, A.,** Bromobenzofurans: a new class of potent, non-peptide antagonists of angiotensin II, *Bioorg. Med. Chem. Lett.,* 1, 711, 1991.
39. **Carini, D. J. and Duncia, J. J. V.,** European Patent Appl. 0,253,310, 1988.
40. **Henning, R., Wagner, A., Gerhards, H., and Scholkens, B.,** German Patent 4,023,215, 1992.
41. **Bouy, P. R., Collins, J. T., Olins, G. M., McMahon, E. G., and Hutton, W. C.,** Conformationally restricted polysubstituted biphenyl derivatives with angiotensin II receptors antagonist properties, *J. Med. Chem.,* 34, 2410, 1991.
42. **Wagner, A., Hennning, R., Gerhards, H., and Scholkens, B.,** European Patent 450,566, 1991.
43. **Buehlmayer, P., Criscione, L., Fukrer, W. Furet, P., de Gasparo, M., Stutz, S., and Whitebread, S.,** Nonpeptide angiotensin II antagonists: synthesis and in vitro activity of a series of novel napthalene and tetrahydronapthalene derivatives, *J. Med. Chem.,* 34, 3105, 1991.
44. **Lin, H.-S., Rampersaud, A. A., Zimmerman, K., Steinberg, M. I., and Boyd, D. B.,** Nonpeptide angiotensin II receptor antagonists: synthetic and computational chemistry of N-[[4-[2-(2H-tetrazole-5-yl)-1-cycloalken-1-yl] phenyl] methyl] imidazole derivatives and their in vitro activity, *J. Med. Chem.,* 35, 2658, 1992.
45. **Sirca, I.,** World Patent 92/06081, 1992.
46. **Hodges, J. C. and Sirca, I.,** U. S. Patent, 5,045,540, 1991.
47. **Carini, D. J., Wong, P. C. B., and Duncia, J. J. V.,** European Patent Appl. 0,324,377, 1989.
48. **Hilditch, A.,** personal communication, 1993.
49. **Keenan, R. M., Weinstock, J., Finkelstein, J. A., Franz, R. G., Gaitanopoulos, D. E., Girard, G. R., Hill, D. T., Morgan, T. M., Samanen, J. M., Peishoff, C. E., Aiyar, N., Brooks, D. P., Fredrickson, T. A., Griffin, E., Ohlstein, E. H., Stack, E. J., Weidley, E. F., and Edwards, R. M.,** Potent nonpeptide angiotensin II receptor antagonists: 1-(carboxybenzyl)imidazole-5-acrylic acids, *J. Med. Chem.,* 36, 1880, 1993.
50. **Carini, D. J., Wells, G. J., and Duncia, J. J. V.,** European Patent Appl. 323, 841, 1989.
51. **Bovy, P. R., O'Neal, J., Collins, J. T., Olins, G. M., Corpus, V. M., Burrows, S. D., McMahon, E. G., Palomo, M., and Koehler, K.,** New cycloheptimidazolones are nonpeptide antagonists for angiotensin II receptors, *Med. Chem. Res.,* 1, 86, 1991.
52. **Allen, E. E., Chakravarty, P. K., Greenlee, W. J., Maccoss, M., Patchett, A. A., and Walsh, T. F.,** European Patent Appl. 490,587, 1992.
53. **Chakravarty, P. K., Mantlo, N. B., Chen, A., Ondeyka, C. D., Camara, V. J., Chang, R. S. L., Lotti, V. J., Siegl, P. K. S., Patchett, A. A., and Greenlee, W. J.,** Presented at 201st American Chemical Society National Meeting, Atlanta, GA, April 14–19, 1991.
54. **Chakravarty, P. K., Patchett, A. A., Schoen, W., Greenlee, W. J., Mantlo, N. B., and Walsh, T. F.,** European Patent 401,030, 1990.
55. **Morimoto, A., Nishikawa, K., and Naka, T.,** European Patent 443,568, 1991.
56. **Herold, P. and Buehlmayer, P.,** European Patent 424,317, 1991.
57. **Oldham, A. A., Allott, C. P., Major, J. S., Pearce, R. J., Roberts, D. A., and Russell, S. T.,** ICI D8731: A novel, potent, and orally effective angiotensin II antagonist, *Br. J. Pharmacol.,* 1992.
58. **Allen, E. E., DeLaszlo, S. E., Greenlee, W. J., Patchett, A. A., Chakravarty, P. K., and Walsh, E. F.,** European Patent 411,766, 1991.
59. **Morimoto, A. and Nishikawa, K.,** European Patent 445,811, 1991.

60. **Wahhab, A., Smith, J., and Moore, G. J.,** Presented at the 203rd American Chemical Society National Meeting, San Francisco, CA, 1992, Medi 167.
61. **Katano, K., Shimizu, M., Shidara, N., Ogino, H., Nagura, J., Kono, K., Fujishima, K., Konno, F., Machinami, T., and Tsuruoka, T.,** Preparation of pyridines as angiotensin II antagonists and pharmaceuticals containing them for treatment of hypertension and heart failure, *Jpn. Kokai Tokkyo Koho JP,* 05, 178, 836 [93, 178, 836], 1993.
62. **Allen, E. E., Greenlee, W. J., Patchett, A. A., Walsh, T. F., and Chakravarty, P. K.,** European Patent Appl. 419,048, 1991.
63. **Herold, P. M. and Buhlmayer, P. H.,** European Patent 0407, 342A2, 1991.
64. **Buehlmayer, P., Ostermayer, F., and Schnidlin, T.,** European Patent 490,820, 1992.
65. **Bernhart, C., Breliene, J. C., Clement, J., Nisato, D., and Perreaut, P.,** World Patent Appl. 91/14679, 1991.
66. **Fortin, M., Frechet, D., Hamon, G., Jouquey, S., and Venert, J. P.,** European Patent Appl. 461,040, 1991.
67. **Franz, R. G. and Weinstock, J.,** World Patent 91/16313, 1991.
68. **Roberts, D. A. and Russell, S. T.,** European Patent 399,732, 1990.
69. **Shibouta, Y., Inada, Y., Ojima, M., Wada, T., Noda, M., Sanada, T., Kubo, K., Kohara, Y., Naka, T., and Nishikawa, K.,** Pharmacological profile of a highly potent and long-acting angiotensin II receptor antagonist CV-11974 and its proding TCV-116, *J. Pharmacol. Exp. Ther.,* 266, 114, 1993.
70. **Yanagisawa, I., Watanabe, T., Kikuchi, K., Tanaka, A., Okazaki, T., Inagaki, O., and Okada, M.,** World Patent 92/04343, 1992.
71. **Entzeroth, M., Grell, W., Haurel, N., Van Meel, J., Narr, B., Ries, U., and Wienen, W.,** German Patent 4,031,635, 1992.
72. **Hauel, N., Narr, B., Ries, U., Van Meel, J., Wienen, W., and Entzeroth, M.,** European Patent 0,502,314-A1, 1992.
73. **Entzeroth, M., Narr, B. Ries, U, Van Meel, J., Wienen, W., Bombard, A., and Haurel, N.,** European Patent 470,543, 1992.
74. **Lifer, S. L., Marshall, W. S., Mohamadi, F., Reel, J. K., Simon, R. L., Steinberg, M. I., and Whitesitt, C. A.,** European Patent 438,869, 1991.
75. **Entzeroth, M., Haurel, N., Narr, B., Ries, U., Van Jacques, M., Wienen, W., and Van Meel, J.,** European Patent 468,470, 1992.
76. **Middlemiss, D., Ross, B., Robertson, M., Eldred, C., Panchal, T., Watson, S., and Hilditch, T.,** Presented at the 203rd American Chemical Society Meeting, San Francisco, CA, April 5, 1992.
77. **Middlemiss, D., Ross, B. C., Eldred, C., Montana, J. G., Shah, P., Hirst, G. C., Watson, S., P., Panchal, T. A., Paton, J. M. S., Hubbard, T., Drew, G. M., Robertson, M. J., Hilditch, A., and Clark, K. L.,** C-linked pyrazole biaryl tetrazoles as antagonists of angiotensin II, *Bioorg. Med. Chem. Lett.,* 2, 1243, 1992.
78. **Ashton, W. T., Hutchins, S. M., Greenlee, W. J., Doss, R. S., Chang, L., Lotti, V. J. Kivlighn, S. D., and Siegl, P. K. S.,** Presented at 203rd American Chemical Society National Meeting, San Francisco, CA, April 5–10, 1992.
79. **Bru-Magniez, N., Micolai, E., and Teulon, J.,** European Patent 449,699, 1991.
80. **Naka, T. and Nishikawa, K.,** European Patent 411,507, 1991.
81. **Cardwell, K. S., Doule, M. D., Eldred, C. D., Jack, T. I. M., Middlemiss, D., Montana, J. G., Ross, B. C., Scopes, D. I. C., and Shah, P.,** World Patent 92/09600-A1, 1992.
82. **Reitz, D. B., Penick, M. A., Brown, M. S., Reinhard, E. J., Olins, G. M., Corpus, V. M., McMahon, E. G., Palomo, M. A., Koepke, J. P., Moore, G. K., Smits, G. J., McGraw, D. E., and Blaine, E. H.,** Presented at the 203rd National Meeting, American Chemical Society, San Francisco, CA, April 5, 1992.
83. **Carini, D. J., Wells, G. J., and Duncia, J. J. V.,** European Patent 323,841, 1989.

84. **Ashton, W. T., Cantone, C. L., Chang, L. L., Strelitz, R. A., MacCoss, M., Chang, R. S. L., Lotti, V. J., Kivlighn, S. D., Sieg, P. K. S.,** Presented at 202nd American Chemical Society National Meeting, New York, August 25, 1991.
85. **Manning, R. E., Reitz, D. B., and Huang, H. C.,** World Patent 91/18888, 1991.
86. **Greenlee, W. J., Patchett, A. A., Hangauer, D., Ashton, W., Fitch, K. J., Walsh, T., Rivero, R. A., and Dhanon, D. S.,** World Patent Appl. 91/11909, 1991.
87. **Ashton, W. T., MacCoss, M., Chang, L. L., Chakravarty, P. K., Cantone, C. L., Greenlee, W. J., Patchett, A. A., and Walsh, T. F.,** European Patent 409,332, 1991.
88. **Reitz, D. B. and Manning, R. E.,** World Patent 92/04335, 1992.
89. **Bovy, P. R., Reitz, D. B., Collins, J. T., Chamberlain, T. S. Olins, G. M., Corpus, V. M., McMahnon, E. G., Palomo, M. A., Koepke, J. P., Smits, G. J., McGraw, D. E., and Gaw, J. F.,** Nonpeptide angiotensin II antagonists: N-phenyl-1H-pyrrolyl-tetrazole derivatives are angiotensin II receptor antagonists, *J. Med. Chem.,* 36, 101, 1993.
90. **Olins, G. M.,** Personal communication, 1993.
91. **Mantlo, N. B., Ondeyka, D., Chang, R. S. L., Lotti, V. J., Kivlighn, S. D., Siegel, P. K. S., Patchett, A. A., Greenlee, W. J.,** Presented at 202nd American Chemical Society National Meeting, New York, August 25, 1991.
92. **Chakravarty, P. K., Maccross, M., Mantlo, N. B., and Walsh, T. F.,** U.S. Patent 5,128,327, 1992.
93. **Inoue, T., Kayahiri, H., Kuroda, A., Oku, T., Saitoh, Y., Satoh, S., Setoi, H., and Tanaka, H.,** Canada Patent 2,052,125, 1992.
94. **Greenlee, W. J., Patchett, A. A., Hangauer, D., Walsh, T., Fitch, K., Rivero, R. A., and Dhanoa, D. S.,** World Patent 91/11999, 1991.
95. **Dhanoa, D. S., Bagley, S. W., Chang, R. S. L., Lotti, V. J., Siegl, P. K. S., Patchett, A. A., and Greenlee, W. J.,** Presented at 204th American Chemical Society National Meeting, Washington, D.C., August 23–28, 1992.
96. **Ries, U., Narr, B., Bombard, A., Haurel, N., Van Meel, J., Wienen, W., and Entzeroth, M.,** European Patent 470,543, 1992.
97. **Jaunin, R.,** European Patent 467,207, 1992.
98. **Morimoto, A. and Nishikawa, K.,** European Patent 430,300, 1991.
99. **Bovy, P. R., Collins, J. T., Chamberlain, T. S., and Gheng, B. K.,** World Patent 92/07852, 1992.
100. **Murray, W. V., Lalan, P., Gill, A., Addo, M. F., Lewis, J. M., Lee, D. K. H., Rampulla, R., Wachter, M. P., Hsi, J. D., and Underwood, D. C.,** Substituted piperidin-2-one biphenyltetrazoles as angiotensin II antagonists, *Bioorg. Med. Chem. Lett.,* 2, 1772, 1992.
101. **Roberts, D. A., Russell, S. T., and Pearce, R. J.,** European Patent 412,848, 1991.
102. **Clemence, F., Fortin, M., and Haesslein, J.,** European Patent 498,721-A1, 1992.
103. **Bash, A. F., Debernardi, J. F., Holleman, W. H., Kerkman, D. J., Kester, J. A., Rosenberg, S. H., Shiosaki, K., Tasker, A. S., Van Geldern, T. W., Winu, M., and Zidowshy, T. M.,** European Patent Appl., 1992.
104. **Bradbury, R. H. and Roberts, D. A.,** European Patent 499,414-A2, 1992.
105. **Edwards, M. P., Allott, C. P., Bradbury, R. H., Major, J. S., Masek, B. B., Oldham, A. A., Pearce, R. J., Roberts, D. A., and Russell, S. T.,** Presented at 203rd American Chemical Society National Meeting, San Francisco, CA, April 5, 1992.
106. **Bradbury, R. N., Pearce, R. J., and Roberts, D. A.,** European Patent Application 487,252, 1992.
107. **Roberts, D. A., Pearce, R. J., and Bradbury, R. H.,** European Patent 487,252, 1992.
108. **Allen, E. E. G., Geenlee, W. J., MacCoss, M., and Patchett, A. A.,** World Patent 91/15209, 1991.
109. **Aletun, M., Daumas, M., Hoornaert, C., and Muller, J.,** European Patent 500,409, 1992.
110. **Herold, P. and Buehlmayer, P.,** European Patent 407,342, 1991.
111. **Naka, T. and Nishikawa, K.,** European Patent 442,473, 1991.

112. **Atwal, K.,** European Patent 481,448, 1992.
113. **Greenlee, W. J., Patchett, A. A., Hangauer, D., Walsh, T., Fitch, K. J., Rivero, R. A., and Dhanon, D. S.,** World Patent 91/12001, 1991.
114. **Allen, E. E., Chakravarty, P. K., Dilaszlo, S. E., Greenlee, W. J., Patchett, A. A., and Walsh, T. F.,** European Patent 481,614, 1992.
115. **Naka, T. and Inada, Y.,** European Patent 483,683, 1992.
116. **Kuhnke, J., Schoellkopf, K., Beckmann, R., and Graf, H.,** German Patent Appl. 4,034,728, 1992.
117. **Greenlee, W. J., Johnston, D. B., MacCoss, M., Mantlo, N. B., Patchett, A. A., Chakravarty, P. K., and Walsh, T. F.,** European Patent 407,102, 1991.
118. **Kuhnke, J., Schoellkopf, K., Buckmann, R., and Graf, M.,** German Patent 4,032,522, 1992.
119. **Naka, T. and Inada, Y.,** European Patent 483,683, 1992.
120. **Nakanishi, M., Uchida, K., Nakano, J., Nagahara, M., Murai, T., and Osada, E.,** World Patent 91/09847, 1991.
121. **Mantlo, N. B., Kim, D., Ondeyka, D., Chang, R. S. L., Kivlighn, S. D., Siegl, P. K. S., and Greenlee, W. J.,** Imidazo[4,5-b]pyridine-based AT_1/AT_2 angiotensin II receptor antagonists, *Bioorg. Med. Chem. Lett.,* 4, 17, 1994.
122. **Kevin, N. J., Rivero, R. A., Greenlee, W. J., Chang, R. S. L., and Chen, T. B.,** Substituted phenylthiophene benzoylsulfonamides with potent binding affinity to angiotensin II AT_1 and AT_2 receptors, *Bioorg. Med. Chem. Lett.,* 4, 189, 1994.
123. **Weinstock, J., Hill, D. T., Keenan, R. M., Franz, R. G., Gaitanopoulos, D. E., Girard, G. R., Eggleston, D. S., Aiyar, N., Griffin, E., Ohlstein, E., Brooks, D. P., Gellai, M., Frederickson, T. A., Weidley, E. F., and Edwards., R. M.,** A potent long-acting imidazole-5-acrylic acid angiotensin II AT-1 receptor antagonist, *Bioorg. Med. Lett.,* 4, 23, 1994.
124. **Cazaubon, C., Gougat, J., Bousquet, F., Guiraudou, P. Gayraud, R., Lacour, C., Roccon, A., Galindo, G., Barthelemy, G., Gautret, B., Bernhart, C., Perreaut, P., Breliere, J., LeFur, G., and Nisato, D.,** Pharmacological characterization of SR 47436, a new nonpeptide AT_1 subtype angiotensin II receptor antagonist, *J. Pharmacol. Exp. Ther.,* 265, 826, 1993.

INDEX

A

B

C

N

O

P

Q

R

S

T

V, W, X, Y, Z